# Sulfate-Reducing
# Bacteria

# BIOTECHNOLOGY HANDBOOKS

Series Editors: Tony Atkinson and Roger F. Sherwood

*Centre for Applied Microbiology and Research*
*Division of Biotechnology*
*Salisbury, Wiltshire, England*

Volume 1    *PENICILLIUM* AND *ACREMONIUM*
            Edited by John F. Peberdy

Volume 2    *BACILLUS*
            Edited by Colin R. Harwood

Volume 3    CLOSTRIDIA
            Edited by Nigel P. Minton and David J. Clarke

Volume 4    *SACCHAROMYCES*
            Edited by Michael F. Tuite and Stephen G. Oliver

Volume 5    METHANE AND METHANOL UTILIZERS
            Edited by J. Colin Murrell and Howard Dalton

Volume 6    PHOTOSYNTHETIC PROKARYOTES
            Edited by Nicholas H. Mann and Noel G. Carr

Volume 7    *ASPERGILLUS*
            Edited by J. E. Smith

Volume 8    SULFATE-REDUCING BACTERIA
            Edited by Larry L. Barton

Volume 9    *THERMUS* SPECIES
            Edited by Richard Sharp and Ralph Williams

A Continuation Order Plan is available for this series. A continuation order will bring delivery of each new volume immediately upon publication. Volumes are billed only upon actual shipment. For further information please contact the publisher.

# Sulfate-Reducing Bacteria

Edited by

## Larry L. Barton
*University of New Mexico*
*Albuquerque, New Mexico*

Plenum Press • New York and London

Library of Congress Cataloging-in-Publication Data

Sulfate-reducing bacteria / edited by Larry L. Barton.
        p.    cm. -- (Biotechnology handbooks ; v. 8)
     Includes bibliographical references and index.
     ISBN 0-306-44857-2
     1. Sulphur bacteria.    I. Barton, Larry.    II. Series.
QR92.S8S84   1995
589.9'6--dc20                                          95-15113
                                                            CIP

ISBN 0-306-44857-2


© 1995 Plenum Press, New York
A Division of Plenum Publishing Corporation
233 Spring Street, New York, N. Y. 10013

10 9 8 7 6 5 4 3 2

Printed in the United States of America

# Contributors

**J. M. Akagi** • Department of Microbiology, The University of Kansas, Lawrence, Kansas 66045

**Larry L. Barton** • Department of Biology, University of New Mexico, Albuquerque, New Mexico 87131

**Liang Chen** • Department of Biochemistry, The University of Georgia, Athens, Georgia 30605

**Heribert Cypionka** • Carl von Ossietzky Universität Oldenburg, Institut für Chemie und Biologies des Meeres, D-26111 Oldenburg, Germany

**Richard Devereux** • United States Environmental Protection Agency, Environmental Research Laboratory, Gulf Breeze, Florida 32561

**Burt D. Ensley** • Phytotech, Newton, Pennsylvania 18940

**Guy D. Fauque** • Centre d'Oceanologie de Marseille, Université d'Aix-Marseille II, F-13288 Marseille Cedex 9, France

**W. Allan Hamilton** • Molecular and Cell Biology, Marischal College University, Aberdeen AB9 1AS, Scotland

**T. R. Jack** • Novacor Research and Technology Corporation, Calgary, Alberta, Canada, T2E 7K7

**Jasper Kunow** • Laboratorium für Mikrobiologie des Fachbereichs Biologie der Philipps-Universität Marburg, and Max-Planck-Institut für Terrestrische Mikrobiologie, Marburg, D-35043 Marburg, Germany

**Whonchee Lee** • Center for Biofilm Engineering, Montana State University, Bozeman, Montana 59717-0398

**Jean Le Gall** • Department of Biochemistry, The University of Georgia, Athens, Georgia 30605

**Ming-Y. Liu** • Department of Biochemistry, The University of Georgia, Athens, Georgia 30605

**Erko Stackebrandt** • Deutsche Sammlung von Mikroorganismen und Zellkulturen, Mascheroder, Weg 1B, 38124 Braunschweig, Germany

**David A. Stahl** • College of Veterinary Medicine, Department of Veterinary Pathobiology, University of Illinois, Urbana, Illinois 61801

**Joseph M. Suflita** • Department of Botany and Microbiology, The University of Oklahoma, Norman, Oklahoma 73019

**Rudolf K. Thauer** • Laboratorium für Mikrobiologie des Fachbereichs Biologie der Philipps-Universität and Max-Planck-Institut für Terrestrische Mikrobiologie, Marburg, D-35043 Marburg, Germany

**Francisco A. Tomei** • Army Environmental Policy Institute, Champaign, Illinois 61821

**Walter M. A. M. van Dongen** • Department of Biochemistry, Agricultural University, Dreyenlaan 3, 6703 HA Wageningen, The Netherlands

**D. W. S. Westlake** • Westec Microbes Ltd., Victoria, British Columbia, Canada V9C 1X4

# Preface

Over the years, sulfate-reducing bacteria (SRB) have been examined from many different perspectives. In several areas research has been highly active and molecular mechanisms have been developed, while research in other areas has progressed at a markedly slower rate. This book was initiated with the objective of summarizing the information available on the SRB with the intent that biology of the bacteria would be linked with biochemistry processes. This approach should enable us to identify and define appropriate questions for future research.

The first chapter provides an overview of the sulfate-reducing eubacteria, while the second chapter summarizes the activities of the sulfate-reducing archaebacteria. Included in these two chapters are an enumeration of key enzymes that are unique to these bacteria and use sulfate as the final electron acceptors. Taxonomic relationships, pathways for sulfate reduction, electron transport, cell energetics, and genetics of these specialized bacteria are addressed in five separate chapters. This information provides an insight into life processes of the SRB. In fact, it is apparent from these chapters that the depth of knowledge of biochemical and physiological processes of the sulfate reducers is so great that these organisms may be used as model systems for understanding anaerobic activities. The broad topics of ecology, biocorrosion, bioremediation, and problems in industry related to SRB are skillfully reviewed in the last four chapters. Many of these highly dynamic processes assume a new dimension in light of the information provided here detailing chemical activities of SRB.

The reader will notice the breadth and complexity of the various topics. Without the cooperation from the authors of these chapters, it would be difficult or perhaps impossible to provide an accurate assessment of the activities of SRB. This book was written for advanced undergraduates, graduate students, engineers, scientists, and other professionals. For readers seeking additional information, a substantial number of references are given to original literature and recent reviews.

Larry L. Barton

*Albuquerque, New Mexico*

# Contents

Chapter 1

**Characteristics and Activities of Sulfate-Reducing Bacteria** .............................................. 1

Larry L. Barton and Francisco A. Tomei

1. Introduction ............................................. 1
2. Diversity of Bacteria ................................... 2
   2.1. Biology ............................................. 2
   2.2. Cell Characteristics ............................... 2
3. Growth ................................................... 3
   3.1. Cultivation ........................................ 3
   3.2. Stress Response .................................... 6
   3.3. Identification ..................................... 7
   3.4. Isolation .......................................... 7
   3.5. Detection .......................................... 8
4. Biochemical Activities .................................. 8
   4.1. Enzymes ............................................ 8
   4.2. Immobilization of Enzymes ......................... 9
   4.3. Metabolic Products ................................ 9
   4.4 Transformation Reaction ........................... 15
   4.5. Energy Technology ................................ 18
5. Interactions with Plants and Animals ................... 20
6. Perspectives ........................................... 22
   References .............................................. 22

Chapter 2

**Sulfate-Reducing Archaea** ............................. 33

Rudolf K. Thauer and Jasper Kunow

1. Introduction: Discovery of *Archaeoglobus* ............. 33
2. Phylogenetic Position among the Archaea ............... 34

3. Morphology and Components of Cellular Structures ........    37
4. Habitats and Growth Requirements ......................    38
5. Coenzymes, Enzymes, and Metabolic Pathways ............    39
   5.1. Coenzymes ........................................    39
   5.2. Enzymes ..........................................    40
   5.3. Metabolic Pathways ...............................    43
6. Assimilatory Sulfate Reduction in Archaea ................    43
   References .............................................    45

Chapter 3

**Taxonomic Relationships** ............................    49

Erko Stackebrandt, David A. Stahl, and Richard Devereux

1. Introduction ...........................................    49
2. Traditional Classification ..............................    51
   2.1. The Mesophilic Gram-Negative Sulfate Reducers ...    53
   2.2. The Thermophilic Gram-Negative Sulfate Reducers    57
   2.3. The Gram-Positive Sulfate Reducers ..............    59
   2.4. The Archaeal Sulfate Reducers ...................    60
3. The Phylogeny of the Mesophilic Gram-Negative Bacterial
   Sulfate-Reducers ......................................    61
   3.1. The Family *"Desulfovibrionaceae"* ...................    63
   3.2. The Family *"Desulfobateriaceae"* ...................    64
4. The Phylogeny of the Thermophilic Gram-Negative
   Bacteria ..............................................    65
5. The Phylogeny of the Gram-Positive Sulfate Reducers .....    66
6. The Phylogeny of the Archaeal Sulfur Reducers ..........    69
7. The Analysis of Genetic Markers other than the 16S rDNA    70
   7.1. Ferredoxin .......................................    71
   7.2. Cytochrome *c* ...................................    72
   7.3. Hydrogenases ....................................    72
8. Determinative Tools and New Developments ..............    73
   8.1. Lipid Biomarkers ...............................    73
   8.2. The Use of Nucleic Acid Hybridization for
        Determinative Studies ..........................    74
   8.3. Gene Probes .....................................    75
9. Origin and Evolution of Sulfate-Reducing Bacteria in
   Relationship to the Biosphere .........................    77
10. Uncultured Novel Sulfate-Reducing Bacteria ..............    78

10.1. Greigite Magnetosomes ............................ 78
10.2. Hyperthermophilic Sulfate-Reducing Bacteria ....... 79
References ............................................. 79

Chapter 4

**Respiratory Sulfate Reduction** .........................  89

J. M. Akagi

1. Introduction ............................................. 89
2. Activation of Sulfate and Its Reduction to Bisulfate ......... 89
   2.1. Adenylyl Sulfate (APS) Formation .................... 89
   2.2. Reduction of APS to AMP and Bisulfate .............. 90
3. Bisulfate Reduction ...................................... 91
   3.1. Detection of Trithionate and Thiosulfate in Reaction
        Mixtures .......................................... 91
   3.2. Bisulfate Reductase ............................... 92
4. The Trithionate Pathway ................................. 94
   4.1. Trithionate Reductase(s) ........................... 98
   4.2. Thiosulfate Reductase ............................. 99
   4.3. Bisulfite Reduction by Cell Extracts and Whole Cells ... 100
   4.4. Role of Membranes in Bisulfite Reduction ............. 101
5. Direct Six-Electron Reduction Mechanism ................. 101
   5.1. Proton Translocation Studies ....................... 103
   5.2. Siroheme and Tetrahydroporphyrin .................. 104
   5.3. Dithionite as a Possible Intermediate ................ 104
   5.4. Is Assimilatory Bisulfate Reductase Involved? .......... 105
6. Concluding Remarks ..................................... 106
   References ............................................. 108

Chapter 5

**Characterization of Electron Transfer Proteins** ...........  113

Liang Chen, Ming-Y. Liu, and Jean Le Gall

1. Introduction ............................................. 113
2. Methods of Characterization ............................. 114
   2.1. Prosthetic Groups ................................. 114

2.2. Apo-Proteins ....................................... 115
2.3. Native, Reconstituted, and Modified Proteins .......... 116
3. Individual Electron Transfer Proteins ..................... 116
   3.1. Non-Heme Iron Proteins ........................... 116
   3.2. Flavoproteins ..................................... 124
   3.3. Cytochromes ...................................... 126
4. Reconstitution of Electron Transfer Chains ............... 129
   4.1. Protein Localization and Reconstitution with Soluble
        and/or Solubilized Proteins ........................ 129
   4.2. Spheroplasts, Membrane Preparations, with or without
        Addition of Soluble Proteins ........................ 134
   4.3. Protein–Protein Interactions: Computer Modeling ..... 134
   4.4. Molecular Biology ................................. 135
   References .............................................. 137

Chapter 6

**Solute Transport and Cell Energetics** ...................   151

Heribert Cypionka

1. Introduction ........................................... 151
   1.1. Thermodynamics of Dissimilatory Sulfate Reduction ... 151
   1.2. Some Relevant Properties of Sulfate, Sulfide,
        and Sulfite ........................................ 152
   1.3. Scalar and Vectorial Processes ...................... 153
   1.4. A First Look at Proton Effects Coupled to Sulfate
        Reduction ......................................... 155
2. Sulfate Transport ...................................... 156
   2.1. Different Feasible Mechanisms of Sulfate Transport .... 156
   2.2. Sulfate Accumulation by Symport with Protons ........ 157
   2.3. Sodium-Dependent Sulfate Accumulation by Marine
        Sulfate Reducers ................................... 158
   2.4. Calculation of Steady State Sulfate Accumulation ...... 158
   2.5. Correlation of Proton Potential and Sulfate
        Accumulation ...................................... 159
   2.6. Stoichiometry of Sulfate Symport with Sodium Ions .... 160
   2.7. Regulation of Sulfate Transport ..................... 160
   2.8. Transport of Thiosulfate and other Sulfate Analogs .... 161
   2.9. Assessment of the Energy Requirement for Sulfate
        Transport ......................................... 162

3. Energetics of Sulfate Activation .......................... 164
   3.1. Energy Coupling of Pyrophosphatase or
        APS Reductase ...................................... 165
4. Energetics of Sulfite Reduction .......................... 166
5. Generation of a Protonmotive Force ...................... 168
   5.1. Hydrogen Cycling .................................. 168
   5.2. Proton Release by Periplasmic Hydrogenase
        and by Translocation .............................. 169
6. Comprehensive Assessment of the Energetics
   of Sulfate Reduction ..................................... 171
7. Energy Conservation by Processes other than Sulfate
   Reduction ............................................... 173
   7.1. Fermentation of Organic Substrates .................. 174
   7.2. Disproportionation of Sulfur Compounds ............. 174
   7.3. Energy Conservation by Reduction of Nitrate .......... 175
   7.4. Energy Conservation by Aerobic Respiration .......... 177
8. Conclusion ............................................... 179
   References ............................................... 179

Chapter 7

**Molecular Biology of Redox-Active Metal Proteins
from *Desulfovibrio*** ...................................... 185

Walter M. A. M. van Dongen

1. Introduction ............................................. 185
2. *Desulfovibrio* Genetics .................................. 187
   2.1. The Genome ...................................... 187
   2.2. Plasmids ......................................... 187
   2.3. Phages ........................................... 188
3. Transfer of Broad-Host-Range Plasmids to *Desulfovibrio*: A
   Basic Step for Genetic Manipulation ...................... 188
   3.1. IncQ Plasmids for Gene Cloning in *Desulfovibrio* ....... 188
   3.2. Applications of Gene Cloning in *Desulfovibrio* .......... 190
4. Molecular Biology as a Tool for Analysis of Redox-Active
   Metal Proteins .......................................... 191
   4.1. Hydrogenases ..................................... 192
   4.2. Proteins with Novel Fe-S Clusters: Prismane Protein and
        Dissimilatory Sulfite Reductase ...................... 202
   4.3. Cytochromes ...................................... 204

5. Perspectives .............................................. 207
References .............................................. 208

Chapter 8

**Ecology of Sulfate-Reducing Bacteria** ................... 217

Guy D. Fauque

1. Introduction ............................................ 217
2. Processes in the Anaerobic Degradation of Organic Matter .. 218
3. The Biological Sulfur Cycle ............................. 218
    3.1. The Sulfuretum ................................... 220
    3.2. Microorganisms Involved in the Sulfur Cycle .......... 221
4. Sulfate Reduction in Natural Habitats .................... 223
    4.1. Gram-Negative Mesophilic Sulfate Reducers ........... 224
    4.2. Gram-Positive Sporeforming Sulfate Reducers ......... 225
    4.3. Gram-Negative Thermophilic Eubacterial Sulfate
        Reducers .......................................... 226
    4.4. Gram-Negative Thermophilic Archaebacterial Sulfate
        Reducers .......................................... 227
5. Effects of Environmental Factors on Growth
   of Sulfate-Reducing Bacteria ............................ 227
    5.1. Effect of pH, Temperature, and Salts ................ 228
    5.2. Aerobic Sulfate Reduction .......................... 230
6. Growth in Phototrophic Associations ..................... 232
7. Interactions with Methanogenic Bacteria .................. 233
    7.1. Synergism: Interspecies Hydrogen Transfer ........... 233
    7.2. Competition ....................................... 234
8. Conclusion .............................................. 234
References .............................................. 235

Chapter 9

**Biocorrosion** ........................................... 243

W. Allan Hamilton and Whonchee Lee

1. Introduction ............................................ 243
2. Corrosion ............................................... 244

2.1. Abiotic Corrosion ................................... 244
2.2. Microbially Influenced Corrosion .................... 247
3. Conclusions ............................................ 260
References .............................................. 262

Chapter 10

**Control in Industrial Settings** ......................... 265

T. R. Jack and D. W. S. Westlake

1. Introduction ........................................... 265
   1.1. *In Situ* Microbial Activity .......................... 265
   1.2. Microbial Activities in Surface Facilities .............. 266
2. Management ........................................... 266
   2.1. Cathodic Protection and Coatings to
        Reduce Corrosion .................................. 266
   2.2. Changes in the Environment ....................... 268
   2.3. Control with Biocides ............................. 270
3. Biocide Application ..................................... 273
   3.1. Bacterial Detection Systems ........................ 273
   3.2. Screening of Biocide Products ...................... 274
   3.3. Biocide Dosages ................................... 276
   3.4. Factors Affecting Biocide Activity ................... 277
   3.5. Field Expectations ................................. 280
   3.6. Improving Biocide Performance ..................... 281
4. Miscellaneous Control Options .......................... 282
   4.1. Heat Shock ....................................... 282
   4.2. Ice Nucleation .................................... 282
   4.3. Manipulation of Microbial Growth Patterns ........... 282
5. Oil Reservoir Souring .................................. 283
6. Future Needs .......................................... 285
   6.1. Correlation of Biological Parameters to
        Corrosion Risk .................................... 285
   6.2. Relating Corrosion Mechanisms and Bacterial
        Populations ....................................... 285
7. Recommendations ...................................... 286
   7.1. Treatment Chemicals ............................... 286
   7.2. Understanding Biofilm Communities ................. 286
8. Summary .............................................. 288
   References .............................................. 289

Chapter 11

**Metabolism of Environmental Contaminants by Mixed
and Pure Cultures of Sulfate-Reducing Bacteria** . . . . . . . . . .    293

Burt D. Ensley and Joseph M. Suflita

1. Introduction . . . . . . . . . . . . . . . . . . . . . . . . . . . . . . . . . . . . . . . . . . . . . . .    293
2. Metabolism of Nonhalogenated Compounds by Mixed
   Culture . . . . . . . . . . . . . . . . . . . . . . . . . . . . . . . . . . . . . . . . . . . . . . . . . .    303
3. Metabolism of Nonhalogenated Compounds by Pure
   Cultures . . . . . . . . . . . . . . . . . . . . . . . . . . . . . . . . . . . . . . . . . . . . . . . . . .    311
4. Metabolism of Halogenated Compounds by Isolates
   and Mixed Cultures . . . . . . . . . . . . . . . . . . . . . . . . . . . . . . . . . . . . . . . .    317
5. Perspectives . . . . . . . . . . . . . . . . . . . . . . . . . . . . . . . . . . . . . . . . . . . . . . .    326
   References . . . . . . . . . . . . . . . . . . . . . . . . . . . . . . . . . . . . . . . . . . . . . . . .    327

**Index** . . . . . . . . . . . . . . . . . . . . . . . . . . . . . . . . . . . . . . . . . . . . . . . . . . . .    333

# Characteristics and Activities of Sulfate-Reducing Bacteria

LARRY L. BARTON and FRANCISCO A. TOMEI

## 1. INTRODUCTION

The sulfate-reducing bacteria (SRB) are a unique physiological group of procaryotes because they have the capability of using sulfate as the final electron acceptor in respiration. Initially, these bacteria were treated as biological curiosities and little research effort was devoted to them. An appreciation of the SRB grew, in part, from an interest in understanding their relationship with other life forms. In the last few decades, the metabolic processes of the SRB have received considerable attention, and from these observations it can be concluded that SRB are markedly similar to other bacteria. A hallmark characteristic that distinguishes SRB is the manner in which sulfate is metabolized. With the demonstration that SRB are broadly distributed on earth, it was recognized that these organisms displayed significant roles in nature by virtue of their potential for numerous interactions (Fig. 1). Recent reports have summarized specific life processes mediated by SRB (Postgate, 1984; Fauque *et al.*, 1991; Odom and Singleton, 1993; Barton, 1993; Peck and LeGall, 1994; Widdel and Hansen, 1992). This chapter provides an insight into the basic activities of the SRB, with special reference to biotechnology.

LARRY L. BARTON • Department of Biology, University of New Mexico, Albuquerque, New Mexico 87131. FRANCISCO A. TOMEI • Army Environmental Policy Institute, 1501 B Interstate Drive, Champaign, Illinois 61821.

*Sulfate-Reducing Bacteria*, edited by Larry L. Barton. Plenum Press, New York, 1995.

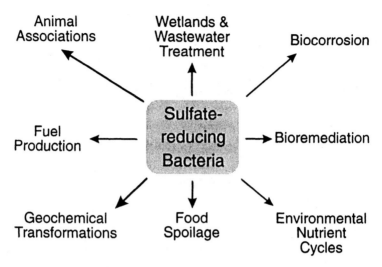

**Figure 1.** Interactions of the sulfate-reducing bacteria.

## 2. DIVERSITY OF BACTERIA

### 2.1. Biology

For many years it was thought that there were only a few species of SRB and that these bacteria used lactate or pyruvate exclusively to support growth. It is now apparent that SRB have considerable capability and diversity in using various compounds for electron donors. The list of substrates supporting growth is approaching one hundred (Fauque *et al.*, 1991; Hansen, 1993). In the absence of sulfate, certain strains of SRB can use a single carbon compound as both an electron donor and an electron acceptor by a process termed dismutation. Various *Desulfovibrio* strains grow with the dismutation of pyruvate, choline, malate, glycerol, and dihydroxyacetone. Recent observations reveal that some strains can grow through disproportionation of sulfite or thiosulfate by a mechanism where the sulfur oxyanions serve as both the electron donor and acceptor. While sulfate reduction was initially the only inorganic reaction linked to these bacteria, it is now apparent that SRB interact with many chemicals in their environment. Figure 2 illustrates some of the oxidation-reduction reactions attributed to various members of the SRB family.

### 2.2. Cell Characteristics

There are three basic cellular groups of SRB: gram-negative eubacteria, gram-positive eubacteria, and archaebacteria. General characteris-

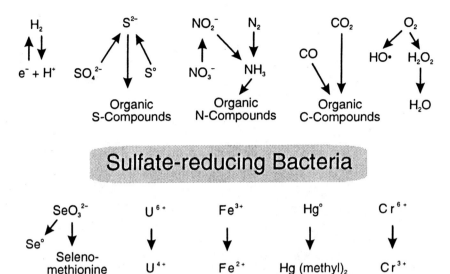

**Figure 2.** Chemical transformations attributed to sulfate-reducing bacteria.

tics of the genera in these three groups are given in Chapter 3. As can be seen in Fig. 3, the cellular characteristics of SRB resemble other bacteria. Morphological characteristics of the cells vary considerably, ranging from long rods to cocci. Bacteriophages have been isolated from *D. vulgaris* NCIB 8303 (Seyedirashti *et al.*, 1992), *D. desulfuricans* ATCC 13541 (Seyedirashti *et al.*, 1991), *D. desulfuricans* ATCC 27774 (Rapp and Wall, 1987), and *D. salexigens* NCIB 8308 (Kamimura and Araki, 1989).

## 3. GROWTH

### 3.1. Cultivation

Various media have been used to grow SRB; details of cultivation are given in the papers that report new isolates. A growth medium that has numerous applications is Postgate's medium "C," which has the following composition: $KH_2PO_4$, 0.5 g; $NH_4Cl$, 1.0 g; $Na_2SO_4$, 4.5 g; $CaCl_2.2H_2O$, 0.06 g; $MgSO_4.7H_2O$, 2.0 g; yeast extract, 1.0 g; $FeSO_4.7H_2O$, 0.004 g; sodium citrate, 0.3 g; and sodium lactate, 3.5 g (or appropriate electron donor) in 1 liter of water. The $E_h$ of the medium must be below $-150$ mV and the reducing environment may be obtained through the use of 1mM $Na_2S$ or 1 mM sodium thioglycolate plus 1mM sodium ascorbate. While most media are adjusted to pH 7.2

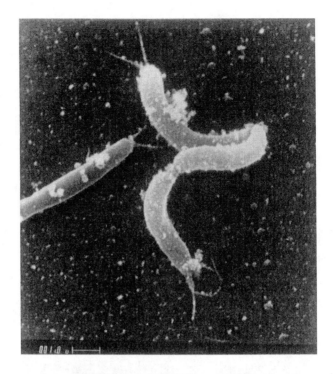

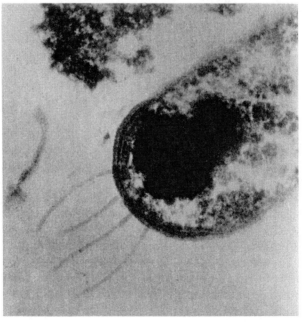

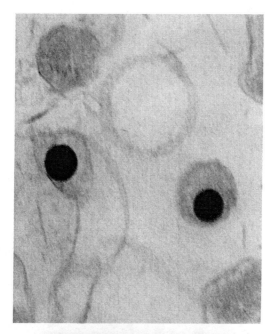

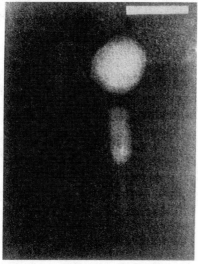

**Figure 3.** Features of bacteria and viruses as observed by electron microscopy. (A) Cells of *D. gigas* observed by scanning electron microscopy; bar = 1 micron (B) Thin section of *D. gigas* showing origin of the flagella; bar = 0.5 micron (C) Thin section of *D. desulfuricans* with internal granules of selenium; bar = 1 micron (D) Negative stain of *D. vulgaris* NCIMB 8303 virus with full head of DNA, bar = 50 nm. Photographs (A) and (B) were provided by J.J. Ruocco, (C) by L.L. Barton and F.A. Tomei, and (D) by E.J. Handley.

for growth, SRB are found in high acid environments of pH 4 (Postgate, 1984) and tolerate alkaline environments of pH 9.5 (Philp *et al.*, 1991). Excellent procedures for enrichment, isolation, and maintenance are available in recent reviews (Widdel and Pfennig, 1984; Campbell and Singleton, 1986; Widdel and Pfennig, 1992; Widdel and Bak, 1992; Widdel, 1992).

Metals are important as cofactors for several enzymes of SRB and thus metals must be present for growth. Nickel and selenium are required for hydrogenase activity (Table I). Calcium is important for increased efficiency in plating (Singleton *et al.*, 1988), which could be attributed to the Ca-binding protein employed in the electron transport system (Chen *et al.*, 1991). Perhaps it was the presence of calcium in Baar's medium that made it ideal for maintenance of SRB. Iron is an essential element for cytochrome production and hydrogenase production (Sadana and Rittenberg, 1964; Czechowski *et al.*, 1990; Bryant *et al.*, 1993).

With respect to nutrition, one of the more interesting features of SRB is the nature of nitrogen compounds used to support their growth. Several strains can fix dinitrogen (see Chapter 8) and use nitrogen from amino acids as nitrogen source. Few strains are capable of growth with nitrate as a final electron acceptor. *D. desulfuricans* 27774 uses nitrate (Chapter 8) and strains of *Desulfovibrio* reduce nitrate to ammonium (McCready *et al.*, 1983). Nitrate and nitrite reductases were not observed in a survey of *Desulfotomaculum* strains, but *Desulfovibrio* isolate FBA could be grown on nitrate (Mitchell *et al.*, 1986). For some time it was considered that nitrite reduction was mediated by the sulfite reductase; however, many *Desulfovibrio* strains possess a specific nitrite reductase that can be important in membrane mediated electron transport activity (Klemm and Barton, 1987; 1989; Barton *et al.*, 1983). Recently, a strain of *Desulfovibrio* was isolated that can use nitrate, nitrite, and 2,4,6-trinitrotoluene (TNT) as a nitrogen source and electron aceptor in the support of growth (Boopathy and Kulpa, 1992). It remains to be seen if future isolates of SRB will be able to use additional nitrogen compounds for respiration or a source of nitrogen.

## 3.2. Stress Response

Unlike many obligate anaerobic bacteria, SRB are capable of limited oxygen metabolism. Laboratory cultures adapted to anaerobic growth cannot tolerate exposure to air; however, isolation of SRB after exposure to oxygen has been accomplished (see Chapter 6). Perhaps tolerance to air can be attributed to the presence of catalase and superoxide dismutase. Both of these enzymes have been demonstrated in *D. vulgaris* (Hatchikian *et al.*, 1977). It will be important to determine the extent to which these enzymes are distributed in other bacterial strains and the nature of their regulation. An oxygen metabolizing enzyme—rubredoxin: oxygen oxidoreductase—has been isolated; it reduces oxygen to water in certain respiratory mechanisms (Chen *et al.*, 1993). Oxygen metabolism by SRB is covered further in Chapters 5 and 6.

While the appropriate chemical environment for the cultivation of SRB has been established, recent studies provide new information about chemical stress on these bacteria. A *Desulfovibrio* isolate from Husainagar Lake is more sensitive to Zn than aerobic bacteria, and Cd had a synergestic effect on Zn inhibition; however, toxicity of both Cd and Zn could be reduced by adjusting the concentrations of Cl, Ca, phosphate, Fe, and Mn in the medium (Radha and Seenayya, 1992). Molybdate is an effective inhibitor even at high phosphate concentrations, and transition metal cations selectively inhibited sulfate reduction in digesters (Clancy *et al.*, 1992). The mechanism of molybdate and selenate inhibition on sulfate metabolism has been discussed (Newport and Nedwell, 1988), and the general activities of Group VI ions have been reviewed (Oremland *et al.*, 1991). Hydrogen sulfide at elevated concentrations (16 mM $H_2S$) markedly inhibits growth of SRB and inhibitory kinetics have been described (Reis *et al.*, 1992). Other parameters of sulfate respiration include temperature and phosphorus concentrations, which have been studied in *D. desulfuricans* (Okabe and Characklis, 1992).

### 3.3. Identification

Conventional identification of isolated bacteria relies on the characteristics described in systematic reviews (Widdel and Pfennig, 1984; Campbell and Singleton, 1986). DNA or RNA probes are becoming useful (see Chapter 3) as well as immune-based reactions (Singleton, 1993). The cellular fatty acids of SRB have been studied, and the significant differences in lipid composition would suggest they could be used in identification (Vainshtein *et al.*, 1992).

### 3.4. Isolation

The standard procedure for isolating SRB has been to incorporate electron donor compounds into media containing sulfate. This has been used extensively when organic compounds are used to isolate bacteria with unique metabolic capability. To screen bacteria for electron donor and electron acceptor combinations, it would be useful to employ an automated assay system previously described by Weimer and Cooling (1991). To obtain a total count of SRB in littoral sediment of Lake Constance, Switzerland, media have been employed that contained $H_2$, lactate, acetate, propionate, butyrate, caprylate, succinate, and benzoate as electron donors (Bak and Pfennig, 1991). After six weeks of incubation, 87% of the cultures grew with $H_2$ as the electron donor, 12% oxidized propionate, 0.3% utilized long-chain fatty acids, and 0.05% utilized acetate. In another study *Desulfotomaculum* sp. were the predominant strains isolated from North Sea oil field waters when lactate, propionate, butyrate or C4-C6 fatty acids were used as carbon and energy sources (Rosnes *et al.*, 1991). An ingenious system was developed for the isolation of specific bacteria by coating immunomagnetic beads with polyclonal antibodies. When using cells of *Thermodesulfobacterium mobile* as the antigen, new isolates of *T. mobile* were captured from the North Sea (Christensen *et al.*, 1992). Since polyclonal antibodies lack absolute specificity for SRB, *Desulfotomaculum* isolates were captured with antibody against *T. mobile*.

### 3.5. Detection

The economic problems resulting from bacterially induced corrosion have stimulated the development of tests indicating the presence of SRB. One approach is the use of a simple, broad-range enzyme assay kit (Costerton *et al.*, 1988) to detect anaerobic microorganisms that cause metal corrosion. Another employs antibodies specific for APS reductase, the enzyme that is unique to SRB (Tatnall *et al.*, 1988). Both of these methods are either commercially available or in final stages of development.

In the area of ecology, the estimation of bacterial biomass in an environment has frequently relied on the uptake of radiolabeled thymidine. Information has been presented to indicate that *D. desulfuricans* and SRB isolated from mesohaline sediments of Chesapeake Bay incorporated about the same amount of $H^3$ as formalin-killed cells (Gilmour *et al.*, 1990). Thus, the use of thymidine uptake to estimate bacterial production in sediments and anoxic waters may provide an underestimation of bacterial population.

### 4. BIOCHEMICAL ACTIVITIES

#### 4.1. Enzymes

The intense study of SRB has resulted in the characterization of several enzymes. A selection of enzymes from the Eubacteria are listed in

Table I, while enzymes isolated or demonstrated in archaebacteria are given in Chapter 2. Enzymes associated with the TCA cycle and other intermediary metabolic processes have not been isolated but are reviewed by Hansen (1993). Clearly, enzymes of the sulfur pathway and electron transport have been extensively studied with in-depth discussions in Chapters 5 and 6, respectively. Molecular characterization of hydrogenase is central to the understanding of cell energetics, and for an excellent review, consult Chapter 7.

### 4.2. Immobilization of Enzymes

Biosensor development to measure $H_2/H^+$ activity requires immobilization of hydrogenase on an appropriate electrode. Several systems have provided considerable information on the immobilization of materials from *Desulfovibrio* sp. (Table II). With immobilized hydrogenase from *D. desulfuricans* strain Norway, responses varied with cytochrome $c_3$ from different bacteria; however, cytochrome $c_3$ from *D. gigas* provided the lowest level of activity (Draoui et al., 1991). In another study, hydrogenase immobilized to a glass electrode surface coupled with methyl viologen or soluble cytochrome $c_3$; however, no stable electrochemical response resulted with cytochrome $c_3$ immobilized on the electrode (Bianco and Haladjian, 1991). This absence of activity with bound cytochrome may reflect denaturation of heme groups of cytochrome $c_3$ in close contact with the electrode surface (Haladjian et al., 1991). In general, hydrogenase activity in cells of immobilized *D. desulfuricans* and *D. vulgaris* was less sensitive to oxygen or temperature denaturation than soluble enzymes.

### 4.3. Metabolic Products

The principle end products of SRB are $H_2S$ and $CO_2$ at a ratio of 1:2, respectively, in cultures where acetate accumulates from lactate catabolism. Several other compounds are produced by these bacteria under special conditions. $H_2$ evolution by *D. vulgaris* NCIB 8303 occurred in all growth phases, with an average amount of 10–15% of total gases while $H_2$ production by *D. gigas* was observed only after maximum growth had occurred (Hatchikian et al., 1976). *D. africanus* did not produce $H_2$ at any phase of growth. During the first few hours of growth, a burst of $H_2$ production has been reported for *D. vulgaris* Miyazaki where two different hydrogenases were implicated in this evolution of $H_2$ (Tsuji and Yagi, 1980). $H_2$ may be produced from the phosphoroclastic reaction and this is discussed elsewhere (Barton, 1994).

**Table I. Selected Enzymes Isolated From Sulfate-Reducing Eubacteria**

| Enzyme | Characteristic[a] | Location[b] | Bacterial strain[c] | Reference |
|---|---|---|---|---|
| *Oxidases/Dehydrogenases* | | | | |
| Aldehyde oxidoreductase | 120 kDa | s | *D. gigas* | Turner *et al.*, 1987 |
| Carbon monoxide dehydrogenase | 180 kDa | s | *D. desulfuricans* | Meyer and Fiebig, 1985 |
| Catalase | 232 kDa | s | *D. vulgaris* NCIB 8303 | Hatchikian *et al.*, 1977 |
| Formate dehydrogenase | p | s | *D. vulgaris* | Yagi, 1969 |
| | 240 kDa | p | *D. gigas* | Riederer-Henderson and Peck, 1986 |
| Fumarate reductase | 134 kDa | m | *D. multispirans* | He *et al.*, 1986 |
| D-Lactate dehydrogenase | p | s | *D. vulgaris* (Miyazaki) | Ogata *et al.*, 1981 |
| | p | m | *D. desulfuricans* | Czechowski and Rossmoore, 1980 |
| | d | m | *D. desulfuricans* | Steenkamp and Peck, 1980 |
| L-Lactate dehydrogenase | d | m | *D. gigas* | Barton and Peck, 1971 |
| | d | m | *D. desulfuricans* | Stams and Hansen, 1982 |
| Malate:NADP oxidoreductase | p | s | *D. gigas* | Hatchikian and Le Gall, 1970 |
| NAD:rubredoxin oxidoreductase | p | s | *D. gigas* | LeGall, 1968 |
| NAD(P)H₂:menadione oxidoreductase | d | s | *D. gigas* | Hatchikian, 1970 |
| Protoporphyrinogen oxidase | 148 kDa | m | *D. gigas* | Klemm and Barton, 1987 |
| Pyruvate:ferredoxin oxidoreductase | d | s | *D. vulgaris* NCIB 8303 | Akagi, 1967 |
| Rubredoxin:oxygen oxidoreductase | 43 kDa | s | *D. gigas* | Chen *et al.*, 1993 |
| Superoxide dismutase | 21.5 kDa | s | *D. vulgaris* NCIB 8303 | Hatchikian *et al.*, 1977 |

| | | | | |
|---|---|---|---|---|
| *Amino acid metabolism* | | | | |
| L-Alanine dehydrogenase | p | s | *D. desulfuricans* ATCC 7757 | Germano and Anderson, 1968 |
| β-Aspartate decarboxylase | p | s | *D. desulfuricans* | Cattanéo-Lacombe *et al.*, 1958 |
| Cysteine synthase | d | s | *D. vulgaris* NCIB 8303 | Gevertz *et al.*, 1980 |
| Serine transacetylase | d | s | *D. vulgaris* NCIB 8303 | Gevertz *et al.*, 1980 |
| L-delta-1-pyroline-5-carboxylate reductase | d | s | *D. desulfuricans* (Norway) | Fons *et al.*, 1991 |
| *Sulfur metabolism* | | | | |
| Adenosine-5'-phosphosulfate (APS) reductase | 220 kDa | s | *D. vulgaris* | Bramlett and Peck, 1975 |
| | 175 kDa | s | *Desulfobulbus propionicus* | Stille and Trüper, 1984 |
| Adenosine triphosphate sulfurylase | p | s | *D. vulgaris* NCIB 8303 | Akagi and Campbell, 1962 |
| | p | s | *Dm. nigrificans* strain 8351 | Akagi and Campbell, 1962 |
| Bisulfite reductase (desulfoviridin) | 200 kDA | s | *D. vulgaris* | Le Gall and Postgate, 1973 |
| Bisulfite reductase (desulforubidin) | 200 kDa | s | *D. gigas* | Lee and Peck, 1971b |
| | 225 kDa | s | *D. desulfuricans* (Norway) | Lee *et al.*, 1973 |
| Bisulfite reductase (p582) | 145 kDa | s | *Dm. nigrificans* NTCC 8351 | Trudinger, 1970 |
| Rhodanese | d | s | *Dm. nigrificans* strain 8351 | Burton and Akagi, 1971 |
| Sulfite reductase | 26.8 kDa | s | *D. vulgaris* | Lee *et al.*, 1971a |
| Sulfite reductase | 40 kDa | s | *D. vulgaris* | Haschke and Campbell, 1968 |
| Trithionate reductase | d | s | *D. vulgaris* | Drake and Akagi, 1977 |
| Thiosulfate-forming enzyme | d | s | *D. vulgaris* NCIB 8303 | Drake and Akagi, 1977 |

*(Continued)*

**Table I.** (*Continued*)

| Enzyme | Characteristic[a] | Location[b] | Bacterial strain[c] | Reference |
|---|---|---|---|---|
| Thiosulfate reductase | 220 kDa | s | *D. gigas* | Hatchikian, 1975 |
| | 16 kDa | s | *D. vulgaris* NCIB 8303 | Haschke and Campbell, 1971 |
| | p | s | *Dm. nigrificans* strain 8351 | Nakatsukasa and Akagi, 1969 |
| *Nitrogen metabolism* | | | | |
| Nitrite reductase | 66 kDa | m | *D. desulfuricans* ATCC 27774 | Liu and Peck, 1981 |
| Nitrogenase | d | m | *D. gigas* | Barton *et al.*, 1983 |
| | d | s | *Desulfovibrio* strains | Chapter 8, this volume |
| *Hydrogen metabolism* | | | | |
| Hydrogenase | 82.5 kDa [NiFe] | c | *D. multispirans* NCIB 12078 | Czechowski *et al.*, 1984 |
| | 56.6 kDa [Fe] | p | *D. vulgaris* NCIB 8303 | Fauque *et al.*, 1988 |
| | 98 kDa [NiFeSe] | p | *D. salexigens* NCIB 8403 | Teixeira *et al.*, 1986 |
| | 89.5 kDa [NiFe] | p | *D. gigas* | Fauque *et al.*, 1988 |
| | 58 kDa [NiFeSe] | m | *D. baculatus* NCIB 8310 | Fauque *et al.*, 1988 |
| | 85 kDa [NiFeSe] | p | *D. baculatus* DSM 8310 | Fauque *et al.*, 1988 |
| | 53.5 kDa [Fe] | p | *D. desulfuricans* ATCC 7757 | Hatchikian *et al.*, 1992 |
| | 89 kDa [Fe] | m | *D. vulgaris* Miyazaki F | Fauque *et al.*, 1988 |
| | [NiFe] | s | *D. africanus* NCIB 8401 | Niviere *et al.*, 1986 |

|  |  |  | Organism | Reference |
|---|---|---|---|---|
|  | d [NiFe] | m | *D. vulgaris* NCIB 8303 | Lissolo *et al.*, 1986 |
|  | 60 kDa | s | *D. desulfuricans* | Yagi *et al.*, 1968 |
|  | p | p | *D. vulgaris* (Miyazaki) | Yagi, 1970 |
| *Nucleic acid metabolism* |  |  |  |  |
| Adenine nucleotide deaminase | p | s | *D. desulfuricans* NCIB 8388 | Yates, 1969 |
| Apurinic endodeoxy-ribonuclease | 18.5 kDa | s | *Dm. nigrificans* IFO 13698 | Sako *et al.*, 1984 |
| Restriction endonuclease | 68.5 kDa | s | *D. desulfuricans* (Norway) | Makula and Meagher, 1980 |
| *Phosphate metabolism* |  |  |  |  |
| Adenosine triphosphatase | d | m | *D. gigas* | Guarraia and Peck, 1971 |
| Inorganic pyrophosphatase | 41.6 kDa | s | *D. desulfuricans* NCIB 8388 | Ware and Postgate, 1971 |
| Pyrophosphatase (rubrerythrin) | 43.8 kDa | s | *D. vulgaris* NCIB 8303 | Vanbeeumen *et al.*, 1991 |

[a] p = purified, d = demonstrated.
[b] s = soluble, p = periplasmic, m = membrane.
[c] *D. multispirans* is now reclassified as *D. desulfuricans* strain multispirans.

**Table II.  Immobilized Enzymes and Cells From Sulfate-Reducing Bacteria**

| Structure | System | Reference |
|---|---|---|
| Cytochrome $c_3$ and hydrogenase (*D. desulfuricans*) | Immobilized on glassy carbon electrodes with carbidiimide | Bianco and Haladjian, 1991 |
| Hydrogenase (*D. desulfuricans*) | Modified pyrolytic graphite electrode | Draoui *et al.*, 1991 |
| Hydrogenase (*D. gigas*) | Amphiphilic biolayer of octadecyltrichlorosilane and octadecylviologen on electrode surface | Parpaleix *et al.*, 1992 |
| Hydrogenase (*D. gigas*) | Glutaraldehyde activated porous silica support | Hatchikian and Monsan, 1980 |
| Hydrogenase (*D. vulgaris*) | Mediated electron transfer with di (n-aminopropyl)-viologen modified glassy carbon electrodes | Hooguliet *et al.*, 1991 |
| Hydrogenase (*D. vulgaris*) | Immobilized in polyvinyl alcohol film | Yagi, 1979 |
| Hydrogenase and cells (*D. desulfuricans*) | Immobilized in radiation polymerized acrylamides | Ziomek *et al.*, 1984 |
| Cells with hydrogenase (*D. vulgaris*) | Immobilized in calcium alginate | Barreto and Cabral, 1991 |

Marine and freshwater strains of SRB produce traces of methane when grown under conditions of pyruvate dismutation (Postgate, 1969). Recently, it has been shown that methane can be produced by *D. desulfuricans* from the metabolism of methylmercury (Baldi *et al.*, 1993). Propionate is produced by *Desulfobulbus propionicus* (Stams *et al.*, 1984; Tasaki *et al.*, 1992). Three strains of *Desulfovibrio* oxidize benzaldehydes to benzoate derivatives (Zellner *et al.*, 1990). With cyanide to inhibit acetyl-CoA oxidation, cells of *Desulfobacterium anilini* produced benzoate from aniline or 4-aminobenzoate (Schnell and Schink, 1991). Although these products may not be important at the current levels of production, advances in molecular biology could prove valuable in the future.

A novel metabolite produced by *D. desulfuricans* was 3-methyl -1,2,3,4-tetrahydroxybutane-1,3-cyclic bisphosphate (Turner *et al.*, 1992). *Desulfotomaculum nigrificans* produces tocopherol as an end product of metabolism (Cinquina, 1968). Internal granules of polyphosphate have been reported to be produced by *D. gigas* (Jones and Chambers, 1975) and β-hydroxy butyrate granules produced by *D. sapovorans* (Nanninga and Gottschal, 1987); granules of elemental selenium have been identified inside cells of *D. desulfuricans* growing in selenate or selenite media (see Fig. 3 and Tomei and Barton, unpublished reports), and polyglucose compounds are produced by several strains of *Desulfovibrio* (Hansen, 1993). Extracellular polymeric substances containing hexoses have been reported for *D. desulfuricans* (Beech *et al.*, 1991) and may be similar to the mannose-rich material reported earlier for the same species (Ochynski and Postgate, 1963).

## 4.4. Transformation Reactions

### 4.4.1. Methylation of Mercury

*D. desulfuricans* was demonstrated by Berman *et al.* (1990) to transform mercury to methylmercury, and this methylation is mediated by cobalamin ($B_{12}$) (Choi and Bartha, 1993). During fermentative growth, *D. desulfuricans* displayed tolerance to mercury with a high level of methylation; however, when cell growth was attributed to sulfate respiration the rate of mercury methylation was low. It will be interesting to learn if SRB can also methylate other heavy metals.

### 4.4.2. Sulfur Compounds

In certain settings, there may be an application to the use of sulfate metabolizing bacteria. The sulfate content in acid mine drainage can be reduced by SRB (Herlihy and Mills, 1985). Sulfate in lignosulfonate was used as an electron acceptor and supported growth of *D. desulfuricans* (Ziomek and Williams, 1989). Following bacterial metabolism of lignin, the polyphenolic backbone and the functional groups of lignin were affected. Additionally, *D. desulfuricans* promoted a twofold increase of sulfur content in Kraft lignin but not lignosulfonate. After bacterial treatment, the metal binding capacity of both Kraft lignin and lignosulfonate was markedly increased.

The utilization of $SO_2$ from flue gas desulfurization process by *D. desulfuricans* appears feasible (Lee and Sublette, 1991) and has economic potential (Sublette and Gwozdz, 1991). *Desulfotomaculum orientis* can be grown on $H_2$, $CO_2$, and $SO_2$ with the production of $H_2S$, while *D. desulfuricans* transforms $SO_2$ to $H_2S$, using minerals and pretreated sewage sludge (Deshmane *et al.*, 1993). Postgate (1965) has reviewed the British project for sulfide production from sewage sludge and the formation of natural sulfur from gypsum. Processes involving production of elemental sulfur may be important when the current supply of native sulfur becomes depleted.

### 4.4.3. Biogenic Metal Sulfides

The production of heavy metal sulfides can result from a chemical reaction between the hydrogen sulfide produced by SRB and a heavy metal cation. With hydrogen sulfide generated by batch cultures of *Desulfovibrio*, biogenic production of lead sulfide (galena), zinc sulfide (wurtzite), antimony sulfide (stibnite), bismuth sulfide (bismuthinite), cobalt sulfide and nickel sulfide (haeszlewoodite) have been demonstrated (Miller, 1950). Mercury sulfide (cinnabar) and copper sulfide (covellite) have been produced by continuous culture of *D. desulfuricans* (Vosjan and van der Hoek, 1972). SRB have been suggested to be involved in the production of hydrotroilite (Berner, 1962), a precursor of iron pyrite, and in special cases in the formation of magnetic iron sulfide (Freke and Tate, 1961).

Because the solubility of these metal sulfides ranges from $10^{-4}$ to $10^{-6}$ grams/100 ml, the removal of toxic heavy metals from solution represents a detoxifying activity. In the use of wetlands systems (bogs) to bioremediate water containing toxic heavy metals, hydrogen sulfide production by SRB can precipitate soluble metal cations. Wetlands and ponds have been reported to remove toxic metals from acid mine drainage in Colorado, a uranium mine in New Mexico, and a lead mine in Missouri (Brierley *et al.*, 1989). When mine and smelter wastes collected in a Canadian lake, biogenic formation of ZnS, CdS, CuS, and FeS by action of SRB in the lake sediments was observed (Jackson, 1978). Nickel removal from mine wastes by SRB has been reported (Hammack and Edenborn, 1992), and bacteria isolated from paddy fields readily precipitate Cu from solution (Panchanadikar and Kar, 1993).

Removal of heavy metals in wastewaters has been studied in pilot-scale reactors (Dvorak *et al.*, 1992). A reactor installed at Budelco's zinc refinery at Budel-Dorplein in the Netherlands removed heavy metals in underlying groundwater (Barnes *et al.*, 1992). Heavy metal sulfides from industrial wastewaters produce a sludge that may be immobilized when mixed with fly-ash, gypsum, or bitumen (Wasay and Das, 1993) prior to recovery of metals and sulfur in a metal refinery process. The biogenic metal sulfides could lead to harvesting of certain metals; however, to become economically feasible new systems must be designed for the removal of these metal sulfide ores from the environment.

### 4.4.4. Reduction by $H_2S$

Hydrogen sulfide is a strong reducing agent and oxyanions may react chemically with it. The Otago Harbor in New Zealand received effluent from a tannery where Cr(VI) was reduced by $H_2S$ produced from SRB, yielding $Cr^{3+}$ (Smillie *et al.*, 1981). Additionally, selenite will react with $H_2S$ to produce elemental selenium. Because $Cr^{3+}$ and $Se^0$ are less toxic than their oxidized counterparts, the reduction by $H_2S$ would be a detoxification system. Decomposition of methylmercury by cultures consisting of SRB and methanogens was reported by Oremland *et al.* (1991). In a controlled experiment by Baldi *et al.* (1993), it was demonstrated that dimethylmercury sulfide was formed as a result of $H_2S$ reaction with methylmercury. Dimethylmercury sulfide is spontaneously converted under anaerobic conditions to metacinnabar with the release of dimethylmercury and methane (Baldi *et al.*, 1993). The use of biogenic hydrogen sulfide production to detoxify metals in aquatic sediments should be used with caution because $H_2S$ levels can exceed approved environmental standards.

### 4.4.5. Reduction of Selenate and Selenite

Cultures of *Desulfovibrio* will reduce selenite and selenate. Selenate reduction to hydrogen selenide has been attributed to respiratory action (Zehr and Oremland, 1987), while the formation of elemental selenium from selenite has been attributed to respiratory activity (Oremland *et al.*, 1989) independent of sulfate-respiration (Oremland *et al.*, 1989) or metabolic reactions in *D. desulfuricans* (Tomei and Barton, unpublished results). The application of SRB to detoxify wastewaters from ore mines containing selenium oxy-anions has been the subject of a recent patent (Baldwin *et al.*, 1985). It is curious that SRB do not reduce elemental selenium to hydrogen selenide, even though many of these bacteria can reduce elemental sulfur to hydrogen sulfide.

### 4.4.6. Uranium and Iron Reduction

Several species of SRB have been shown to reduce Fe(III) to magnitite ($Fe_3O_4$) or siderite ($FeCO_3$) and to reduce U(VI) to uraninite ($UO_2$) (Lovley *et al.*, 1993b). SRB cannot grow with the reduction of these two metals, because there appear to be no mechanisms of energy conservation. In *D. vulgaris* NCIB 8303, uranium-reducing activity is mediated by cytochrome $c_3$ and another c-type cytochrome (Lovley *et al.*, 1993c). $H_2$ oxidation catalyzed by hydrogenase is coupled to uranium reduction in the presence of $c_3$. This newly acquired knowledge of the enzymatic reduction of uranium can lead to new bioremediation techniques (Lovley and Phillips, 1992a,b). Uranyl carbonate, the prevalent form in oxic groundwaters, can be readily removed by reduction and precipitation to the insoluble uraninite form.

Growth with uranium and iron as electron acceptors has been demonstrated in the newly identified bacterium, *Geobacter metallireducens* (Lovley *et al.*, 1993a). This bacterium is taxonomically related to *Desulfuromonas acetoxidans*, a bacterium that uses elemental sulfur but not sulfate as the terminal electron acceptor. *Desulfuromonas acetoxidans* has been shown to utilize Fe(III) as the terminal electron acceptor, with the production of magnetite and siderite as products, as in the case of SRB (Roden and Lovley, 1993). Sulfate reducers can also deposit internal iron granules when cultured in the laboratory (Tomei and Barton, unpublished results). The exact chemical nature of these granules is not known, but magnetite and phosphate precipitates can be discounted.

### 4.5. Energy Technology

### 4.5.1. Hydrogen Production by Biophotolysis

The production of molecular hydrogen has been demonstrated in an *in vitro* biophotolysis system containing spinach chloroplasts, bacterial hydrogenase and bacterial ferredoxin. In this system, chlorophyll absorbs the photons, and the Mn-enzyme of photosystem II in the chloroplast catalyzes the photolysis of water to molecular oxygen plus protons and electrons (Fig. 4). Ferredoxin transfers electrons to hydrogenase

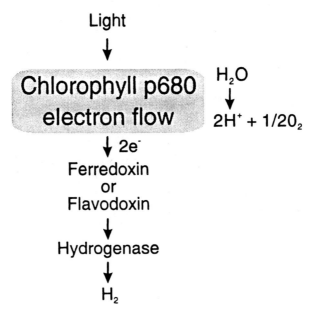

**Figure 4.** Biophotolysis of water with the production of molecular hydrogen.

where the electrons and protons are united to produce molecular hydrogen. The rate and amount of hydrogen generated by this system has received considerable attention (Rao *et al.*, 1976; Fry *et al.*, 1977). A difficulty with the *in vitro* system is that bacterial hydrogenases and ferredoxins are sensitive to molecular oxygen, and generation of $O_2$ in the biophotolysis reaction would not exceed 50 micromoles of $H_2$/mg chlorophyll/hr. Kinetic analysis of the $H_2$ production from a system consisting of hydrogenase from *D. desulfuricans* and cytochrome $c_3$ revealed that efficiency varies with the use of proflavin, 5-deazariboflavin or Zn-tetraphenyl porphyrin sulfonate as mediators (Eng *et al.*, 1993). While the hydrogenase from *D. desulfuricans* exhibits remarkable oxygen and thermal stability when cell-associated (Barreto and Cabral, 1991), oxygen sensitivity of hydrogenase from most sulfate reducers limits its usefulness in biophotolysis reactions. Of great potential is the use of flavodoxin from *D. desulfuricans* NCIB 27774 with hydrogenase from *Nocardia opaca* lb (Lissolo *et al.*, 1983).

### 4.5.2. Oil and Tar Sands

The injection of SRB into oil wells has been used in an attempt to increase oil production (Beck, 1947; Updegraff and Wren, 1954; La RiviEre, 1955; ZoBell, 1957). Since the production of surface active substances is important in release of oil from oil-containing sands, it would appear important to assess the activity of "mucin" secreted by *D. desulfuricans* (Ochynski and Postgate, 1963) on secondary oil recovery. Production of extracellular polysaccharides by *D. desulfuricans* has been stimulated in cultures forming biofilms (Beech *et al.*, 1991). Provocative reports suggest that SRB are involved in oil formation (Jankowski and ZoBell, 1944; Novelli and ZoBell, 1944; Voroshilov and Dianova, 1950). Recently it was reported that *D. desulfuricans* strain 2198 produced aliphatic hydrocarbons with a carbon chain length of 14-25 (Belyaeva *et al.*, 1992). Also, evidence has been produced to suggest that SRB contributed to the formation of the PR Spring tar and deposit in Utah (Mason and Kirchner, 1992).

### 4.5.3. Waste Treatment and Commercial Processes

The treatment of commercial effluents frequently results in a system that encourages the growth of SRB. SRB have been used to treat wastes from distilleries in India and Czechoslovakia (Postgate, 1965), sour whey (Zellner *et al.*, 1989), a starch producing factory (Nanninga and Gottschal, 1987), fish canning industries (Sato *et al.*, 1991), methanol wastes (Yamaguchi *et al.*, 1991), furfural containing wastewater (Boopathy and Daniels, 1991) and molasses wastewater (Boopathy and Tilche, 1992). An advantage in treatment of industrial wastes with SRB is that the resulting sludge is lower in water content than sludge derived from methane fermentation.

Several reports enumerate problems in commercial processing that have been attributed to SRB. In several instances, the paper industry has reported product modification by the SRB (Russel, 1961; Postgate, 1965). Lignin modification has been explored in a recent publication (Ziomek and Williams, 1989). The deterioration of cutting emulsions (Bennett, 1957) and the spoilage of olive brines (Levin *et al.*, 1959), canned corn (Werkman and Weaver, 1927), and various canned vegetables (Werkman, 1929) have been attributed to mesophilic or thermophilic SRB. *Desulfotomaculum* continue to be of considerable importance in food processing (Doores, 1983).

SRB have an important role in biocorrosion of iron (see Chapter 9), various metal alloys (Little *et al.*, 1991) and concrete (Mori *et al.*, 1991). As discussed in Chapter 10, control of these biocorrosion activities is a problem of considerable magnitude.

## 5. INTERACTIONS WITH PLANTS AND ANIMALS

Interactions between *Oscillotoria* sp., *Nostoc* sp. and *D. desulfuricans* in high-sand content golf greens results in the formation of black subsurface globules (Hodges 1992a). This growth of cyanobacteria and SRB has been implicated in promoting the pathogenicity of *Pythium torulosum* on the roots of creeping bent grass (Hodges, 1992b).

In addition to the demonstration of SRB in the rumen of cattle, these bacteria also are associated with many animals, including humans (Table III). Fecal samples from human populations in the United Kingdom and rural South Africa were examined: 70% of the people from the UK and 15% of the South Africans were found to have feces containing SRB. Isolates were of the genera *Desulfovibrio, Desulfotomaculum, Desulfobacter, Desulfomonas,* and *Desulfobulbus* (Gibson *et al.*, 1988). In a study of sudden-death victims, SRB were present in the ascending and traverse colons, with an average of $10^9$ viable cells/g dry weight in the colon contents (Macfarlane *et al.*, 1992). Mucin, a sulfated polysaccharide, can serve as the source of sulfate for the bacteria in the human large gut (Gibson *et al.*, 1988); the release of sulfate from mucin for the growth of SRB is attributed to enzymes from *Bacteroides* residing in the gut (Tsai *et al.*, 1992). Diet strongly influences the type of bacteria in the human large intestine: when sulfate is available SRB are abundant,

**Table III. Association of Sulfate-Reducing Bacteria with Animals**

| Source | Genus present | Reference |
|---|---|---|
| Human colon (healthy and with ulcerative colitis) | *Desulfovibrio* *Desulfotomaculum* *Desulfobulbus* *Desulfomonas* *Desulfobacter* | Gibson *et al.*, 1991 |
| Human colon | *Desulfomonas pigra* | Widdal and Pfennig, 1984 |
| Human bacteremia | *Desulfovibrio* sp. | Widdal and Pfennig, 1984 |
| Sheep rumen fluid | *Desulfotomaculum ruminis* | Widdal and Pfennig, 1984 |
| Bovine rumen fluid | *Desulfobulbus propionicus* | Widdal and Pfennig, 1984 |
| Rumen fluid, feces of higher animals | *Desulfotomaculum acetoxidans* | Campbell and Singleton, 1986 |
| Termite hindgut | *D. termitidis* | Trinkerl *et al.*, 1990 |
| Termite gut | Isolate STp (*D. desulfuricans*) Isolate STg (*D. giganteus*) | Bauman *et al.*, 1990 |
| Ectosymbiont on marine anaerobic ciliates | Undetermined | Fenchel and Ramsing, 1992 |
| Epizoic on clam shells | Undetermined | Bussmann and Reichardt, 1991 |

but when sulfate is limiting, methanogens are abundant (Christl *et al.*, 1992). Species of bacteria belonging to the genera of *Desulfovibrio* and *Desulfobulbus* account for consumption of $H_2$ in the intestine and represent 66% and 16%, respectively, of all colony-producing SRB (Gibson *et al.*, 1993).

The presence of SRB in gut contents of healthy persons and patients with ulcerative colitis has been examined (Gibson *et al.*, 1991). It was learned that about 50% of people have significant numbers of SRB in their feces and up to 92% of the isolates were of the genus *Desulfovibrio*. Notably, Gibson and colleagues observed that the production of $H_2S$ was greater in patients with ulcerative colitis than in healthy individuals.

Two *Desulfovibrio* species were isolated from the soil-feeding termite, *Cubitermes speciosus,* but their contribution to the biology of the termite is unresolved (Bauman *et al.*, 1990). Through the use of dye-conjugated probes to 16S ribosomal RNA, SRB were detected as ectosymbiots on the surface of free-living anaerobic ciliates of the following species: *Metopus contortus, Caenomorpha levanderi,* and *Parablepharisma* sp. (Fenchel and Ramsing, 1992). Bacteria that would react with the probe specific for *Desulfobacter* were not present. Shells of the clam *Artica islandica* from sediments were colonized by epizoic SRB (Bussmann and Reichardt, 1991). These clams can survive anoxia and it was suggested that bound bacteria had high biogeochemical activity.

## 6. PERSPECTIVES

Bacteria displaying sulfate respiration are a highly diverse group with broad metabolic capabilities. The presence of SRB throughout our environment indicates a genetic reservoir of considerable size to account for successful competition and cooperative growth with other organisms. Rather than considering that the SRB would have strange metabolic details because of their unique respiratory habits, one should consider that the genes for sulfate respiration may have been broadly distributed thoughout bacteria. Thus, the nuances of metabolic regulation and molecular biology that function in aerobic bacteria may be expected to be present in these sulfate-respiring bacteria.

Numerous biotechnology applications involving SRB (Chapter 11) are either characterized or may be expected to be developed, just as these exist for aerobic or fermentative bacteria. While the primary focus to date has been on fresh water eubacterial isolates of SRB, the research has only started with archaebacterial types. Clearly, the SRB are a more important life form than Winogradsky and other early workers could have ever envisioned.

## REFERENCES

Akagi, J. M., 1967, Electron carriers for the phosphoroclastic reaction of *Desulfovibrio desulfuricans, J. Biol. Chem.* **242:**2478–2483.
Akagi, J. M., and Campbell, L. L., 1962, Studies on thermophilic sulfate-reducing bacteria III. Adenosine triphosphate sulfurylase of *Clostridium nigrificans* and *Desulfovibrio desulfuricans, J. Bacteriol.* **84:**1194–1201.
Bak, F., and Pfennig, N., 1991, Sulfate-reducing bacteria in littoral sediment of Lake Constance, *FEMS Microbiol. Ecol.* **85:**43–52.
Baldi, F., Pepi, M., and Filippelli, M., 1993, Methymercury resistance in *Desulfovibrio desulfuricans* strains in relation to methylmercury degradation, *Appl. Environ. Microbiol.* **59:**2479–2485.
Baldwin, R. A., Sauter, J. C., Kaufman, J. W., and Laughlin, W. C., 1985, in: United States Patent number, 4,519,913.
Barnes, L. J., Janssen, F. J., Scheeren, P. J. H., Versteegh, J. H., and Koch, R. O., 1992, Simultaneous microbial removal of sulfate and heavy metals from waste water, *Trans-*

actions of the Institution of Mining and Metallurgy Section C Mineral Processing and Extractive Metallurgy, **101**:183–189.

Barreto M. C., and Cabral, J. M. S., 1991, Immobilization of *Desulfovibrio vulgaris* with hydrogenase activity, *J. Chem. Technol. Biotechnol.* **50**:563–570.

Barton, L. L., 1994, The pyruvic acid phosphoroclastic reaction in, *Inorganic Microbiol Sulfur Metabolism, Methods in Enzymology* (H. D. Peck, Jr., and J. Le Gall, eds), Academic Press, San Diego, In Press.

Barton, L. L. 1993, Sulfur metabolism, in: *The Encyclopedia of Microbiology*, Vol. 4 (J. Lederberg, ed.), Academic Press, San Diego, pp. 135–150.

Barton, L. L., and Peck, H. D., Jr., 1971, Phosphorylation coupled to electron transfer between lactate and fumarate in cell-free extracts of the sulfate-reducing anaeobe, *Desulfovibrio gigas, Bacteriol. Proc.*:149.

Barton, L. L., Odom, M., LeGall, J., and Peck, H. D., Jr., 1983, Energy coupling to nitrite respiration in the sulfate-reducing bacterium *Desulfovibrio gigas, J. Bacteriol.* **153**:867–871.

Bauman, A., Koenig, J. F., Dutreix, J., and Garcia, J. L., 1990, Characterization of 2 sulfate-reducing bacteria from the gut of the soil-feeding termite, *Curitermes speciosus, J. Gen. Mol. Biol.* **58**:271–275.

Beech, I. B., Gaylarde, C. C., Smith, J. J., and Geesey, G. G., 1991, Extracellular polysaccharides from *Desulfovibrio desulfuricans* and *Pseudomonas fluorescens* in the presence of mild and stainless steel, *Appl. Microbiol. Biotechnol.* **35**:65–71.

Beck, J. V., 1947, Penn grade progress on use of bacteria for releasing oil from sands, *Producers Monthly* **11**:13–19.

Belyaeva, M. I., Mukhitova, F. K., Zolotukhina, L. M., Kiyachko, S. V., Bagaeva, T. V., and Karpilova, I. Y., 1992, Extracellular metabolic products of sulfate-reducing bacteria of the genus *Desulfovibrio, Microbiol.* **61**:122–126.

Bennett, E. O., 1957, The role of sulfate-reducing bacteria in the deterioration of cutting emulsions, *Lubrication Engr.* **13**:215–219.

Berman, M., Chase, T., Jr., and Bartha, R., 1990, Carbon flow in mercury biomethylation by *Desulfovibrio desulfuricans, Appl. Environ. Microbiol.* **56**:298–300.

Berner, R. A., 1962, Experimental studies of the formation of sedimentary iron sulfides, in: *Biogeochemistry of Sulfur Isotopes* (M. L. Jensen, ed.) Yale University Press, New Haven, pp. 156–172.

Bianco, P., and Haladjian, J., 1991, Electrocatalysis at hydrogenase or cytochrome $c$-3 modified glassy electrodes, *Electroanalysis* **3**:973–977.

Boopathy, R., and Daniels, L., 1991, Isolation and characterization of a furfural degrading sulfate-reducing bacterium from an anaerobic digester, *Current Microbiol.* **23**:327–332.

Boopathy, R., and Tilche, A., 1992, Pelletization of biomass in a hybrid anaerobic baffled reactor (HABR) treating acidified wastewater, *Bioresource Technol.* **40**:101–107.

Boopathy, R., and Kulpa, C. F., 1992, Trinitrotoluene (TNT) as a sole nitrogen source for a sulfate-reducing bacterium *Desulfovibrio* sp. (B strain) isolated from an anaerobic digester, *Current Microbiol.* **25**:235–241.

Brierley, C. L., Brierley, J. A., and Davidson, M. S., 1989, Applied microbial processes for metal recovery and removal from wastewater, in: *Metal Ions and Bacteria* (T. J. Beveridge and R. J. Doyle, eds.) J. Wiley and Sons, New York, pp. 359–382.

Bramlett, R. N., and Peck, H. D., Jr., 1975, Some physical and kinetic properties of adenylyl sulfate reductase from *Desulfovibrio vulgaris J. Biol. Chem.* **250**:2979–2986.

Bryant, R. D., Van Ommen Kloeke, K. F., and Laishley, E. J., 1993, Regulation of the periplasmic Fe hydrogenase by ferrous iron in *Desulfovibrio vulgaris* (Hildenborough), *Appl. Environ. Microbiol.* **59**:491–595.

Burton, C. P., and Akagi, J. M., 1971, Observations of the rhodanese activity of *Desulfotomaculum nigrificans J. Bacteriol.* **107**:375–376.

Bussmann, I., and Reichardt, W., 1991, Sulfate-reducing bacteria in temporarily toxic sediments with bivalves, *Marine Ecol. Progress Series* **78**:97–102.

Campbell, L. L., and Singleton, R., Jr., 1986, Genus IV. *Desulfotomaculum*, in: *Bergey's Manual of Systematic Bacteriology* (P.H.A. Sneath, N. S. Mair, M. E. Sharpe and J. G. Holt, eds.) Williams & Wilkins, Baltimore, pp. 1200–1205.

CattanEo-Lacombe, Jr., Senez, J. C., and Beaumont, P., 1958, Sur la purification de la 4 aspartique d'Ecarboylase de *Desulfovibrio desulfuricans*, *Biochim. et Biophys. Acta* **30**:458–465.

Chen, L., Liu, M.-Y., and Le Gall, J., 1991, Calcium is required for the reduction of sulfite from hydrogen in a reconstituted electron transfer chain from the sulfate-reducing bacterium, *Desulfovibrio gigas*, *Biochem. Biophys. Res. Commun.* **180**:238–242.

Chen, L., Liu, M.-Y, Le Gall, J., Fareleira, P., Santos, H., and Xavier, A. V., 1993, Rubredoxin oxidase, a new flavo-hemo-protein, is the site of oxygen reduction to water by the "strict anaerobe" *Desulfovibrio gigas*, *Biochem. Biophys. Res. Commun.*

Choi, S. C., and Bartha, R., 1993, Cobalamin-mediated mercury methylation by *Desulfovibrio desulfuricans* LS, *Appl. Environ. Microbiol.* **59**:290–295.

Christensen, B., Torsvik, T., and Lien, T., 1992, Immunomagnetically captured thermophilic sulfate-reducing bacteria from the North Sea oil field waters, *Appl. Environ. Microbiol.* **58**:1244–1248.

Cinquina, C. L., 1968, Isolation of tocopherol from *Desulfotomaculum nigrificans, J. Bacteriol.* **95**:2436–2438.

Christl, S. U., Gibson, G. R., and Cummings, J. H., 1992, Role of dietary sulphate in the regulation of methanogenesis in the human large intestine, *Gut* **33**:1234–1238.

Clancy, P. B., Venkataraman, N., and Lynd, L. R., 1992, Biochemical inhibition of sulfate reduction in batch and continuous anaerobic digesters, *Water Sci. and Technol.* **25**:51–60.

Costerton, J. W., Boivin, J. W., Laishley, E. J., and Bryant, R. D., 1988, A new test for microbial corrosion, in: *The 6th Asian-Pacific Corrosion Control Conference*, Corrosion Association of Singapore Asian-Pacific Materials and Corrosion Association, Singapore, pp. 20–25.

Czechowski, M. H., Chatelus, C., Fauque, G., Libertcoquempot, M. F., Lespinat, P. A., Berlier, Y., and Le Gall, Jr., 1990, Utilization of cathodically-produced hydrogen from mild steel by *Desulfovibrio* species with different types of hydrogenases, *J. Industrial Microbiol.* **6**:227–233.

Czechowski, M. H., and Rossmoore, H. W., 1980, Factors affecting *Desulfovibrio desulfuricans* lactate dehydrogenase, *Dev. Ind. Microbiol.* **21**:349–356.

Czechowski, M. H., He, S. H., Nacro, M., DerVartanian, D. V., Peck, H. D., Jr. and Le Gall, J., 1984, A cytoplasmic nickel-iron hydrogenase with high specific activity from *Desulfovibrio multispirans* sp. n.: a new species of sulfate-reducing bacterium, *Biochim. Biophys. Res. Commun.* **125**:1025–1032.

Deshmane, V., Lee, C. M., and Sublette, K. L., 1993, Microbial reduction of sulfur dioxide with pretreated sewage sludge and elemental hydrogen as electron donors, *Appl. Biochem. Biotech.* **39**:739–752.

Doores, S., 1983, Bacterial spore resistance—species of emerging importance, *Food Technol.* **37**:127–134.

Draoui, K., Bianco, D., Haladjian, J., Guerlesquin, F., and Bruschi, M., 1991, Electrochemical investigation of intermolecular electron transfer between 2 physiological partners - cytochrome $c_3$ - and immobilized hydrogenase from *Desulfovibrio* desulfuricans Norway, *J. Electroanalytical Chem. and Interfacial Electrochem.* **313**:201–214.

Dvorak, D. H., Hedin, R. S., and Edenborn, H. M., 1992, Treatment of metal-containing water using bacterial sulfate reduction: results from pilot-scale reactors, *Biotechnol. Bioengr.* **40**:609–616.

Drake, H. L., and Akagi, J. M., 1977, Characterization of a novel thiosulfate-forming enzyme isolated from *Desulfovibrio vulgaris, J. Bacteriol.* **132**:132–138.

Eng, L. H., Lewin, M.B.M., and Neujahr, H. Y., 1993, Kinetic properties of the periplasmic hydrogenase from *Desulfovibrio desulfuricans* NCIMB 8372 and use in photosensitized $H_2$ production, *J. Chemical. Technol. and Biotechnol.* **56**:317–325.

Fauque, G., Le Gall, J., and Barton, L. L., 1991, Sulfate-reducing and sulfur-reducing bacteria, in: *Variations in Autotrophic Life* (J. M. Shively and L. L. Barton, eds.) Academic Press, London, pp. 271–338.

Fauque, G., Peck, H. D., Jr., Moura, J. J. G., Huynh, B. H., Berlier, Y., DerVartanian, D. V., Teixeira, M., Przybyla, A. E., Lespinat, P. A., Moura, I., and Le Gall, J., 1988, The three classes of hydrogenases from sulfate-reducing bacteria of the genus *Desulfovibrio, FEMS Microbiol. Rev.* **54**:299–344.

Fenchel, T., and Ramsing, N. B., 1992, Identification of sulfate-reducing ectosymbiotic bacteria from anaerobic ciliates using 16S binding oligonucleotide probes, *Arch. Microbiol.* **158**:394–397.

Fons, M., Cami, B., and Chippaux, M., 1991, Possible involvement of a L-delta-1-pyroline-5-carboxylate (P5C) reductase in the synthesis of proline in *Desulfovibrio desulfuricans* Norway, *Biochem. Biophys. Res. Commun.* **179**:1088–1094.

Freke, A. M., and Tate, D., 1961, The formation of magnetic iron sulphide by bacterial reduction of iron solutions. *J. Biochem. Microbiol. Technol. Engr.* **3**:29–39.

Fry, I., Papageorgiou, G., Tel-Or, E., and Packer, L., 1977, Reconstruction of a system for $H_2$ evolution with chloroplasts, ferredoaxin, and hydrogenase, *Z. Naturforsch* **32c**:110–117.

Gevertz, D., Amelunxen, R., and Akagi, J. M., 1980, Cysteine synthesis by *Desulfovibrio vulgaris* extracts, *J. Bacteriol.* **141**:1460–1462.

Germano, G. J., and Anderson, K. E., 1968, Purification and properties of L-alanine dehydrogenase from *Desulfovibrio desulfuricans, J. Bacteriol.* **96**:55–60.

Gibson, G. R., Cummings, J. H., and Macfarlane, G. T., 1988, Use of a three-stage continuous culture system to study the effect of mucin on dissimilatory sulfate reduction and methanogenesis by mixed populations of human gut bacteria. *Appl. Environ. Microbiol.* **54**:2750–2755.

Gibson, G. R., Cummings, J. H., Macfarlane, G. T., 1991, Growth and activities of sulphate-reducing bacteria in gut contents of healthy subjects and patients with ulcerative colitis. *FEMS Microbiol. Ecol.* **86**:1003–111.

Gibson, G. R., Macfarlane, G. T., and Cummings, J. H., 1993, Sulphate reducing bacteria and hydrogen metabolism in the human large intestine, *Gut* **34**:437–439.

Gibson, G. R., Macfarlane, G. T., and Cummings, J. H., 1988, Occurance of sulphate-reducing bacteria in human feces and the relationship of dissimilatory sulphate reduction to methanogenesis. *J. Appl. Bacteriol.* **65**:103–111.

Gilmour, C. C., Leavitt, M. E., and Shiaris, M. P., 1990, Evidence against incorporation of exogenous thymidine by sulfate-reducing bacteria, *Linmol. Oceanography* **35**:1401–1409.

Guarraia, L., and Peck, H. D., Jr., 1971, Dinitrophonol-stimulated adenosine triphosphatase activity in extracts of *Desulfovibrio gigas, J. Bacteriol.* **106**:890–895.

Haladjian, J., Draoui, K., Bianco, P., 1991. Electron transfer reaction of cytochrome c-3 at pyrolytic graphite electrodes, *Electrochimica Acta* **36**:1423–1426.

Hammack, R. W., and Edenborn, H. M., 1992, The removal of nickel from mine waters using bacterial sulfate reduction, *Appl. Microbiol. Biotechnol.* **37:**674–678.

Hansen, T. A., 1993, Carbon metabolism of sulfate-reducing bacteria, in: *The Sulfate-Reducing Bacteria: Contempary Perspectives* (J. M. Odom and R. Singleton, Jr., eds.) Springer-Verlag, Berlin, pp. 21–40.

Haschke, R. H., and Campbell, L. L., 1971, Thiosulfate reductase of *Desulfovibrio vulgaris, J. Bacteriol.* **106:**603–607.

Haschke, R., and Campbell, L. L., 1968, Partial resolution of the sulfite reducing system of *Desulfovibrio vulgaris, Fedn. Proc. Fedn. Am. Socs. Expl. Biol.* **27:**390.

Hatchikian, C. E., 1970, Menadione reductase from *Desulfovibrio gigas, Biochim. Biophys. Acta* **212:**353–355.

Hatchikian, E. C., 1975, Purifican and properties of thiosulfate reductase from *Desulfovibrio gigas, Arch. Microbiol.* **105:**249–256.

Hatchikian, E. C., Le Gall, J., and Bell, G. R., 1977, Significance of superoxide dismutase and catalase activities in the strict anaerobes, sulfate-reducing bacteria, *Superoxide and Superoxide Dismutases* (A. M. Michelson, J. M. McCard and I. Fridovich, eds.) Academic Press, London, pp. 159–172.

Hatchikian, E. C., Forget, N., Fernandez, V. M., Willliams, R., and Cammarack, R., 1992, Further characterization of the 'Fe' hydrogenase from *Desulfovibrio desulfuricans* ATCC 7757. *Eur. J. Biochem.* **209:**357–365.

Hatchikian, E. C., and Le Gall, J., 1970, Étude du Métabolisme des acides dicarboxyliques et du private chez les bactéries sulfato-réductases. 1. Étude de l'oxidation enzymatique du fumarate en acétate, *Annls. Inst. Pasteur,* Paris **118:**125–142.

Hatchikian, E. C., Chaigneau, M., and Le Gall, J., 1976, Analysis of gas production by growing cultures of three species of sulfate-reducing bacteria, in: *Microbial Production and Utilization of Gases* (H. G. Schlegel, G. Gottschalk and J. N. Pfenning, eds.) K. B. Goltze, Gottingen, pp. 389–402.

Hatchikian, E. C., and Monsan, P., 1980, Highly active immobilized hydrogenase from *Desulfovibrio gigas, Biochem. Biophys Res. Commun.* **92:**1091–1096.

He, S. H., DeVartanian, D. V., and Le Gall, J., 1986, Isolation of fumarate reductase from *Desulfovibrio multispirans* a sulfate- reducing bacteria, *Biochem. Biophys. Res. Commun.* **135:**1000–1007.

Herlihy, A. T., and Mills, A. L., 1985, Sulfate reduction in freshwater sediments receiving acid mine drainage, *Appl. Environ. Microbiol.* **49:**179–186.

Hodges, C. F., 1992a. Interaction of cyanobacteria and sulfate-reducing bacteria in subsurface black-layer forming in high sand content golf greens, *Soil Biol. Biochem.* **24:**15–20.

Hodges, C. F., 1992b. Pathogenicity of *Pythium torulosum* to roots of *Agrsotis palustrus* in black-layered sand produced by the interaction of the cyanobacterium species *Lyngbya, phormidium,* and *Nostoc* with *Desulfovibrio desulfuricans, Can. J. Botany,* **70:**2193–2197.

Hooguliet, J. C., Vanos, P. J. H. J., Vandermark, E. J., and VanBennekom, W. P., 1991, Modification of glassy carbon electrode surfaces with mediations and bridge molecules, *Biosensors and Bioelectronics,* **6:**413–423.

Jackson, T. A., 1978, The biochemistry of heavy metals in polluted lakes and streams at Fin Lion, Canada, and a proposed method for limiting heavy metal pollution of natural waters, *Inviron. Geol.* **2:**173–189.

Jankowski, G. J., and ZoBell, C. E., 1944, Hydrocarbon production by sulfate-reducing bacteria. *J. Bacteriol.* **47:**447.

Jones, H. E., and Chambers, L. A., 1975, Localization of intracellular polyphosphate formation by *Desulfovibrio gigas, J. General. Microbiol.* **89:**67–72.

Kamimura, K., and Araki, M., 1989, Isolation and characterization of a bacteriophage lytic for *Desulfovibrio salexigens*, a salt-requiring sulfate-reducing bacterium, *Appl. Environ. Microbiol.* **55**:645–648.

Klemm, D., and Barton, L. L., 1987, Purification and properties of protoporphyrinogen odixase from an anaerobic bacterium, *Desulfovibrio gigas*, *J. Bacteriol.* **169**:5209–5215.

Klemm, D., and Barton, L. L., 1989, Protoporphyrinogen oxidation coupled to nitrite reduction with membranes from *Desulfovibrio gigas*, *FEMS Microbiol. Lett.* **61**:61–64.

La RiviEre, J. W. Q. M., 1955, The production of surface active compounds by microorganisms and its possible significance in oil recovery. 2. On the release of oil from oil-sand mixtures with the aid of sulfate-reducing bacteria. *Antonine van Leeuwenhoek, J. Mikrobiol. Serol.* **21**:9–27.

Lee, J.-P., Le Gall, J., and Peck, H. D., Jr., 1971a, Purification of an assimilatory type of sulfite reductase from *Desulfovibrio vulgaris*, *Fedn. Proc. Fedn. Am. Soc. Exp. Biol.* **30**:1202.

Lee, J.-P., and Peck, H. D., Jr., 1971b, Purification of the enzyme reducing bisulfite to trithionate from *Desulfovibrio gigas* and its identification as desulfoviridin, *Biochem Biopohys. Res. Commun.* **45**:583–589.

Lee, J.-P., Yi, C.-S, Le Gall, J., and Peck, H. D., Jr., (1973), Isolation of a new pigment, desulforubidin, from *Desulfovibrio desulfuricans* (Norway strain) and its role in sulfite reduction, *J. Bacteriol.* **115**:453–455.

Lee, K. H., and Sublette, K. L. 1991, Simultaneous combined microbial removal of sulfur dioxide and nitric oxide from a gas stream, *Appl. Biochem and Biotech.* **28–29**:623–634.

Levin, R. E., Ng, H., Nagel, C. W., and Vaughn, R. H., 1959, Desulfovibrios associated with hydrogen sulfide formation in olive brines, *Bacteriol. Proc.* p. 7.

Le Gall, J., 1968, Purification particle et étude de la NAD: rubredoxine oxydo-réductase de *D. gigas*, *Annls. Inst. Pasteur*, Paris, 114:109–115.

Le Gall, J., and Postgate, J. R., 1973, The physiology of sulphate-reducing bacteria, *Adv. Microbial. Physiol.* **10**:81–133.

Lissolo, T., Choi, E. S., Le Gall, J., and Peck, H. D., Jr., 1986, The presence of multiple intrinsic membrane nickel-containing hydrogenases in *Desulfovibrio vulgaris* (Hildenborough), *Biochem. Biophys. Res. Commun.* **139**:701–708.

Lissolo, T., Cocquempot, M.-F., Thomas, D., Le Gall, J., Schneider, K., and Schlegel, H. G., 1983, Hydrogen production using chloroplast membranes without oxygen scavengers: an assay with hydrogenases from *Desulfovibrio* sp., *Eur. J. Appl. Microbiol. Biotechnol.* **17**:158–162.

Little, B., Wagner, P., and Mansfeld, F., 1991, Microbiologically influenced corrosion of metals and alloys. *International Materials Rev.* **36**:253–272.

Liu, M. C., and Peck, H. D., Jr., 1981, The isolation of a hexaheme cytochrome from *Desulfovibrio desulfuricans* and its identification as a new type of nitrite reductase. *J. Biol. Chem.* **256**:13159–13164.

Lovley, D. R., and Phillips, E. J. P., 1992a, Reduction of uranium by *Desulfovibrio desulfuricans*, *Appl. Environ. Microbiol.* **58**:850–856.

Lovley, D. R., and Phillips, E. J. P., 1992b, Bioremediation of uranium contamination with enzymatic uranium reduction, *Environ. Sci. and Technol.* **26**:2228–2234.

Lovley, D. R., Widman, P. K., Woodward, J. C., and Phillips, E. J. P., 1993c, Reduction of uranium by cytochrome $c_3$ of *Desulfovibrio vulgaris*, *Appl. Environ. Microbiol.* **59**:3572–3576.

Lovley, D. R., Roden, E. E., Phillips, E. J. P., and Woodward, J. C., 1993b, Enzymatic iron and uranium reduction by sulfate-reducing bacteria, *Marine Geol.* **113**:41–53.

Lovley, D. R., Giovannoni, S. J., White, D. C., Champine, J. E., Phillips, E. J. P., Gorby, Y. A., and Goodwin, S., 1993a, *Geobacter metalireducens* gen. nov. sp. nov. a microorganism capable of coupling the complete oxidation of organic compounds to the reduction of iron and other metals. *Arch. Microbiol.* **159**:336–344.

Macfarlane, G. T., Gibson, G. R., and Cummings, J. H., 1992, Comparison of fermentation reactions in different regions of the human colon, *J. Appl. Bacteriol.* **72**:57–62.

McCready, R. G. L., Gould, W. D., and Barendregt, R. W., 1983, Nitrogen isotope fractionation during the reduction of $NO_3^-$ to $NH_4^+$ by *Desulfovibrio* sp., *Can. J. Microbiol.* **29**:231–234.

Makula, R. A., and Meagher, R. B., 1980, A new restriction endonuclease from the anaerobic bacterium, *Desulfovibrio desulfuricans*, Norway, *Nucleic Acids Res.* **8**, 3125–3131.

Mason, G. M., and Kirchner, G., 1992, Authentic pyrite-evidence for a microbial origin of tar sand, *Fuel* **71**:1403–1405.

Meyer, O., and Fiebig, K., 1985, Enzymes oxidizing carbon monoxide, in: *Gas Enzymology* (H. Degn, R. P. Cox, H. Toftlundeds, eds.) D. Reidel Publishing Co. Boston, pp. 147–168.

Miller, L. P., 1950, Formation of metal sulphides through the activities of sulphate-reducing bacteria. *Contr. Boyce Thomson Inst.* **16**:85–89.

Mitchell, G. J., Jones, J. G., and Cole, J. A., 1986, Distribution and regulation of nitrate and nitrite reduction by *Desulfovibrio* and *Desulfotomaculum* species, *Arch. Microbiol.* **144**:35–40.

Mori, T., Koga, M., Hikosaka, Y., Nonaka, T., Mishina, F., Sakai, Y., and Koizumi, J., 1991, Microbial corrosion of concrete sewer pipes, $H_2S$ production from sediments and determination of corrosion rate, *Water Sci. Technol.* **23**:1275–1282.

Nakatsukasa, W., and Akagi, J. M., 1969, Thiosulfate reductase isolated from *Desulfotomaculum nigrificans, J. Bacteriol.* **98**:429–433.

Nanninga, H. J., and Gottschal, J. C., 1987, Properties of *Desulfovibrio carbinolicus* sp. nov. and other sulfate-reducing bacteria isolated from an anaerobic-purification plant, *Appl. Environ. Microbiol.* **53**:802–809.

Niviere, V., Forget, N., Gayda, J. P., and Hatchikian, E. C., 1986, Characterization of the soluble hydrogenase from *Desulfovibrio africanus*, *Biochem. Biophys. Res. Commun.* **139**:658–665.

Novelli, G. D., and ZoBell, C. E., 1944, Assimilation of petroleum hydrocarbons by sulfate-reducing bacteria, *J. Bacteriol.* **47**:447–448.

Ochynski, F. W., and Postgate, J. R., 1963, Some biochemical differences between fresh water and salt water strains of sulfate-reducing bacteria, in: *Symposium on Marine Microbiology* (C. H. Oppenheimer, ed.) C. C. Thomas Publisher, Springfield, IL, pp. 426–441.

Odom, J. M., and Singleton, R., Jr., 1993, *The Sulfate-Reducing Bacteria: Contemporary Perspectives*, Springer-Verlag, New York, p. 289.

Ogata, M., Arihara, K., and Yagi, T., 1981, D-Lactate dehydrogenase of *Desulfovibrio vulgaris, J. Biochem.* Tokyo, **89**:1423–1431.

Okabe, S., and Characklis, W. G., 1992, Effects of temperature and phosphorus concentrations on microbiol sulfate reduction by *Desulfuricans desulfovibrio, Biotechnol. Bioengr.* **39**:1031–1042.

Oremland, R. S., Culbertson, C. W., and Winfrey, M. R., 1991, Methylmercury decomposition in sediments and bacterial cultures - involvement of methanogens and sulfate reducers in oxidative demethylation, *Appl. Environ. Microbiol.* **57**:130–137.

Oremland, R. S., Hollibaugh, J. T., and Maest, A. S. 1989, Selenate reduction to elemental selenium by anaerobic bacteria in sediments and culture: biochemical significance of a novel, sulfate-independent respiration, *Appl. Environ. Microbiol.* **55**:2333–2343.

Panchanadikar, V. V., and Kar, R. N., 1993, Precipitation of copper using *Desulfovibrio* sp., *World J. Microbiol. Biotechnol.* **9:**280–281.

Parpaleix, T., Laval, J. M., Majda, M., and Bourdillon, C., 1992, Potentiometric and voltammetric investigations of $H_2/H^+$ catalysis by periplasmic hydrogenase from *Desulfovibrio gigas* immobilized at the electrode surface in an amphiphilic bilayer assembly, *Anal. Chem.* **64:**641–646.

Peck, H. D., Jr., and Le Gall, J., 1994, *Inorganic Microbiol Metabolism, Methods in Enzymology*, vol. 243, Academic Press, Inc., San Diego, CA, pp. 682.

Philp, J. C., Taylor, K. J., and Christofi, N., 1991, Consequences of sulphate-reducing bacterial growth in a lab-simulated waste disposal regime, *Experimentia* **47:**553–559.

Postgate, J. R., 1965, Recent advances in the study of the sulfate-reducing bacteria. *Bacteriol. Rev.* **29:**425–441.

Postgate, J. R., 1969, Methane as a minor product of pyruvate metabolism by sulphate-reducing and other bacteria, *J. Gen. Microbiol.* **57:**293–302.

Postgate, J. R., 1984, *The Sulphate-Reducing Bacteria*, Cambridge, University Press, Cambridge, pp. 208.

Radha, S., and Seenayya, G., 1992, Environmental factors affecting the bioavailability and toxicity of Cd and Zn to an anaerobic bacterium *Desulfovibrio, Sci. of the Total Environ.* **125:**123–136.

Rao, K. K., Rosa, L., and Hall, H. O., 1976, Prolonged production of hydrogen gas by a chloroplast biocatalytic system, *Biochem. Biophys. Res. Commun.* **68:**21–27.

Rapp, B. J., and Wall, J. D., 1987, Genetic transfer in *Desulfovibrio desulfuricans, Proc. National Acad. Sciences of the USA,* **8:**9128–9130.

Reis, M. A. M., Almeida, J. S., Lemos, P. C., Carrondo, M. J. T., 1992, Effect of hydrogen sulfide on growth of sulfate-reducing bacteria. *Biotechnol. Bioengr.,* **40:**593–600.

Riederer-Henderson, M. A., and Peck, H. D., Jr., 1986, Properties of formate dehydrogenase from *Desulfovibrio gigas, Can. J. Microbiol.* **32:**430–435.

Roden, E. E., and Lovley, D. R., 1993, Dissimilatory FE(III) reduction by the marine microorganism *Desulfuromonas acetoxidans, Appl. Environ. Microbiol.* **59:**734–742.

Rosnes, J. T., Torsvik, T., and Lien, T., 1991, Spore-forming thermophilic sulfate-reducing bacteria isolated from the North Sea oil field waters, *Appl. Environ. Microbiol.* **57:**2302–2307.

Russell, P., 1961, Microbiological studies in relation to moist groundwood pulp, *Chem. Ind.* (London) 642–649.

Sadana, J. C., and Rittenberg, D., 1964, Iron requirement for the hydrogenase of *Desulfovibrio desulfuricans, Arch. Biochem. Biophys.* **108:**255–257.

Sako, Y., Uchida, A., and Kadota, H., 1984, Isolation and characterization of an apurinic endodeoxyribonuclease from the anaerobic thermophile *Desulfotomaculum nigrificans, J. Gen. Microbiol.* **130:**1524–1534.

Sato, M., Mendez, R., and Lemma, J. M., 1991, Biodegradability and toxicity in the anaerobic treatment of fish canning wastewaters, *Environ. Technol.* **12:**669–677.

Schnell, S., and Schink, B., 1991, Anaerobic aniline degradation via redictive deamination of 4-aminobenzoyl-CoA in *Desulfobacterium anilini, Arch. Microbiol.* **155:**183–190.

Seyedirashti, S., Wood, C., and Akagi, J. M., 1991, Induction and partial purification of bacteriophages form *Desulfovibrio vulgaris* (Hildenborough) and *Desulfovibrio desulfuricans* ATCC 13541, *J. Gen. Microbiol.* **137:**1545–1549.

Seyedirashti, S., Wood, C., and Akagi, J. M., 1992, Molecular characterization of two bacteriophages isolated from *Desulfovibrio vulgaris* NCIMB 8303 (Hildenborough), *J. Gen. Microbiol.* **138:**1393–1397.

Singleton, R., Jr., 1993, The sulfate-reducing bacteria: An overview, in: *The Sulfate-*

*Reducing Bacteria: Contempory Perspectives* (J. M. Odom and R. Singleton, Jr., eds.) Springer-Verlag, Berlin, pp. 1–21.

Singleton, R., Jr., Ketcham, R. B., and Campbell, L. L., 1988, Effect of calcium on plating efficiency of the sulfate-reducing bacterium *Desulfovibrio vulgaris*, *Appl. Environ. Microbiol.* **54**:2318–2319.

Smillie, R. H., Hunter, K., and Louitit, M., 1981, Reduction of chromium(VI) by bacterially produced hydrogen sulphide in a marine environment, *Water Res.* **15**:1351–1354.

Stams, A. J. M., and Hansen, T. A., 1982, Oxygen labile L(+)lactate dehydrogenase activity in *Desulfovibrio desulfuricans*, *FEMS Microbiol. Lett.* **13**:389–394.

Stams, A. J. M., Kremer, D. R., Nicolay, K., Weenk, G. H., and Hansen, T. A., 1984, Pathway of propionate formation in *Desulfovibrio propionicus*, *Arch. Microbiol.* **139**:167–173.

Steenkamp, D. J., and Peck, H. D., Jr., 1980, The association of hydrogenase and dithionite reductase activities with the nitrite reductase of *Desulfovibrio desulfuricans*, *Biochem. Biophys. Res. Commun.* **94**:41–48.

Stille, W., and Trüper, H. G., 1984, Adenylylsulfate reductase in some new sulfate-reducing bacteria. *Arch. Microbiol.* **137**:145–150.

Sublette, K. L., and Gwozdz, K. J., 1991, An economic analysis of microbial removal of sulfur dioxide as a means of byproduct recovery from regenerable processes for flue gas desulfurization, *Appl. Biochem. Biotech.* **28**:635–646.

Tasaki, M., Kamagata, Y., Nakamura, K., and Mikami, E., 1992, Propionate formation from alcohols or aldehydes by *Desulfobulbus propionicus* in the absence of sulfate, *J. Ferment. and Bioengr.* **73**:329–331.

Tatnall, R. E., Stanton, K. M., and Ebersole, R. C., 1988, Methods for testing the presence of sulfate-reducing bacteria, in: *Corrosion 88 Conference.* pp. 1–34. NACE Publication Department, Houston, Texas.

Teixeira, M., Moura, I., Fauque, G., Czechowski, M., Berlier, Y., Lespinat, P. A., Le Gall, J., Xavier, A. V., and Moura, J. J. G., 1986, Redox properties and activity studies on a nickel-containing hydrogenase isolated from a halophilic sulfate reducer *Desulfovibrio salexigens, Biochimie* **68**:75–84.

Trinkerl, M., Breunig, A., Schauder, R., and Konig, H., 1990, *Desulfovibrio termitidis* sp. nov., a carbohydrate-degrading sulfate-reducing bacterium from the hindgut of a termite, *Syst. and Appl. Microbiol.* **13**:372–377.

Trudinger, P. A., 1970, Carbon monoxide-reacting pigment from *Desulfotomaculum nigrificans* and its possible relevance to sulfite reduction, *J. Bacteriol.* **104**:158–170.

Tsai, H. H., Sunderland, D., Gibson, R. G., Hart, C. A., and Rhodes, J. M., 1992, A novel mucin sulphatase from human faeces: its identification, purification and characterization, *Clin. Sci.* **82**:447–454.

Tsjui, K., and Yagi, T., 1980, Significance of hydrogen burst from growing cultures of *Desulfovibrio vulgaris* Miyazaki, and the role of hydrogenase and Cytochrome $c_3$ in energy production system, *Arch. Microbiol.* **125**:35–42.

Turner, N., Barata, B., Bray, R. C., Deistung, J., Le Gall, J., and Moura, J. J. G., 1987, The molybdenum iron-sulfur protein from *Desulfovibrio gigas* as a form of aldehyde oxidase, *Biochem. J.* **243**:755–761.

Turner, D. L., Santos, H., Fareleira, P., Pacheco, I., Le Gall, J., and Xavier, A. V., 1992, Structure determination of a novel cyclic phosphocompound isolated from *Desulfovibrio desulfuricans*, *Biochem, J.* **285**:387–390.

Updegraff, D. M., and Wren, G. B., 1954, The release of oil from petroleum-bearing materials by sulfate-reducing bacteria, *Appl. Microbiol.* **2**:309–322.

Vainshtein, M., Hippe, H., Kroppenstedt, R. M., 1992, Cellular fatty acid composition of *Desulfovibrio* species and its use in classification of sulfate-reducing bacteria. *Systematic and Appl. Microbiol.* **15**:554–566.

Vanbeeumen, J. J., Vandriessche, G., Liu, M. Y., and Le Gall, J., 1991, The primary structure of rubrerythrin, a protein with inorganic pyrophosphatase activity from *Desulfovibrio vulgaris* - comparison with hemerythrin and rubredoxin, *J. Biol. Chem.* **266**:20645–20653.

Voroshilov, A. A., and Dianova, E. V., 1950, Concerning the bacterial oxidation of petroleum and its migration in connate waters, *Mikrobiologiya* **19**:203–210.

Vosjan, J. H., and van der Hoek, G. J., 1972, A continuous culture of *Desulfovibrio* on a medium containing mercury and copper ions, *Netherlands J. Sea Res.* **5**:440–444.

Wasay, S., and Das, H. A., 1993, Immobilization of chromium and mercury from industrial wastes. *J. of Environ. Sci. and Health Part A Environ. Sci. and Engr.* **28**:285–297.

Ware, D. A. and Postgate, J. R., 1971, Physiological and chemical properties of a reductant-activated inorganic pyrophosphatase from *Desulfovibrio desulfuricans, J. Gen microbiol.* **67**:145–160.

Weimer, P. J., and Cooling, F. B., 1991, Automated Screening of inhibitors of bacterial dissimilatory sulfate reduction, *Appl. Microbiol. Biotechnol.* **35**:297–300.

Werkman, D. H., 1929, Bacteriological studies on sulfide spoilage of canned vegetables. *Iowa State Col. Agr. Exp. Sta. Res. Bull.* **117**:161–180.

Werkman, C. H., and Weaver, H. J., 1927, Studies in the bacteriology of sulfur stinker spoilage of canned sweet corn, *Iowa State Col. J. Sci.* **2**:57–67.

Widdel, F., 1992, The genus *Thermodesulfobacterium*, in: *The Prokaryotes*, Vol. IV. (A. Balows, H. G. Trüper, M. Dworkin, W. Harder, and K.-H. Schleifer, eds.) Springer-Verlag, Berlin, pp. 3390–3392.

Widdel, F., and Pfennig, N., 1984, Dissimmilatory sulfate- or sulfur-reducing bacteria, in: *Bergey's Manual of Systematic Bacteriology* (N. R. Krieg and J. G. Holt, eds.) Williams & Wilkins, Baltimore, pp. 663–679.

Widdel, F., and Bak, F., 1992, Gram-negative mesophilic sulfate-reducing bacteria. in: *The Prokaryotes*, Vol. IV. (A. Balows, H. G. Trüper, M. Dworkin, W. Harder, and K.-H. Schleifer, eds.) Springer-Verlag, Berlin, pp. 3352–3378.

Widdel, F., and Pfennig, N., 1992, The genus *Desulfuromonas* and other Gram-negative sulfur-reducing eubacteria, in: *The Prokaryotes*, Vol. IV. (A. Balows, H. G. Trüper, M. Dworkin, W. Harder, and K.-H. Schleifer, eds.) Springer-Verlag, Berlin, pp. 3380–3389.

Widdel, F., and Hansen, T. A., 1992, The dissimilatory sulfate-and sulfur-reducing bacteria, in: *The Prokaryotes*, Vol I (A. Ballows, H. G. Trüper, M. Dworkin, W. Harder, and K.-H. Schleifer, eds.) Springer-Verlag, Berlin, pp. 583–624.

Yagi, T., 1969, Formate: cytochrome oxidoreductase of *Desulfovibrio vulgaris, J. Biochem.* Tokyo, **66**:473–478.

Yagi, T., 1979, Preparation of hydrogenase immobilized in polyvinyl alcohol film, *J. Appl. Biochem.* **1**:448–454.

Yagi, T., 1970, Solubilization, purification and properties of particulate hydrogenase from *Desulfovibrio vulgaris, J. Biochem.* **68**:649–657.

Yagi, T., Honya, M., and Tamiya, N., 1968, Purification and properties of hydrogenases of different origins, *Biochim. Biophys. Acta* 153:699–705.

Yamaguchi, M., Hake, J., Tanimoto, Y., Naritomi, T., Okamura, K., and Minami, K., 1991, Enzyme activity for monitoring the stability in a thermophilic anaerobic digestion of wastewater containing methanol, *J. Fermentation and Bioengineering* **71**:264–269.

Yates, M. G., 1969, A nonspecific adenine nucleotide deaminase from *Desulfovibrio desulfuricans, Biochim. Biophys. Acta* **171**:299–310.

Zehr, J. P., and Oremland, R. S., 1987, Reduction of selenate to selenide by sulfate-respiring bacteria: experiments with cell suspensions and estuary sediments, *Appl. Environ. Microbiol.* 53:1365–1369.

Zellner, G., Kneifel, H., and Winter, J., 1990, Oxidation of benzaldehydes to benzoic acid derivatives by three *Desulfovibrio* strains, *Appl. Environ. Microbiol.* **56:**2228–2233.

Zellner, G., Messner, P., Kneifel, H., and Winter, J., 1989, *Desulfovibrio simplex* spec. nov., a new sulfate-reducing bacterium from a sour whey digester, *Arch. Microbiol.* **152:**329–334.

Ziomek, E., and Williams, R. E., 1989, Modification of lignins by growing cells of the sulfate-reducing anaerobe *Desulfovibrio desulfuricans, Appl. Environ. Microbiol.* **55:**2262–2266.

Ziomek, E., Martin, W. G., and Williams, R. E., 1984, Immobilization of isolated and cellular hydrogenase of *D. desulfuricans* in radiation-polymerized polyacrylamides, *Appl. Biochem. Biotechnol.* **9:**57–64.

ZoBell, C. E., 1957, Ecology of sulfate-reducing bacteria, in: *Sulfate-Reducing Bacteria-Their Relation to the Secondary Recovery of Oil,* Science Symposium, St. Bonaventure University, pp. 1–24.

# Sulfate-Reducing Archaea 2

## RUDOLF K. THAUER and JASPER KUNOW

## 1. INTRODUCTION: DISCOVERY OF *ARCHAEOGLOBUS*

Hyperthermophilic sulfate-reducing microorganisms were first isolated from anaerobic submarine hydrothermal systems at Vulcano and Stufe di Nerone, Italy, by Stetter *et al.* (1987). The isolates were identified as Archaea by 16S-rRNA sequence comparisons and by characteristic features such as the presence of phytanyl ether lipids, lack of a peptidoglycan cell wall, the possession of a rifampicin- and streptolydigin-resistant multicomponent RNA polymerase, the inhibition of DNA synthesis by aphidicolin, and the demonstration of an NAD-dependent ADP ribosylation of a soluble protein catalyzed by diphtheria toxin. Evidence for dissimilatory sulfate reduction was obtained by the demonstration of growth dependence on sulfate and of the formation of large amounts of $H_2S$. The isolates could grow on molecular hydrogen (or formate) and sulfate as sole energy sources, indicating that sulfate reduction is coupled with energy conservation (Thauer *et al.*, 1977). These results established without doubt the existence of Archaea, which can use sulfate as external electron acceptor for anaerobic respiration, a physiological trait previously thought to be restricted to the domain bacteria (Widdel, 1988).

So far all isolated archaeal sulfate reducers belong to only one genus named *Archaeoglobus*, of which two species have been described: *A. fulgidus* (Stetter, 1988; Zellner *et al.*, 1989) and *A. profundus* (Burggraf *et al.*, 1990). Characteristic features distinguishing the two species are summarized in Table I. Most of what we know about these organisms comes from studies of *A. fulgidus* strain VC-16 (DSM 4304) (Stetter, 1988). Generalization is probably possible but may not always be correct.

RUDOLF K. THAUER and JASPER KUNOW • Laboratorium für Mikrobiologie des Fachbereichs Biologie der Philipps-Universität Marburg and Max-Planck-Institut für Terrestrische Mikrobiologie, Marburg, D-35043 Marburg, Germany.

*Sulfate-Reducing Bacteria*, edited by Larry L. Barton. Plenum Press, New York, 1995.

**Table I. Characteristic Features Distinguishing *Archaeoglobus fulgidus*
and *A. profundus***

| Feature | *A. fulgidus* | *A. profundus* |
|---|---|---|
| GC-content of DNA (mol %) | 46 | 41 |
| Flagella | Monopolar polytrichous | None |
| Nutrition type | Chemolithoautotroph and chemoorganoheterotroph | Obligately chemolithoheterotroph |
| Growth temperature (°C) | 64–92 (optimum 83) | 65–90 (optimum 82) |
| pH of growth | 5.5–7.5 (optimum 6.5) | 4.5–7.5 (optimum 6) |
| NaCl requirement (%) | 1.0–3.5 (optimum 1.5) | 0.9–3.6 (optimum 1.8) |
| Minimethanogenesis | + | − |

## 2. PHYLOGENETIC POSITION AMONG THE ARCHAEA

The domain of Archaea comprises two kingdoms, the Crenarchaeota and the Euryarchaeota (Woese *et al.*, 1990; Wheelis *et al.*, 1992). The phylogenetic position of the Archaeoglobus species among the Archaea has been determined by using both 16S-rRNA and 23S-rRNA sequences. The analysis shows clearly that the *Archaeoglobus* lineage groups with the unit that comprises the *Methanomicrobiales* and extreme halophiles in the kingdom of Euryarchaeota (Fig. 1) (Woese *et al.*, 1991). The previously reported deep branching very near the base of the euryarchaeal side of the archaeal tree (Achenbach-Richter *et al.*, 1987) is now considered incorrect since it was obtained from sequence data not corrected for branching-order artifacts resulting from compositional disparities, which artificially cluster sequences of similar composition. The simplest evolutionary interpretation of the poylogenetic data is that the archaeal sulfate reducers have evolved from methanogens of the *Methanomicrobiales* lineage. It is interesting that this lineage, besides the sulfate reducers and extreme halophiles, has also spawned a variety of methanogenic phenotypes not encountered on the other methanogenic lineages. For example, methane formation from acetate, methanol, or from trimethylamine and its production under halophilic and alkalophilic conditions are confined to the various species of *Methanomicrobiales*.

That the phylogenetic relationship of *Archaeoglobus* is closer to the *Methanomicrobiales* than to the other orders of methanogens is probably reflected in some of the following similarities: only members of the "*Archaeoglobales*" and members of the aceticlastic *Methanomicrobiales*

Crenarchaeota                    Euryarchaeota

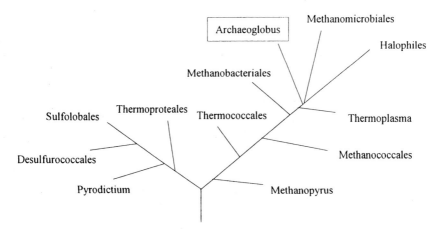

**Figure 1.** The phylogenetic position of *Archaeoglobus* among the Archea. Branching orders are based upon rRNA sequence comparisons (Woese *et al.*, 1990, 1991; Burggraf *et al.*, 1991). The line lengths do not correspond to phylogenetic distances.

(*Methanosarcina* and *Methanothrix*) can degrade acetyl-CoA via decarbonylation (Thauer, 1988; Thauer *et al.*, 1989; Möller-Zinkhan and Thauer, 1990); only they synthesize 2-oxoglutarate from oxaloacetate via citrate and isocitrate (Weimar and Zeikus, 1979; Ekiel *et al.*, 1985b; Möller-Zinkhan *et al.*, 1989) (all nonaceticlastic methanogens synthesize 2-oxoglutarate from oxaloacetate by reduction via malate, fumarate, succinate, and succinyl-CoA; Zeikus *et al.*, 1977; Ekiel *et al.*, 1983, 1985a); and only they appear to contain methanofuran b with a tetraglutamate side chain (White, 1988) and coenzyme $F_{420}$-5 with five glutamate residues in the side chain (Gorris *et al.*, 1991) (for structures see Fig. 2). Also, some properties of the key enzymes, present both in *Archaeoglobus* and in methanogens, are more similar between *Archaeoglobus* and the aceticlastic *Methanomicrobiales* than between the archaeal sulfate reducers and the other methanogens (Schwörer *et al.*, 1993).

   Immunological cross-reactivity of protein synthesis elongation factors was found to grossly correlate with the phylogenetic relatedness of *Archaeoglobus* to the other Archaea as inferred from similarities in rRNA-sequences (Cammarano *et al.*, 1989; Tiboni *et al.*, 1989). The exact position of the archaeal sulfate reducers among the Euryarchaeota was, however, not deducible from these data.

   The structures of 7S-RNA from *Archaeoglobus* was determined and

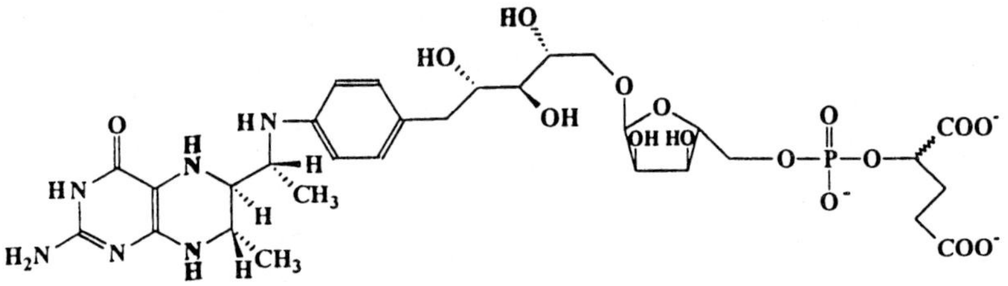

**Tetrahydromethanopterin (H₄MPT)**

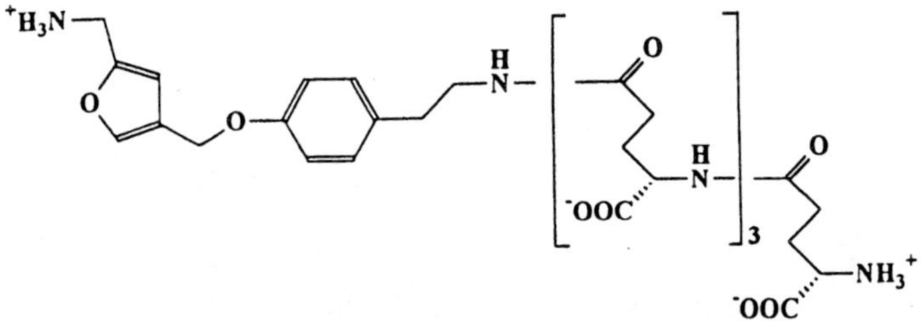

**Methanofuran b (MFR)**

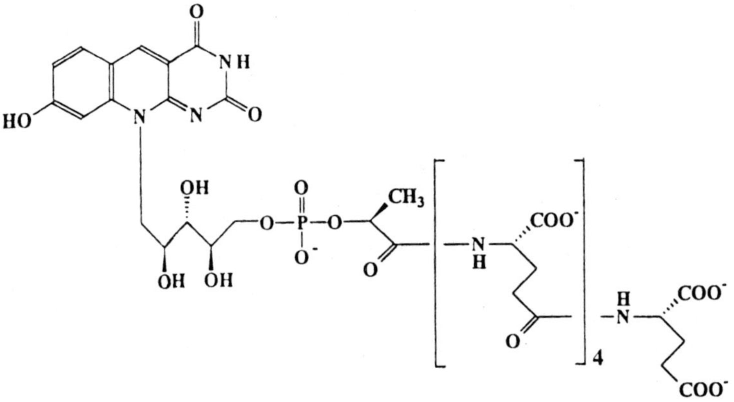

**Coenzyme F₄₂₀-5 (F₄₂₀)**

**Figure 2.** Structures of tetrahydromethanopterin, of methanofuran b, and of coenzyme F₄₂₀-5 found in *Archaeoglobus*.

compared with the 7S-RNA from other Archaea (Kaine, 1990). The secondary structure was found to be highly conserved. The data clearly indicate that *Archaeoglobus* belongs to the domain of Archaea but do not allow a more precise phylogenetic placement.

The phylogenetic position of *Archaeoglobus* within the Archaea has been investigated by comparison of the DNA-dependent RNA-polymerase (Stetter *et al.*, 1987). Within the Archaea two distinct types of RNA polymerase structures are known that parallel the two phylogenetic branches (Pühler *et al.*, 1989; Klenk *et al.*, 1992): the BAC-type of the Crenarchaeota and the AB'B"C-type of the Euryarchaeota. The basis of this difference is the occurrence of two polypetides, B' and B", in the AB'B"C-type that show homology to component B of the other type. The RNA polymerase purified from strain VC-16 exhibited eight identifiable components each having a different molecular mass. The three heaviest of these showed serological cross-reaction with components A, B', and B" of the enzyme from methanogens, indicating homology to the AB'B"C-structure. Surprisingly, antibodies directed against the fourth largest component (C) of the methanogen enzyme exhibited a strong cross-reaction with the heaviest component (A) of the VC-16 RNA polymerase, demonstrating that the largest component of the *Archaeoglobus* RNA polymerase contains structural elements of components A and C of the methanogen enzymes. *Archaeoglobus* thus exhibits a previously unknown third type of archaeal RNA polymerase, (A+C)B'B", which is phylogenetically removed from the euryarchaeal branch (Stetter *et al.*, 1987). The finding was interpreted to support a deep branching of the *Archaeoglobus* lineage. This interpretation is now questionable.

## 3. MORPHOLOGY AND COMPONENTS OF CELLULAR STRUCTURES

Cells of members of the genus *Archaeoglobus* are regular to irregular cocci (sometimes triangular), occurring singularly or in pairs. They may be immotile or motile by means of monopolar polytrichous flagella. They form greenish-black colonies on agar and show a blue-green fluorescence at 420 nm previously thought to be characteristic of methanogenic Archaea (Stetter, 1992).

The cell envelope of archaeal sulfate reducers consists of a single surface layer of glycoprotein subunits, which are closely associated with the outside of the cytoplasmic membrane (König, 1988). There is no rigid sacculus and pseudomurein is absent. The S-layer is composed of subunits arranged as hexagonal lattice with a center-to-center spacing of

17.5 nm. The three-dimensional structure reveals the morphological unit to be dome-shaped with a wide opening to the cell membrane and a narrow opening to the exterior of the apex (Kessel *et al.*, 1990).

The lipid phase of the cytoplasmic membrane consists of phytanyl ($C_{40}H_{72-80}$) ether lipids with core lipids based on 2,3-di-$O$-phytanyl-sn-glycerol and tetraether. The former is present only in trace amounts in some of the isolates. *A. fulgidus* and *A. profundus* differ in their lipid composition as judged by staining of thin layer chromatograms. In *A. fulgidus* two phosphoglycolipids at Rf 0.10 and 0.25, one phospholipid at Rf 0.30, and one glycolipid at Rf 0.60 are present as major complex lipids. In *A. profundus* the major complex lipids are two phosphoglycolipids at Rf 0.10 and 0.13 and four glycolipids at Rf 0.40, 0.45, 0.60, and 0.65. In all isolates examined so far aminolipids appear to be absent (Trincone *et al.*, 1992). For the composition of ether polar lipids of methanogenic Archaea see Koga *et al.* (1993).

## 4. HABITATS AND GROWTH REQUIREMENTS

*Archaeoglobus* species have been isolated from the following habitats: hydrothermally heated shallow marine sediments at the beach of Vulcano and Naples, Italy; a shallow (depth 10 m) hydrothermal system at Ribeira Quente, Azores; a submarine hydrothermal system (depth 103 m) at the Kolbeinsey Ridge, north of Iceland; within the abyssal hot sediments at the Guaymas trench, Mexico; and the crater of an active Polynesian submarine volcano and its submarine plume of eruption in the open ocean (10,000 viable cells per l sea water; Huber *et al.*, 1990). Finding the archaeal sulfate reducers only in these habitats can probably be explained by their requirement for high temperature (optimum 82 °C–85 °C) and salt (optimum 1.5% NaCl–1.8% NaCl). The relatively high salt requirement is probably also the reason why *Archaeoglobus* appears to be absent in continental sulphataric fields, which usually have a low salt content.

The upper temperature limit for growth of the *Archaeoglobus* isolates is 92 °C and the lower temperature limit is 64 °C (Table 1). Indirect evidence is available, however, for the existence of sulfate reducers that can grow at higher temperatures, since radiotracer studies revealed that in some submarine hydrothermic areas sulfate reduction can occur at temperatures up to 110 °C with an optimal rate at 103 °C to 106 °C (Jørgensen *et al.*, 1992). It remains to be shown whether the organisms responsible for this activity are phylogenetically close to *Archaeoglobus*.

*A. fulgidus* is able to grow chemolithoautotrophically on $H_2$, $CO_2$, and thiosulfate, or, less efficiently, on $H_2$, $CO_2$, and sulfate. It also grows chemoorganoheterophically on formate, formamide, D- and L-lactate, pyruvate, glucose, starch, peptone, gelatine, caseine, and meat- or yeast-extract with sulfate, thiosulfate or sulfite, but not S°, as electron acceptor. In the presence of yeast extract and peptone, growth is stimulated by methanol, ethanol, 1-propanol, 2,3-butanediol, and fumarate (Zellner *et al.*, 1989). $H_2S$, $CO_2$, and trace amounts of methane are formed during growth.

A. *profundus* grows chemolithoheterotrophically on $H_2$ and sulfate, thiosulfate, or sulfite as energy sources and organic compounds such as acetate, lactate, pyruvate, yeast extracts or peptone as carbon sources. Growth of these species in the absence of $H_2$ is not possible. A. *profundus* differs from A. *fulgidus* in that it is an obligate chemolithotroph and that it cannot grow autotrophically.

Growth of A. *fulgidus* and of A. *profundus* is not very rapid. The doubling time of both species is four hours under optimal growth conditions.

For a selected enrichment of the two Archaeoglobus species it can be exploited that A. *fulgidus* rather than A. *profundus* can grow on lactate and sulfate as sole carbon and energy sources and that A. *profundus* rather than A. *fulgidus* is dependent on both acetate and $CO_2$ for growth on $H_2$ and sulfate (Stetter, 1992).

## 5. COENZYMES, ENZYMES, AND METABOLIC PATHWAYS

### 5.1. Coenzymes

*Archaeoglobus fulgidus* contains two coenzymes previously thought to occur only in methanogens: tetrahydromethanopterin and methanofuran (Stetter *et al.*, 1987) (Fig. 2). The methanopterin from this organism (Gorris *et al.*, 1991) has been shown to be structurally identical with the cofactor isolated from *Methanobacterium thermoautotrophicum* (Schleucher *et al.*, 1992) and the methanofuran to be identical to that isolated from *Methanosarcina barkeri* (White, 1988). The archaeal sulfate reducer also contains the 5-deazaflavin coenzyme $F_{420}$ (Fig. 2) in high concentrations, which are otherwise only found in methanogens (Gorris *et al.*, 1991). The presence of this cofactor is responsible for the blue-green fluorescence at 420 nm exhibited by the cells when irradiated with long wave UV light. *Archaeoglobus* has been shown to have at least three distinct coenzymes $F_{420}$. One of these is identical with $F_{420}$-5 containing five glutamate residues (Fig. 2) while the other two are as yet unknown analogues of $F_{420}$ (Gorris *et al.*, 1991).

*Archaeoglobus* does not appear to contain coenzyme M (2-mercaptoethanesulfonate) and coenzyme $F_{430}$, which is a nickel porphinoid. Also the genes for methyl-coenzyme M reductase were not found (Stetter, 1992). The small amount of methane reported to be formed by *A. fulgidus* during growth on lactate and sulfate is thus probably not generated by methyl-coenzyme M reduction as in methanogens (Weiss and Thauer, 1993) but by mini-methanogenesis known to be associated with acetyl-CoA decarbonylation (Schauder *et al.*, 1986). Interestingly, mini-methanogenesis is absent in *A. profundus*, which is incapable of acetyl-CoA decarbonylation (Burggraf *et al.*, 1990; Vorholt *et al.*, 1994).

*Archaeoglobus* contains relatively high concentrations of the corrinoid 5-methylbenzimidazolyl cobamide. This corrinoid is also found in the bacterial sulfate reducers *Desulfobacterium autotrophicum* and *Desulfobulbus propionicum*, and in the sulfur metabolizing Archaeon *Thermoplasma acidophilus*. It is not found in methanogens, which generally contain 5-hydroxybenzimidazolyl cobamide (Kräutler *et al.*, 1988).

Bacterial sulfate reducers all contain a naphthoquinone as lipid diffusible electron carrier (Collins and Widdel, 1986), whereas methanogens are devoid of this lipoquinone. In *Archaeoglobus fulgidus* a novel menaquinone was found with a fully saturated heptaprenyl side chain ($MK-7H_{14}$) (Tindall *et al.*, 1989). The feature of a fully saturated side chain is shared by the respiratory lipoquinones of *Thermoproteus tenax* and of *Sulfolobus*, which both belong to the archaeal branch of Crenarchaeota. The isoprenoid chain in the bacterial sulfate reducers is either fully unsaturated or dihydrogenated in the terminal isoprenoid unit (Tindall *et al.*, 1989).

The levels of FAD and FMN (Gorris *et al.*, 1991) and of riboflavin, biotin, pantothenic acid, nicotinic acid, pyridoxin, and lipoic acid (Noll and Barber, 1988) in *Archaeoglobus fulgidus* VC-16 have been determined. The levels of pyridoxin, nicotinic acid, and lipoic acid were only in the order of 1% of that found in *Escherichia coli*. The concentrations of the other components were found to be in the range comparable to what has been found in *E. coli*.

## 5.2. Enzymes

Cells of *A. fulgidus* VC-16 grown on lactate plus sulfate have been analyzed for the presence of catabolic and anabolic enzymes. These cells catalyze the oxidation of lactate with sulfate to $CO_2$ and $H_2S$ (reaction a) and use lactate as the sole carbon source for biosyntheses.

(a) Lactate $^-$ + 1.5 $SO_4^{2-}$ + 4 $H^+$ → 3 $CO_2$ + 1.5 $H_2S$ + 3 $H_2O$

$$\Delta G^{o'} = -151 \text{ kJ/mol lactate}$$

The following enzymes involved in lactate oxidation to 3 $CO_2$ (reaction a) have been found: lactate dehydrogenase catalyzing the oxidation of lactate to pyruvate, probably with menaquinone (MK) as electron acceptor (reaction b); pyruvate:ferredoxin oxidoreductase catalyzing the oxidation of pyruvate to acetyl-CoA and $CO_2$ with ferredoxin (Fd) as electron acceptor (reaction c); acetyl-CoA decarbonylating carbon monoxide dehydrogenase catalyzing the conversion of acetyl-CoA and tetrahydromethanopterin ($H_4MPT$) to $N^5$-methyl-$H_4MPT$ ($CH_3$-$H_4MPT$), $CO_2$, and CoA probably with ferredoxin as electron acceptor (reaction d); $N^5$, $N^{10}$-methylene-$H_4MPT$ reductase catalyzing the oxidation of $N^5$-methyl-$H_4MPT$ to $N^5$, $N^{10}$-methylene-$H_4MPT$ ($CH_2$=$H_4MPT$) with coenzyme $F_{420}$ ($F_{420}$) as electron acceptor (reaction e); $N^5,N^{10}$-methylene-$H_4MPT$ dehydrogenase catalyzing the oxidation of $N^5,N^{10}$-methylene-$H_4MPT$ to $N^5,N^{10}$-methenyl-$H_4MPT$ ($CH\equiv H_4MPT^+$) with coenzyme $F_{420}$ as electron acceptor (reaction f); $N^5,N^{10}$-methenyl-$H_4MPT$ cyclohydrolase catalyzing the hydrolysis of $N^5,N^{10}$-methenyl-$H_4MPT$ to $N^5$-formyl-$H_4MPT$ ($CHO$-$H_4MPT$) (reaction g); N-formylmethanofuran:$H_4MPT$ formyltransferase catalyzing the formation of N-formylmethanofuran (CHO-MFR) from $N^5$-formyl-$H_4MPT$ and methanofuran (MFR) (reaction h); and N-formylmethanofuran dehydrogenase catalyzing the conversion of N-formylmethanofuran to $CO_2$ and methanofuran with a yet to be identified electron acceptor (reaction i) (Möller-Zinkhan *et al.*, 1989; Möller-Zinkhan and Thauer, 1990; Schmitz *et al.*, 1991).

(b) Lactate + MMK → Pyruvate + $MKH_2$

(c) Pyruvate + CoA + 2 $Fd_{ox}$ → Acetyl-CoA + $CO_2$ + 2 $Fd_{red}$

(d) Acetyl-CoA + $H_4MPT$ + 2 $Fd_{ox}$ → $CH_3$-$H_4MPT$ + $CO_2$ + CoA + 2 $Fd_{red}$

(e) $CH_3$-$H_4MPT$ + $F_{420}$ → $CH_2$ = $H_4MPT$ + $F_{420}H_2$

(f) $CH_2$ = $H_4MPT$ + $F_{420}$ + $H^+$ → $CH \equiv H_4MPT^+$ + $F_{420}H_2$

(g) $CH \equiv H_4MPT^+$ + $H_2O$ → $CHO-H_4MPT$ + $H^+$

(h) $CHO-H_4MPT$ + MFR → $CHO-MFR$ + $H_4MPT$

(i) $CHO-MFR$ + X + $H_2O$ → $CO_2$ + MFR + $XH_2$

Lactate dehydrogenase catalyzing reaction *b* is an integral membrane protein. The carbon monoxide dehydrogenase catalyzing reaction d appears to be membrane associated. All the other enzymes are recovered in the soluble cell fraction.

Of these enzymes, the following have been purified and characterized in detail: pyruvate: ferredoxin oxidoreductase catalyzing reaction $c$ (Kunow *et al.*, 1994b); methylene-$H_4$MPT reductase catalyzing reaction $e$ (Schmitz *et al.*, 1991); methylene-$H_4$MPT dehydrogenase catalyzing reaction $f$ (Schwörer *et al.*, 1993); methenyl-$H_4$MPT cyclohydrolase catalyzing reaction $g$ (Klein *et al.*, 1993a); N-formylmethanofuran:$H_4$MPT formyltransferase catalyzing reaction $h$ (Schwörer *et al.*, 1993); and N-formylmethanofuran dehydrogenase catalyzing reaction $i$ (Schmitz *et al.*, 1991). These enzymes also occur in methanogenic Archaea and are there involved in methanogenesis from $CO_2$ or in methanol oxidation to $CO_2$. Comparisons of the N-terminal amino acid sequences of the eyzmes from *Archaeoglobus fulgidus* and from several methanogens revealed that the respective proteins are phylogentically related (Schwörer *et al.*, 1993; Klein *et al.*, 1993b). The stereospecificity of the two coenzyme $F_{420}$-dependent enzymes was found to be $Si$-face with respect to C5 of the 5-deazaflavin (Kunow *et al.*, 1993a).

The oxidation of lactate to 3 $CO_2$ via reactions $b$-$i$ is associated with the reduction of ferredoxin, coenzyme $F_{420}$, menaquinone and an unknown electron acceptor X. The oxidation of these electron carriers is achieved via sulfate reduction to $H_2S$. The following enzymes involved in this process have been identified: ATP sulfurylase catalyzing the formation of adenylyl sulfate (APS) from ATP and sulfate (reaction j); APS reductase catalyzing the reduction of APS to $SO_3^{2-}$ and AMP, probably with reduced menaquinone as electron donor (reaction k); sulfite reductase catalyzing the reduction of sulfite to $H_2S$ with a not yet identified electron donor (reaction l); inorganic pyrophosphatase catalyzing the hydrolysis of inorganic pyrophosphate ($PP_i$) to 2 inorganic phosphate ($P_i$) (reaction m); and $F_{420}H_2$ dehydrogenase catalyzing the oxidation of $F_{420}H_2$ probably with menaquinone as electron acceptor (reaction n). An enzyme catalyzing the oxidation of reduced ferredoxin has not yet been found.

$$( j )\ SO_4^{2-} + ATP^{4-} \rightarrow APS^{2-} + PP_i^{4-}$$
$$( k )\ APS^{2-} + MKH_2 \rightarrow SO_3^{2-} + MK + AMP^{2-} + 2\ H^+$$
$$( l )\ SO_3^{2-} + 6[H] + 2\ H^+ \rightarrow H_2S + 3\ H_2O$$
$$(m)\ PP_i^{4-} + H_2O \rightarrow 2\ P_i^{2-}$$
$$( n )\ F_{420}H_2 + MK \rightarrow F_{420} + MKH_2$$

The enzyme catalyzing reaction $n$ is a membrane-bound multisubunit complex containing FAD and iron-sulfur clusters (Kunow *et al.*, 1994a). The other enzymes are recovered in the soluble cell fraction. The ATP sulfurylase, the APS reductase and the sulfite reductase have been purified and characterized (Speich and Trüper, 1988; Dahl *et al.*, 1990; Lampreia *et al.*, 1991; Dahl *et al.*, 1993; Speich *et al.*, 1994). Their properties are very similar to those of the respective enzymes from bacterial sulfate reducers. The amino acid sequence deduced from the DNA primary structure of the respective archaeal and bacterial enzymes show a high degree of homology, indicating that the proteins involved in dissimilatory sulfate reduction in Bacteria and Archaea have one phylogenetic origin (Speich, 1991; Dahl *et al.*, 1993).

The following enzymes involved in biosyntheses have been demonstrated in cell extracts of *A. fulgidus:* citrate synthase; aconitase; NADP-dependent isocitrate dehydrogenase; NAD-dependent glutamate dehydrogenase; NAD-dependent malate dehydrogenase; fumarate reductase; NADP-dependent glyceraldehyde phosphate dehydrogenase; and NAD-dependent glycerol phosphate dehydrogenase (Möller-Zinkhan *et al.* 1989; Kunow *et al.*, 1993b). 2-Oxoglutarate:ferredoxin oxidoreductase present in most methanogens, except in the aceticlastic *Methanomicrobiales*, was not found.

*A. fulgidus* also contains a $F_{420}H_2$:NADP oxidoreductase connecting catabolic and anabolic redox reactions. The latter enzyme has been purified and characterized. It is very similar to the respective enzyme found in methanogens, in halobacteria, and in some bacteria (Kunow *et al.*, 1993b).

## 5.3. Metabolic Pathways

From enzymes and coenzymes that have been found in *Archaeoglobus fulgidus* VC-16 it can be deduced that in this archaeal sulfate reducer lactate is oxidized to 3 $CO_2$ via the pathway depicted in Fig. 3. The pathway is similar to that found in the bacterial sulfate reducers *Desulfotomaculum acetoxidans* and *Desulfobacterium autotrophicum*. It differs, however, from that found in the bacteria in that tetrahydromethanopterin rather than tetrahydrofolate is involved as $C_1$-carrier and in that formylmethanofuran rather than free formate is an intermediate (Thauer *et al.*, 1989).

## 6. ASSIMILATORY SULFATE REDUCTION IN ARCHAEA

Very little is known about the ability of Archaea to use sulfate as sulfur source. Growth of most methanogens is dependent on more reduced sulfur compounds such as $H_2S$, thiosulfate or sulfite (Rönnow and Gunnarsson, 1981; Mazumder *et al.*, 1986). Only *Methanococcus thermolithotrophicus* has been reported to perform assimilatory sulfate reduction (Daniels *et al.*, 1986). A sulfite reductase catalyzing the 6-electron reduction of sulfite to $H_2S$ was isolated from *Methanosarcina barkeri*. The

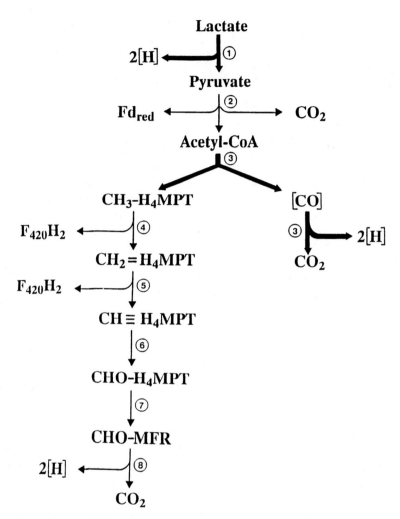

**Figure 3.** Pathway of lactate oxidation to 3 $CO_2$ in *Archaeoglobus fulgidus*. Numbers in circles refer to the enzymes involved: ① lactate dehydrogenase; ② pyruvate:ferredoxin oxidoreductase; ③ carbon monoxide dehydrogenase; ④ $N^5,N^{10}$-methylene-$H_4$MPT reductase; ⑤ $N^5,N^{10}$-methylene-$H_4$MPT dehydrogenase; ⑥ $N^5,N^{10}$-methenyl-$H_4$MPT cyclohydrolase; ⑦ $N$-formylmethanofuran:$H_4$MPT formyltransferase; ⑧ $N$-formylmethanofuran dehydrogenase. Heavy arrows indicate reactions catalyzed by the membrane fraction of cell extracts, thin arrows indicate reactions catalyzed by the soluble fraction of cell extracts. Abbreviations: $CH_3$-$H_4$MPT, $N^5$-methyltetrahydromethanopterin; $CH \equiv H_4$MPT, $N^5,N^{10}$-methylenetetrahydromethanopterin; $CH \equiv H_4$MPT, $N^5,N^{10}$-methenyltetrahydromethanopterin; CHO-$H_4$MPT, $N^5$-formyltetrahydromethanopterin; CHO-MFR, formylmethanofuran; $Fd_{red}$, reduced ferredoxin; $F_{420}H_2$, reduced coenzyme $F_{420}$; [CO], CO bound to carbon monoxide dehydrogenase.

spectral properties of the enzyme were similar to the siroheme containing sulfite reductases from bacteria (Moura *et al.*, 1982).

ACKNOWLEDGMENTS.    This work was supported by a grant from the *Deutsche Forschungsgemeinschaft* and by the *Fonds der Chemischen Industrie*.

## REFERENCES

Achenbach-Richter, L., Stetter, K. O., and Woese, C. R., 1987, A possible biochemical missing link among archaebacteria. *Nature* **327:**348–349.

Burggraf, S., Jannasch, H. W., Nicolaus, B., and Stetter, K. O., 1990, *Archaeoglobus profundus* sp.nov. represents a new species within the sulfate-reducing archaebacteria, *System. Appl. Microbiol.* **13:**24–28.

Burggraf, S., Stetter, K. O., Rouvière, P., and Woese, C. R., 1991, *Methanopyrus kandleri:* An archaeal methanogen unrelated to all other known methanogens, *System Appl. Microbiol.* **14;**346–351.

Cammarano, P., Tiboni, O., and Sanangelantoni, A. M., 1989, Phylogenetic conservation of antigenic determinants in archaebacterial elongation factors (Tu proteins), *Can. J. Microbiol.* **35:**2–10.

Collins, M. D., and Widdel, F., 1986, Respiratory quinones of sulphate-reducing and sulphur-reducing bacteria: a systematic investigation, *System. Appl. Microbiol.* **8:**8–18.

Dahl, C., Koch, H.-G., Keuken, O., and Trüper, H. G., 1990, Purification and characterization of ATP sulfurylase from the extremely thermophilic archaebacterial sulfate-reducer, *Archaeoglobus fulgidus*, *FEMS microbiol. Lett.* **67:**27–32.

Dahl, C., Kredich, N. M., Deutzmann, R., and Trüper, H. G., 1993, Dissimilatory sulphite-reductase from *Archaeoglobus fulgidus:* physicochemical properties of the enzyme and cloning, sequencing and analysis of the reductase genes, *J. Gen. Microbiol.* **139:**1817–1828.

Daniels, L., Belay, N., and Rajagopal, B. S., 1986, Assimilatory reduction of sulfate and sulfite by methanogenic bacteria, *Appl. Environm. Microbiol.* **51:**703–709.

Ekiel, I., Smith, I. C. P., and Sprott, G. D., 1983, Biosynthetic pathways in *Methanospirillum hungatei* as determined by $^{13}$C nuclear magnetic resonance, *J. Bacteriol.* **156:**316–326.

Ekiel, I., Jarrell, K. F., and Sprott, G. D., 1985a, Amino acid biosynthesis and sodium-dependent transport in *Methanococcus voltae*, as revealed by $^{13}$C NMR, *Eur. J. Biochem.* **149:**437–444.

Ekiel, I., Sprott, G. D., and Patel, G. B., 1985b, Acetate and $CO_2$ assimilation by *Methanothrix concilii, J. Bacteriol.* **162:**905–908.

Gorris, L. G. M., Voet, A. C. W. A., and van der Drift, C., 1991, Structural characteristics of methanogenic cofactors in the non-methanogenic archaebacterium *Archaeoglobus fulgidus, BioFactors* **3:**29–35.

Huber, R., Stoffers, P., Cheminee, J. L., Richnow, H. H., and Stetter, K. O., 1990, Hyperthermophilic archaebacteria within the crater and open-sea plume of erupting Macdonald seamount, *Nature* **345:**179–182.

Jørgensen, B. B., Isaksen, M. F., and Jannasch, H. W., 1992, Bacterial sulfate reduction above 100°C in deep-sea hydrothermal vent sediments, *Science* **258:**1756–1757.

Kaine, B. P., 1990, Structure of the archaebacterial 7S RNA molecule, *Mol. Gen. Genet.* **221:**315–321.

Kessel, M., Volker, S., Santarius, U., Huber, R., and Baumeister, W., 1990, Three-dimensional reconstruction of the surface protein of the extremely thermophilic archaebacterium *Archaeoglobus fulgidus, System. Appl. Microbiol.* **13:**207–213.

Klein, A. R., Breitung, J., Linder, D., Stetter, K. O., and Thauer, R K., 1993a, $N^5,N^{10}$-Methenyltetrahydromethanopterin cyclohydrolase from the extremely thermophilic sulfate reducing *Archaeoglobus fulgidus:* comparison of its properties with those of the cyclohydrolase from the extremely thermophilic *Methanopyrus kandleri, Arch. Microbiol.* **159:**213–219.

Klein, A. R., Koch, J., Stetter, K. O., and Thauer, R. K., 1993b, Two $N^5,N^{10}$-methylenetetrahydromethanopterin dehydrogenases in the extreme thermophile *Methanopyrus kandleri:* characterization of the coenzyme $F_{420}$-dependent enzyme, *Arch. Microbiol.* **160:**186–192.

Klenk, H.-P., Palm, P., Lottspeich, F., and Zillig, W., 1992, Component H of the DNA-dependent RNA polymerases of Archaea is homologous to a subunit shared by the three eucaryal nuclear RNA polymerases, *Proc. Natl. Acad. Sci. USA,* **89:**407–410.

Koga, Y., Nishihara, M., Morii, H. and Akagawa-Matsushita, M., 1993, Ether polar lipids of methanogenic bacteria: structures, comparative aspects, and biosyntheses, *Microbiol. Rev.* **57:**164–182.

König, H., 1988, Archaeobacterial cell envelopes, *Can. J. Microbiol.* **34:**395–406.

Kräutler, B., Kohler, H.-P. E., and Stupperich, E., 1988, 5′-Methylbenzimidazolyl-cobamides are the corrinoids from some sulfate-reducing and sulfur-metabolizing bacteria, *Eur. J. Biochem.* **176:**461–469.

Kunow, J., Schwörer, B., Setzke, E., and Thauer, R. K., 1993a, *Si*-face stereospecificity at C5 of coenzyme $F_{420}$ for $F_{420}$-dependent $N^5,N^{10}$–methylenetetrahydromethanopterin dehydrogenase, $F_{420}$-dependent $N^5,N^{10}$–methylenetetrahydromethanopterin reductase and $F_{420}H_2$: dimethylnaphthoquinone oxidoreductase, *Eur. J. Biochem.* **214:**641–646.

Kunow, J., Schwörer, B., Stetter, K. O., and Thauer, R. K., 1993b, A $F_{420}$-dependent NADP reductase in the extremely thermophilic sulfate-reducing *Archaeoglobus fulgidus, Arch. Microbiol.,* **160:**199–205.

Kunow, J., Linder, D., Stetter, K. O., and Thauer, R. K., 1994a, $F_{420}H_2$:quinone oxidoreductase from *Archaeoglobus fulgidus:* characterization of a membrane bound multisubunit complex containing FAD and iron-sulfur clusters, *Eur. J. Biochem.* **233:**503–511.

Kunow, J., Stetter, K. O. und Thauer, R. K., 1994b, Pyruvate: ferredoxin oxidoreductase from the sulfate-reducing *Archaeoglobus fulgidus:* molecular composition, catalytic properties, and sequence alignments, *Arch. Microbiol.,* in press.

Lampreia, J., Fauque, G., Speich, N., Dahl, C., Moura, I., Trüper, H. G., and Moura, J. J. G., 1991, Spectroscopic studies on APS reductase isolated from the hyperthermophilic sulfate-reducing archaebacterium *Archaeoglobus fulgidus, Biochem. Biophys. Res. Commun.* **181:**342–347.

Mazumder, T. K., Nishio, N., Fukuzaki, S., and Nagai, S., 1986, Effect of sulfur-containing compounds on growth of *Methanosarcina barkeri* in defined medium, *Appl. Environm. Microbiol.* **52:**617–622.

Möller-Zinkhan, D., and Thauer, R. K., 1990, Anaerobic lactate oxidation to 3 $CO_2$ by *Archaeoglobus fulgidus* via the carbon monoxide dehydrogenase pathway: demonstration of the acetyl-CoA carbon-carbon cleavage reaction in cell extracts, *Arch. Microbiol.* **15:**215–218.

Möller-Zinkhan, D., Börner, G., and Thauer, R. K., 1989, Function of methanofuran, tetrahydromethanopterin, and coenzyme $F_{420}$ in *Archaeoglobus fulgidus, Arch. Microbiol.* **152:**362–368.

Moura, J. J. G., Moura, I., Santos, H., Xavier, A. V., Scandellari, M., and Le Gall, J., 1982, Isolation of $P_{590}$ from *Methanosarcina barkeri:* evidence for the presence of sulfite reductase activity, *Biochem. Biophys. Res. Commun.* **108:**1002–1009.

Noll, K. M., and Barber, T. S., 1988, Vitamin contents of archaebacteria, *J. Bacteriol.* 170:4315–4321.

Pühler, G., Leffers, H., Gropp, F., Palm, P., Klenk, H.-P., Lottspeich, F., Garrett, R. A., and Zillig, W., 1989, Archaebacterial DNA-dependent RNA polymerases testify to the evolution of the eukaryotic nuclear genome. *Proc. Natl. Acad. Sci. USA* 86:4569–4573.

Rönnow, P. H., and Gunnarsson, L. A. H., 1981, Sulfide-dependent methane production and growth of a thermophilic methanogenic bacterium, *Appl. Environm. Microbiol.* 42:580–584.

Schauder, R., Eikmanns, B., Thauer, R. K., Widdel, F., and Fuchs, G., 1986, Acetate oxidation to $CO_2$ in anaerobic bacteria via a novel pathway not involving reactions of the citric acid cycle. *Arch. Microbiol.* 145:162–172.

Schleucher, J., Schwörer, B., Zirngibl, C., Koch, U., Weber, W., Egert, E., Thauer, R. K., and Griesinger, C., 1992, Determination of the relative configuration of 5,6,7,8-tetrahydromethanopterin by two-dimensional NMR spectroscopy, *FEBS Lett.* 314:440–444.

Schmitz, R. A., Linder, D., Stetter, K. O., and Thauer, R. K., 1991 $N^5,N^{10}$-Methylenetetrahydromethanopterin reductase (coenzyme $F_{420}$-dependent) and formylmethanofuran dehydrogenase from the hyperthermophile *Archaeoglobus fulgidus*, *Arch. Microbiol.* 156:427–434.

Schwörer, B., Breitung, J., Klein, A. R., Stetter, K. O., and Thauer, R. K., 1993, Formylmethanofuran:tetrahydromethanopterin formyltransferase and $N^5,N^{10}$-methylenetetrahydromethanopterin dehydrogenase from the sulfate-reducing *Archaeoglobus fulgidus*: similarities with the enzymes from methanogenic Archaea, *Arch. Microbiol.* 159:225–232.

Speich, N., and Trüper, H. G., 1988, Adenylylsulphate reductase in a dissimilatory sulphate-reducing Archaebacterium, *J. Gen. Microbiol.* 134:1419–1425.

Speich, N., 1991, Enzymologische und molekularbiologische Charakterisierung der Adenosin-5'-Phosphosulfat-Reduktase aus *Archaeoglobus fulgidus*. Dissertation, Universität Bonn, Germany.

Speich, N., Dahl, N., and Heisig, P., Klein, A., Lottspeich, F., Stetter, K. O., and Trüper, H. G., 1994, Adenylylsulfate reductase from the sulfate reducing archaeon *Archaeoglobus fulgidus*: cloning and characterization of the genes and comparison of the enzyme with other iron-sulphur flavoproteins, *Microbiology* 140:1273–1284.

Stetter, K. O., 1988, *Archaeoglobus fulgidus* gen. nov., sp. nov.: a new taxon of extremely thermophilic archaebacteria, *System. Appl. Microbiol.* 10:172–173.

Stetter, K. O., 1992, The genus *Archaeoglobus*, in: *The Procaryotes*, 2nd ed., vol. 1, (Balows, A., Trüper, H. G., Dworkin, M., Harder, W., and Schleifer, K.-H., eds), Springer-Verlag: New York, Berlin, pp. 707–711.

Stetter, K. O., Lauerer, G., Thomm, M., and Neuner, A., 1987, Isolation of extremely thermophilic sulfate reducers: evidence for a novel branch of Archaebacteria, *Science* 236:822–824.

Thauer, R. K., 1988, Citric-acid cycle, 50 years on. Modifications and an alternative pathway in anaerobic bacteria, *Eur. J. Biochem.* 176:497–508.

Thauer, R. K., Jungermann, K., and Decker, K., 1977, Energy conservation in chemotrophic anaerobic bacteria, *Bacteriol. Rev.* 41:100–180.

Thauer, R. K., Möller-Zinkhan, D., and Spormann, A. M., 1989, Biochemistry of acetate catabolism in anaerobic chemotrophic bacteria, *Annu. Rev. Microbiol.* 43:43–67.

Tiboni, O;, Sanangelantoni, A. M., di Pasquale, G., and Cammarano, P., 1989, Immunochemical cross-reactivities of protein synthesis elongation factors (EF-Tu and EF-1α proteins) support the phylogenetic coherence of archaebacteria, *System. Appl. Microbiol.* **12**:237–243.

Tindall, B. J., Stetter, K. O., and Collins, M. D., 1989, A novel, fully saturated menaquinone from the thermophilic sulphate-reducing Archaebacterium *Archaeoglobus fulgidus, J. Gen. Microbiol.* **135**:693–696.

Trincone, A., Nicolaus, B., Palmieri, G., de Rosa, M., Huber, R., Huber, G., Stetter, K. O., and Gambacorta, A., 1992, Distribution of complex and core lipids within new hyperthermophilic members of the *Archaea* domain, *System. Appl. Microbiol.* **15**:11–17.

Vorholt, J., Kunow, J., Stetter, K. O., and Thauer, R. K., 1994, Enzymes and coenzymes of the carbon monoxide dehydrogenase pathway for autotrophic $CO_2$ fixation in *Archaeoglobus lithotrophicus* and the lack of carbon monoxide dehydrogenase in the heterotrophic *A. profundus, Arch. Microbiol.*, in press.

Weimer, P. J. and Zeikus, J. G., 1979, Acetate assimilation pathway of *Methanosarcina barkeri, J. Bacteriol.* **137**:332–339.

Weiss, D. S., and Thauer, R. K., 1993, Methanogenesis and the unity of biochemistry, *Cell* **72**:819–822.

Wheelis, M. L., Kandler, O., and Woese, C. R., 1992, On the nature of global classification, *Proc. Natl. Acad. Sci. USA* **89**:2930–2934.

White, R. H., 1988, Structural diversity among methanofurans from different methanogenic bacteria, *J. Bacteriol.* **170**:4594–4597.

Widdel, F., 1988, Microbiology and ecology of sulfate- and sulfur-reducing bacteria, in: *Biology of Anaerobic Microorganisms*, (A. J. B. Zehnder, ed.), John Wiley & Sons, Inc., pp. 469–585.

Woese, C. R., Achenbach, L., Rouvière, P., and Mandelco, L., 1991, Archaeal phylogeny: reexamination of the phylogenetic position of *Archaeoglobus fulgidus* in light of certain composition-induced artifacts, *System. Appl. Microbiol.* **14**:364–371.

Woese, C. R., Kandler, O., and Wheelis, M. L., 1990, Towards a natural system of organisms: proposal for the domains Archaea, Bacteria, and Eucarya, *Proc. Natl. Acad. Sci. USA* **87**:4576–4579.

Zeikus, J. G., Fuchs, G., Kenealy, W., and Thauer, R. K., 1977, Oxidoreductases involved in cell carbon synthesis of *Methanobacterium thermoautotrophicum, J. Bacteriol.* **132**:604–613.

Zellner, G., Stackebrandt, E., Kneifer, H., Messner, P., Sleytr, U. B., Conway de Macario, E., Zabel, H.-P., Stetter, K. O., and Winter, J., 1989, Isolation and characterization of a thermophilic, sulfate reducing archaebacterium, *Archaeoglobus fulgidus* strain Z, *System. Appl. Microbiol.* **11**:151–160.

# Taxonomic Relationships     3

ERKO STACKEBRANDT, DAVID A. STAHL,
and RICHARD DEVEREUX

## 1. INTRODUCTION

The last several years have seen a notable increase in basic and applied research into the dissimilatory reduction of sulfate by microorganisms. The most significant change in our knowledge of this functionally defined assemblage has been the recognition of far greater evolutionary, genetic, and metabolic capability than previously appreciated. Sulfate-reducing bacteria (SRB) are now known to directly or indirectly participate in the degradation of a wide variety of substrates, including saturated hydrocarbons and a variety of aromatic and xenobiotic compounds (see Chapter 11, this volume).

The range of electron acceptors utilized by these organisms has been extended to include iron, chlorinated aromatics, and oxygen. The observation of high rates of sulfate reduction coincident with oxygenic photosynthesis in microbial mats, and similar reports of sulfate reduction extending to the oxic zone of aquatic sediments, have altered the conventional wisdom that SRB are obligate anaerobes (Canfield and Des Marais, 1991). In fact, we may be in the midst of a paradigm shift; SRB are increasingly perceived as microaerophiles, rather than obligate anaerobes (Marschall et al., 1993). These fundamental changes in our perception of the SRB can not be separated from systematics. The increasingly routine use of molecular comparisons in microbial taxonomy has

ERKO STACKEBRANDT • Deutsche Sammlung von Mikroorganismen und Zellkulturen, GmbH, 38124 Braunschweig, Germany.     DAVID A. STAHL • College of Veterinary Medicine, Department of Veterinary Pathobiology, University of Illinois, Urbana, IL 61801.     RICHARD DEVEREUX • United States Environmental Protection Agency, Environmental Research Laboratory, Gulf Breeze, Florida 32561–5299.

*Sulfate-Reducing Bacteria*, edited by Larry L. Barton. Plenum Press, New York, 1995.

greatly altered how we perceive and work with the SRB. A natural classification of the SRB, primarily based on comparative 16S rRNA sequencing, is now in place, and has served as the foundation for determinative studies. Today's renaissance in the microbiology of sulfate reduction is inseparable from recent advances in their classification.

What has been most notable in many recent studies is that molecular phylogenetic studies (based primarily on 16S rRNA sequence comparisons) have served as the basis for identifying previously unrecognized metabolic capabilities. Many of the most exciting insights into novel metabolism among characterized SRB, or among organisms initially not recognized to be SRB, have been gained through phylogenetic studies. For example, the recognition that SRB are capable of reducing iron and uranium (Lovley and Phillips, 1992; Lovley *et al.*, 1993) was a direct consequence of the phylogenetic placement of *Geobacter metallireducens* (previously classified as GS-15, an obligate anaerobe capable of using ferric iron as a terminal electron acceptor) within the SRB (Lovley *et al.*, 1993). Thus the view that sulfate and iron reduction are mediated by restricted sets of mutually exclusive populations must be reevaluated. Recognition that an organism capable of using chlorinated aromatics as terminal electron acceptors was an SRB was primarily gained via the phylogenetic insight provided by comparative 16S sequence analysis (DeWeerd *et al.*, 1990). Recognition that the etiologic agent associated with enteritis in swine is related to the *Desulfovibrio* was inferred using comparative 16S rRNA sequence analysis (Gebhart *et al.*, 1993).

The isolation of novel organisms with novel metabolic capacities continues to rapidly expand the number of representatives of this group. Now that this diversity is cast within a phylogenetic framework, the phylogeny is also making for a more systematic exploration of environmental diversity. DNA probes, specifically targeting the 16S rRNAs, are now being used for direct quantification in the environment. Their range of habitats must now be extended to include subsurface aquifer systems and the deep subsurface. Their possible use in bioremediation has received greater attention with the recent isolation of various SRB capable of the complete mineralization of toluene (Rabus *et al.*, 1993), phenol, catechol, benzoate, p- and m-cresol, benzyl alcohol, and vanillate (Kuever *et al.*, 1993), benzoate and 4-hydroxybenzoate (Dryzyzga *et al.*, 1993), and 3-aminobenzoate (Schnell and Schink 1993) coupled to sulfate reduction. The diversity of organisms capable of these anaerobic transformations is virtually unexplored. Thus, a robust classification is essential to defining this diversity and as a foundation for the rational manipulation of natural populations for bioremediation.

Our intention in this chapter is to provide the reader with both a historical and contemporary perspective on the classification and phylogeny of the SRB. The historical chronicle is a prelude to discussion of more recent molecular studies and provides an essential overview of the defining physiological attributes of these bacteria. In turn, the developing molecular phylogeny serves to frame the principal physiological features within a natural classification and provide a foundation for the development of DNA probes used for the direct exploration of environmental diversity and studies of the SRB ecology. It is the latter that remains the greatest challenge. Although we have greatly expanded our appreciation of the genetic diversity of SRB, a complete understanding of that diversity can only come from an understanding of their ecology in relationship to their evolution.

## 2. TRADITIONAL CLASSIFICATION

The early era in the systematics of SRB, from the description of *Spirillum desulfuricans* (Beijerinck, 1895) to the establishment of the genera *Desulfovibrio* (Kluyver and van Niel 1936) and *Desulfotomaculum* (Campbell and Postgate, 1965) and the emendation of *Desulfovibrio* (Postgate and Campbell 1966), is a reflection of the problems with which taxonomists were confronted for more than the first six decades of this century. Lack of knowledge about the phylogenetic significance of morphological and superficial phenotypic characters, and the absence of techniques to determine chemotaxonomic and genetic properties, led to changes in nomenclature that depended more upon the subjective assessment of taxonomists than on objective evaluation of characters.

The main problems in the systematics of these two genera of sulfate reducers originate from the opinion that all members stain gram-negative, and on reports on the transformability of strains. The first thermophilic SRB was isolated by Elion (1925) and described as *Vibrio thermodesulfuricans*. Baars (1930) reisolated this organism and compared it with *Vibrio desulfuricans*, the lactate and ethanol oxidizing organisms described earlier as *Spirillum desulfuricans* by Beijerinck (1895). The author described the possibility of adapting *V. desulfuricans* to grow at temperatures of up to 55 °C, by gradually increasing the incubation temperature. The reverse phenomenon was also observed with *V. thermodesulfuricans* and the authors considered this organism to be a temperature adapted strain of *V. desulfuricans*. A similar phenomenon was observed by Starkey (1933), who described successful transformations of monotrichous short and asporogenous sulfate-reducing vibrios to thermophilic, peritrichous, and sporogenous, large, slightly curved rods for which Starkey (1933) proposed the name *Sporovibrio desulfuricans*. The genus *Desulfovibrio* with *D. desulfuricans* as the type species was established (Kluyver and Van Niel, 1936) for an organism hitherto classified as *Spirillum* (Beijerinck, 1895), *Microspira* (Migula, 1900), *Vibrio* (Baars, 1930), and *Sporovibrio* (Starkey, 1933). A gram-negative (although originally described as gram-positive), thermophilic, sulfate reducing and sporulating bacterium, *Clostridium (C.) nigrivans* resembled *Sporovibrio desulfuricans* and subsequently a comparative study on these two species and the asporogenous mesophilic *Desulfovibrio desulfuricans* was performed (Campbell *et al.*, 1957). The results showed that spore-forming thermophilic sulfate reducers were not temperature induced variants of *D. desulfuricans*. Immunological studies, the absence of pigments, morphological, and biochemical properties, also revealed that *C. nigrificans* was identical to *Sporovibrio desulfuricans* but distinctly different from *Desulfovibrio desulfuricans*. *C. nigrificans* retained priority over *Sporovibrio desulfuricans* and this strain was excluded from *Desulfovibrio*.

Careful examination of unpublished results provided to Butlin and Bonker by Campbell and Postgate (1965), supplemented with additional studies by the latter authors supported the results of Campbell *et al.* (1957); no mesophile could be acclimatized to growth above 50 °C. Based on spore formation, *C. nigrificans, Desulfovibrio orientis* (Adams and Postgate, 1959), and a strain from sheep rumen (Coleman, 1960) were not only separated from *Desulfovibrio*, for example, nonsporing organisms that possessed cytochrome *c* and desulfoviridin (Postgate, 1956), but also from the gram-positive clostridia. Consequently, the name *Desulfotomaculum (Dm)* was proposed for these three gram-negative staining sausage (tomaculum)-like organisms, containing the species *Dm. nigrificans, Dm. orientis,* and *Dm. ruminis,* respectively.

Sulfate reduction, however, is not restricted to members of these two genera but has been found as a energy-yielding mode in other genera as well. In some cases, individual sulfate-reducing species are present in genera that otherwise do not contain sulfate-reducers, such as *Spirillum, Pseudomonas,* and *Campylobacter*. In other cases new genera have been described that are exclusively defined by a few species of sulfate reducers such as *Thermodesulfotobacterium* (Zeikus *et al.*, 1983) and the *Archaea Archaeoglobus* (Stetter, 1988). The following is a brief summary of the historical development of the concepts leading to the establishment of the main genera of sulfate-reducers. The information on species is restricted to gross morphological, chemotaxonomic, and genetic properties and on predominant electron donors. Detailed information on growth temperature, cell size, and lipid composition for *Desulfovibrios* and related taxa, electron donors and carbon sources and medium used for the enrichment of species of *Desulfovibrio, Desulfotomaculum,* and other sulfate-reducers are not discussed. This information has been published comprehensively in previous publications (Campbell and Postgate, 1965, 1969; Pfennig *et al.,* 1981; Pfennig and Biebl, 1981; Postgate, 1984a,b; Widdel and Pfennig, 1984; Widdel and Hansen, 1991; Widdel and Bak, 1992; Devereux and Stahl, 1993; see also Tables I and II). Except for the composition of cellular fatty acids (Vainshtein *et al.,* 1992) and menaquinones (Collins and Widdel, 1986) of *Desulfovibrio* species and some related taxa, the chemotaxonomic characterization of species is still fragmentary.

## 2.1. The Mesophilic Gram-Negative Sulfate-Reducers

The genus *Desulfovibrio* was emended by Postgate and Campbell (1966) to include five nonsporulating, polarly flagellated, mesophilic gram-negative sulfate-reducing species. The authors clearly stated their intention to treat the presented classification (at that time) as a working classification and that further taxonomic revisions might be necessary when new data become available. For example, strain Norway 4 was listed as a strain of *D. desulfuricans* although it does not produce desulfoviridin (to date strain Norway 4 is classified as strain of *Desulfomicrobium [Dmi] baculatus*). Strain Hildenborough, on the other hand, was not included in *D. desulfuricans* because it did not match the growth requirements of the type species and was assigned to *D. vulgaris*. Today this strain is the type strain of *D. vulgaris* subsp. *vulgaris*. The first indication that the genera *Desulfovibrio* and *Desulfotomaculum* are not close phylogenetic neighbors resulted from DNA/rRNA similarity studies (Pace and Campbell, 1971), in which four *Desulfovibrio* species were demonstrated to be as unrelated to *Desulfotomaculum nigrificans* as they were to species of *Vibrio, Escherichia,* and *Spirillum*.

Unlike the apparent homogeneous metabolic and morphological properties of *Desulfotomaculum* species (see below), the gram-negative dissimilatory sulfate reducers use a much wider range of oxidizable substrates, and demonstrate a greater variety of morphologies. As a result, the number of species increased steadily from five in 1965, to nine in 1984 (Postgate, 1984b), to 15 in 1994 (Approved Lists of Bacterial Names and Validation Lists). Until recently, analyses that could determine the genetic validity of the species were limited. Early determination of rRNA cistron similarities revealed that the reference rRNA of *D. vulgaris* Hildenborough was highly similar (if not identical) to the rDNA of *D. desulfuricans* VC NCIB 9467 (the former *Vibrio cholinicus*), while values obtained

**Table I. Key Characters in the Classification of Gram-Negative Dissimilatory Sulfur-Reducing Bacteria[a]**

| Taxon | Oxidation | Form | Motility | Desulfoviridin | Cytochrome | Mole % G+C | Major menaquinone | Sat. Even | Sat. Odd | Branched Iso | Branched Anteiso | Unsaturated | H₂ | For | L | E | A | FA | F | M | B |
|---|---|---|---|---|---|---|---|---|---|---|---|---|---|---|---|---|---|---|---|---|---|
| *Desulfovibrio* | | | | | | | | | | | | | | | | | | | | | |
| *africanus* | I[c] | vibrio | lophotrichous | + | $c_3$ | 65 | MK-6(H₂) | + | – | ++ | ++ | +++ | + | + | + | + | – | – | – | + | nr |
| *"alcoholovorans"* | nr[d] | nr | nr | nr | nr | nr | nr | + | – | ++ | +++ | ++ | nr | nr | nr | nr | nr | nr | nr | nr | nr |
| *"carbinolicus"* | I | rod | – | + | nr | 65 | nr | + | – | + | +++ | + | + | + | + | + | + | + | + | + | nr |
| *desulfuricans* T | I | vibrio | single, polar | + | $c_3$ | 59 | MK-6 | + | – | +++ | – | +++ | + | + | + | (+) | – | – | + | + | nr |
| *fructosovorans* | nr | vibrio | single, polar | + | nr | 64 | nr | + | – | + | +++ | – | nr | + | + | (+) | – | – | + | + | nr |
| *furfuralis* | I | vibrio | single, polar | + | $c_3$ | 61 | nr | nr | nr | nr | nr | nr | nr | nr | + | + | – | – | + | + | nr |
| *giganteus* | I | vibrio/rod | single, polar | + | nr | 56 | nr | – | – | ++ | +++ | + | + | – | + | + | – | – | + | – | – |
| *gigas* | I | spirilloid | lophotrichous | + | $c_3$ | 65 | MK-6 | + | – | ++ | +++ | + | + | + | + | (+) | – | – | – | – | – |
| *halophilus* | nr | vibrio | single, polar | + | nr | 61 | nr | + | – | +++ | ++ | +++ | + | + | + | + | + | nr | – | – | nr |
| *longus* | nr | flexible rod | single, polar | + | nr | 62 | nr | nr | nr | nr | nr | nr | + | + | + | + | – | – | – | – | – |
| *salexigens* | I | vibrio | single, polar | + | $c_3$ | 49 | MK-6(H₂) | ++ | – | +++ | + | +++ | + | + | + | + | – | – | – | + | – |
| *simplex* | I | vibrio | single, polar | + | nr | 48 | nr | + | – | +++ | – | ++ | + | + | + | + | – | – | – | + | – |
| *sulfodismutans* | I | vibrio | + | + | nr | 64 | nr | + | – | ++ | ++ | + | (+) | – | + | + | + | – | – | – | – |
| *termitidis* | nr | curved rod | single, polar | + | nr | 67 | nr | ++ | – | +++ | + | ++ | – | + | + | nr | – | nr | nr | + | nr |
| *vulgaris* | I | vibrio | single, polar | + | $c_3$ | 65 | MK-6 | – | – | +++ | + | ++ | + | + | + | (+) | – | – | + | + | – |
| *Desulfobacter* | | | | | | | | | | | | | | | | | | | | | |
| *curvatus* | C | vibrio | + | – | nr | 46 | MK-7[d] | +++ | – | + | – | + | + | – | – | + | + | – | – | – | – |
| *hydrogenophilus* | C | rod | – | – | nr | 45 | MK-7[d] | nr | nr | nr | nr | nr | + | – | – | (+) | + | – | – | – | – |
| *latus* | C | large oval rod | – | – | nr | 44 | nr | nr | nr | nr | nr | nr | – | – | – | – | + | – | – | – | – |
| *postgatei* T | C | ellipsoidal rod | variable | | $b,c$ | 46 | MK-7 | nr | nr | nr | nr | nr | nr | – | – | – | + | – | – | – | – |
| *Desulfobacterium* | | | | | | | | | | | | | | | | | | | | | |
| *anilini* | C | oval | – | – | $c$ | 59 | nr | nr | nr | nr | nr | nr | + | + | nr | nr | + | 3-18 | nr | nr | nr |
| *autotrophicum* T | C | oval | single, polar | – | nr | 48 | MK-7[e] | + | + | – | – | +++ | (+) | (+) | (+) | (+) | (+) | (3-16) | + | + | – |
| *catecholicum* | C | lemon | single, polar | – | nr | 52 | nr | nr | nr | nr | nr | nr | – | (+) | (+) | (+) | (+) | (3-20) | (+) | (+) | + |
| *indolicum* | C | oval rod | single, polar | – | nr | 47 | MK-7(H₂) | nr | nr | nr | nr | nr | (+) | – | – | (+) | – | (3) | (+) | (+) | – |
| *macestii* | nr | rod | single, polar | – | nr | 58 | nr | nr | nr | nr | nr | nr | + | + | + | + | – | nr | nr | – | – |

| Species | | Shape | Flagella | Desulfo-viridin | Cyt. | mol% G+C | Quinone | | | | | | | | | | | | | | |
|---|---|---|---|---|---|---|---|---|---|---|---|---|---|---|---|---|---|---|---|---|---|
| "naceni" | C | irregular sphere | − | − | nr | 46 | MK-7 | nr | nr | nr | nr | nr | + | + | − | + | (+) | (3)-16 | + | + | − |
| phenolicum | C | oval/curved rod | single, polar | − | nr | 41 | MK-7(H2) | nr | nr | nr | nr | nr | − | (+) | − | (+) | (+) | (4) | (+) | (+) | + |
| "vacuolatum" | C | oval/sphere | − | − | nr | 45 | MK-7(H2) | nr | nr | nr | nr | nr | + | + | + | (+) | (+) | (3)-16 | + | + | − |
| Desulfobulbus | | | | | | | | | | | | | | | | | | | | | |
| elongatus | I | Rod | variable | − | nr | 59 | MK-5(H2) | nr | nr | nr | nr | nr | + | + | + | − | − | 3 | − | nr | nr |
| "marinus" | I | oval | single, polar | − | nr | nr | MK-5(H2) | nr | nr | nr | nr | nr | + | + | + | − | nr | 3 | nr | nr | nr |
| propionicus [T] | I | lemon/onion shape | variable | − | b,c | 60 | MK-5(H2) | nr | nr | nr | nr | nr | + | − | − | + | − | 3 | − | − | nr |
| Desulfococcus | | | | | | | | | | | | | | | | | | | | | |
| biacutus | C | lemon shape | − | + | b,c | 57 | nr | nr | nr | nr | nr | nr | nr | − | − | − | − | 3-6 | − | − | − |
| multivorans [T] | C | spheres | variable | + | b,c | 57 | MK-7 | nr | nr | nr | nr | nr | + | + | (+) | + | − | 3-16 | − | − | − |
| Desulfohalobium | | | | | | | | | | | | | | | | | | | | | |
| retbaense [T] | I | curved rod | polar flagella | desulfi-rubidin | c₃ | 57 | nr | ++ | ++ | + | + | ++ | + | + | + | + | − | nr | − | nr | nr |
| Desulfomonas | | | | | | | | | | | | | | | | | | | | | |
| pigra [T] | I | rod | − | + | nr | 66 | MK-6 | + | +++ | +++ | − | ++ | + | + | + | − | − | (4) | − | nr | nr |
| Desulfomonile | | | | | | | | | | | | | | | | | | | | | |
| tiedje [T] | nr | rod | − | + | nr | 49 | nr | nr | nr | nr | nr | nr | + | − | − | − | − | − | − | nr | + |
| Desulfonema | | | | | | | | | | | | | | | | | | | | | |
| limicola [T] | C | filament | gliding | + | b,c | 35 | MK-7 | nr | nr | nr | nr | nr | + | + | (+) | + | + | 3-14 | + | − | − |
| magnum | C | filament | gliding/rolling | − | b,c | 42 | MK-9 | nr | nr | nr | nr | nr | + | − | (+) | + | + | 3-10 | + | (+) | + |
| Desulfosarcina | | | | | | | | | | | | | | | | | | | | | |
| variabilis [T] | C | irregular packages | variable | − | nr | 51 | MK-7 | nr | nr | nr | nr | nr | + | + | + | + | (+) | 3-14 | + | + | + |
| "Desulfoarculus" | | | | | | | | | | | | | | | | | | | | | |
| baarsii | C | vibrio | single, polar | − | nr | 66 | MK-7(H2) | +++ | ++ | ++ | + | − | + | − | − | − | + | (3)-18 | − | − | − |
| "Desulfobotulus" | | | | | | | | | | | | | | | | | | | | | |
| sapovorans | I | vibrio | single, polar | − | b,c | 53 | MK-7 | ++ | + | + | ++ | ++ | − | − | + | − | − | 4-16 | − | − | − |
| Desulfomicrobium | | | | | | | | | | | | | | | | | | | | | |
| apsheronum | nr | rod | single, polar | − | nr | 52 | nr | + | ++ | ++ | + | +++ | + | + | + | + | − | − | + | + | nr |
| baculatus [T] | nr | short rod | single, polar | − | b,c | 57 | nr | + | +++ | +++ | + | +++ | + | + | + | (+) | + | − | − | + | nr |

[a] The data were compiled from Postgate (1984), Widdel (1988), Gogotava and Vainshtein (1989), Schnell et al. (1989), Platen et al. (1990), Devereux and Stahl (1993), Ollivier et al. (1991).

[b] Data were taken from Vainshtein et al. (1992) and Ollivier et al. (1991).

[c] Abbreviations: nr, not reported; +, utilized; (+), poorly utilized; −, not utilized; C, complete; I, incomplete; For, formiate; L, lactate; E, ethanol; A, acetate; FA, fatty acids; Fum, fumarate; M, malate; B, benzoate.

[d] Listed as MK-7 (H₂) in Widdel (1988) and as MK-7 in Devereux and Stahl (1993).

[e] Listed as a species of Desulfobacter in Validation Lists No. 26 (IJSB 38:328).

with rDNA of *D. salexigens* and *D. africanus* were clearly greater than the values obtained with outgroup organisms (Pace and Campbell, 1971). DNA similarity studies lead to the recognition of the unrelatedness of *D. baculatus* and *D. thermophilus* to the other species of the genus and their subsequent description as *Desulfomicrobium baculatus* (Rozanova *et al.*, 1988), and *Thermodesulfomicrobium mobile* (Rozanova and Pivovarova, 1988), respectively. Furthermore, the DNA base composition within *Desulfovibrio* ranged between 49 and 65 mole%, which indicated either that this genus is phylogenetically not homogeneous or alternatively, if homogeneous, of significant phylogenetic depth (see below).

As Postgate (1984a) pointed out, the small number of diagnostic properties was the reason for the imperfect taxonomy of desulfovibrios. Phenotypic instabilities in many experiments, such as change in morphology, loss of motility, the presence of lophotrichous or double-flagellated cells, and change in cultural behavior had been observed. In particular, morphology did not appear to be an appropriate guide to the classification of *Desulfovibrio* since cells were prone to pleomorphism in old cultures and cell shape was anticipated to vary with growth conditions. Even the desulfoviridin test, originally used to separate the *Desulfovibrios* (positive) from the strains of *Desulfotomaculum* (negative), was negative for certain members of the former genus, such as the rod-shaped *D. desulfuricans* strain Norway, *D. baculatus*, *D. sapovorans*, and *D. baarsii*. As will be shown below, these organisms are no longer considered *Desulfovibrio* species.

The isolation of organisms that differ from *Desulfovibrio* by a combination of key characters (Table I), has led to an increase in the number of genera of nonspore-forming dissimilatory sulfate reducers; from one in 1965 to 10 in 1994 (Approved List). Extensive keys to the classification and properties of *Desulfovibrios* and related taxa, including an extensive listing of size, growth temperature, habitats, and electron donors for sulfate reduction, have been published in the last decades (see above) and we refer the reader to these references rather than reiterate these compilations. Table I is a limited listing of taxonomically important properties extracted from published information, supplemented with data for recently described species. Some taxonomic problems have recently been introduced by the use of taxa not on the Approved Lists of Bacterial Names (Skerman *et al.*, 1980), which have not been validated since then. Although the designation of these novel taxa is legitimate from a taxonomic and phylogenetic point of view (see below), no formal description has been published or included in the lists of validly described taxa since 1980 in the International Journal of Systematic Bacteriology to comply with taxonomic rules. Hence, as of this writing the following taxa have no standing in taxonomy and should be listed in quotes:

- The genera "*Desulfoarculus*" and "*Desulfoarculus baarsii.*" This species is still a member of *Desulfovibrio, D. baarsi* (Widdel, 1980; Approved Lists No. 7).
- The genera "*Desulfobotulus*" and "*Desulfobotulus sapovorans.*" This species is still a member of *Desulfovibrio, D. sapovorans* (Widdel, 1980; Approved Lists No. 7).
- The species "*Desulfobacterium niacini*" (Imhoff-Stuckle and Pfennig, 1983) and "*Desulfobacterium vacuolatum*" (Widdel, 1988).
- The species "*Desulfobulbus marinus,*" a suggested name for a marine isolate (Widdel and Pfennig, 1982).
- The species "*Desulfovibrio piger.*" This species is still a valid member of *Desulfomonas, Desulfomonas pigra* (Moore *et al.,* 1976).
- The species "*Desulfovibrio alcoholovorans*" and "*Desulfovibrio carbionolicus,*" listed in Vainshtein *et al.* (1992).

## 2.2. The Thermophilic Gram-Negative Sulfate-Reducers

The genus *Thermodesulfobacterium* contains two species, *T. commune* and *T. mobilis.* The type species *T. commune* was described by Zeikus *et al.* (1983) for a small (0.3 × 0.9 micron) extremely thermophilic, nonmotile, nonspore-forming, gram-negative staining bacterium from thermal muds and algal mats in Yellowstone National Park (USA) (see Widdel, 1992b).

The maximum growth temperature (Tmax) of 85 °C for this organism is among the highest yet observed for members of the class Bacteria, relative to members of the *Thermotoga* group (Tmax 90 °C) and *Aquifex pyrophilus* (Tmax 95 °C) (Huber *et al.,* 1992). It is characterized by a low DNA G+C content of 34 mole%, no growth on lactate and pyruvate without sulfate, and possession of cytochrome *c3.* This organism contains no desulfoviridin, desulforubidin or 582-type bisulfite reductase, but a hitherto unknown dissimilatory bisulfite reductase, desulfofuscidin (Hatchikian and Zeikus, 1983). In contrast to members of the class Bacteria that are mainly (but not exclusively so) defined by ester-linked fatty acids, *T. commune* contains ether-linked lipids, mainly phospholipids (Langworthy *et al.,* 1983). In contrast to the ether-linked constituents of *Archaea,* the lipids are alkyl chains with one subterminal methyl branch (1,2-di-O-anteisoheptadecyl-sn-glycerol). The extremely thermophilic, rod-shaped species *D. thermophilus* (Rozanova and Khudakova, 1974) also lacks desulfoviridin and has a substantially lower DNA mole% G+C than

**Table II. Key Characters in the Classification of *Desulfotomaculum* Species**[a]

| Taxon | Morphology | Flagellation | Cyto-chrome | Mena-quinone | Temperature optimum (°C) |
|---|---|---|---|---|---|
| *Desulfotomaculum* | | | | | |
| *acetoxidans* | Straight or curved rod | Single, polar | b | MK-7 | 34–36 |
| *antarcticum* | Rod | Peritrichous | b | nr | 20–30 |
| *australicum* | Rod | Motile | nr | nr | 68 |
| *geothermicum* | Rod | At least 2 | c | nr | 54 |
| *guttoideum* | Rod, drop-shaped | Peritrichous | c | nr | 31 |
| *kuznetsovii* | Rod | Peritrichous | nr | nr | 60–65 |
| *nigrificans* | Rod | Peritrichous | b | MK-7 | 55 |
| *orientis* | Straight or curved rod | Peritrichous | b | MK-7 | 37 |
| *ruminis* | Rod | Peritrichous | b | MK-7 | 37 |
| *sapomandens* | Rod | Motile | nr | nr | 38 |
| *thermobenzoicum* | Rod | Motile | nr | nr | 62 |
| *"thermoacetoxidans"* | Straight or curved rod | Motile | nr | nr | 55–60 |

*Desulfovibrio* species. Very low DNA similarities between *D. thermophilus* and two other *Desulfovibrio* species (Nazina *et al.*, 1987) lead Rozanova and Pivovarova (1988) to transfer this species into *Thermodesulfobacterium* as *T. mobilis*. Although the change in specific epithet from *thermophilus* to *mobile* is invalid according to the International Code of Nomenclature of Bacteria, this species has been validated as the only species of the (invalid) genus *Thermodesulfobacterium*. Nevertheless, the transfer appears to be justified based on similarities between *T. mobilis* (*D. thermophilus*) and *T. commune* in growth temperature, morphology, lack of spores, isoprenoid composition (MK-7), and fatty acid composition (predominantly branched iso- and anteiso fatty acids). The enzyme bisulfite reductase from *T. mobilis* (investigated as *D. thermophilus*) and *T. commune* was shown to share some homologous properties (Fauque *et al.*, 1990) but the importance of this finding is restricted by the lack of samples from a phylogenetically broader group of organisms.

## 2.3. The Gram-Positive Sulfate-Reducers

During the 25 years following the original description of three *Desulfotomaculum* species the number was extended by another three meso

**Table II.**  (*Continued*)

| DNA G+C content (mol%) | Oxida-tion[b] | Electron donors[b] | | | | | | | | |
|---|---|---|---|---|---|---|---|---|---|---|
| | | H | For | L | E | A | FA | Fum | M | B |
| 38 | C | − | − | − | + | + | 3–5 | − | − | − |
| nr | I | nr | − | + | nr | − | − | nr | nr | nr |
| 48 | nr | + | − | + | + | + | 3–4 | − | − | + |
| 50 | C | + | + | + | + | − | 3–18 | − | nr | − |
| 52 | I | + | − | + | − | − | nr | nr | − | nr |
| 49 | C | + | + | + | + | + | 3–18 | + | + | − |
| 49 | I | + | + | + | + | − | − | − | − | − |
| 45 | I | + | + | + | + | − | − | − | − | − |
| 49 | I | − | + | + | + | − | − | − | − | − |
| 48 | C | − | + | − | + | (+) | 4–18 | (+) | (+) | + |
| 53 | C | + | + | + | + | − | 3–4 | + | + | + |
| 50 | C | + | + | + | − | + | 3–5 | nr | + | − |

[a] Data taken from Widdel (1992a), Widdel and Hansen (1992), Daumas *et al.* (1988), Patel *et al.* (1993), Tasaki *et al.* (1991) and Min and Zinder (1990).

[b] Abbreviations: nr, not reported; +, utilized; (+), poorly utilized; −, not utilized; C, complete; I, incomplete; For, formiate; L, lactate; E, ethanol; A, acetate; FA, fatty acids; Fum, fumarate; M, malate; B, benzoate.

philic species, *Dm. acetoxidans, Dm. antarcticum,* and *Dm. guttoideum.* While the latter two species resemble the first three species in the incomplete oxidation of organic substrates and peritrichous flagellation (Campbell and Singleton, 1986), *Dm. acetoxidans* (Widdel and Pfennig, 1977) oxidized organic compounds completely, possesses a single polar flagellum and has a significantly lower mole% G+C of only 38%. Today, the eleven validly described species also include thermophilic representatives and the species are mainly defined by metabolic properties and growth requirements (see Table II). The species "*Dm. thermoacetooxidans*" (Min and Zinder, 1990) has not yet been validated.

One of the most unexpected findings of electron microscope studies was that, in contrast to their original description as gram-negative, *Desulfotomaculum* strains stained gram-negative but had a gram-positive cell wall ultrastructure (Nazina and Pivovarova, 1979; Sleytr *et al.,* 1969). This was confirmed by subsequent phylogenetic analysis (see below).

All species are defined by the presence of spores, but the shape (spherical to oval) and location of spores (central, subterminal, terminal) varies. *Desulfotomaculum* species are versatile with respect to electron donors for sulfate reduction (see Widdel, 1992a). Some species are autotrophic and some species grow by fermentation of glucose and other organic substrates. Some species perform homoacetogenesis by converting a few substrates, including $H_2$ and $CO_2$, to acetate. This finding is especially interesting from a phylogenetic standpoint since it may indicate that these species are more closely related to homoacetogenic clostridia than to other members of *Desulfotomaculum*. Desulfoviridin has never been found in any *Desulfotomaculum* species, but sulfite reductase P 582 is present. In addition to b-type cytochromes, c-type cytochromes have been detected (Daumas *et al.*, 1988). The phospholipid pattern does not differ greatly from that of *Desulfovibrio* and related taxa, in that saturated, unbranched, even-numbered (16:0, 18:0) and iso- and anteiso-branched (16:1, 18:1) fatty acids dominate (Ueki and Suto, 1979). The two thermophilic species, *Dm. nigrificans* and *Dm. australicum*, possess high amounts of isosaturated branched (i-15:0.,i-17:0) fatty acids, which may reach up to 87% of total fatty acids in the latter species. These compounds appear to be caused by adaptation to the thermophilic niche, as similar high amounts of these fatty acids have been found in other thermophilic bacteria, for example, thermi and certain clostridia (Patel *et al.*, 1991).

## 2.4. The Archaeal Sulfate Reducers

*Archaeoglobus (A.) fulgidus* (Stetter, 1988) and *A. profundus* (Burggraf *et al.*, 1990), isolated from anaerobic submarine hydrothermal areas, are the only two dissimilatory sulfate reducing members of the domain Archaea so far described. The species, *A. fulgidus*, has been examined in detail with respect to physiology, biochemistry, and phylogeny (see below). Based on initial phylogenetic analyses, this species was initially thought to be a biochemical "missing link" between sulfur metabolizing and methanogenic archaeas (Achenbach-Richter *et al.*, 1987). Cells of both species are regular to irregular coccoid, require at least 1% salt for growth, and are extremely thermophilic (Tmax 90 °C). They show a similar blue-green fluorescence at 420 nm, use sulfate, sulfite, and thiosulfate as electron acceptors, and their growth is inhibited by elemental sulfur. As demonstrated in *A. fulgidus*, ATP sulfurylase, adenylyl sulfate, and bisulfite reductase activities are present (Speich and Trüper, 1988; Dahl *et al.*, 1990), and lactate is oxidized via a pathway normally only used by methanogenic bacteria (Möller-Zinkhan *et al.*, 1989). In addition the cell walls of both species lack peptidoglycan but contain acyclic C40 tetraether and C20 diether lipids. However, while *A. fulgidus* produces small amounts of methane and possesses the cofactors methanofuran and tetrahydromethanopterin, *A. profundus* does not produce methane and these cofactors have not been reported. Further differences between *A. fulgidus* and *A. profundus* exist in the DNA base composition (46 versus 41 mole%, respectively) and in nutrition type (facultatively chemolithoauto-trophic versus obligately chemolithoheterotrophic, respectively).

## 3. THE PHYLOGENY OF THE MESOPHILIC GRAM-NEGATIVE BACTERIAL SULFATE REDUCERS

A listing of the species of SRB isolated and described to date reveals the majority of them to be gram-negative (Widdel and Bak, 1992). Comparative 16S rRNA sequence analyses of these species has placed them into two distantly related phylogenetic lines. One line is composed of thermophilic species, such as *Thermodesulfobacterium*-like spp. (Fig. 1), and the other lineage contains the mesophilic species (Fig. 2). Within each of these lines, considerable phylogenetic diversity has been demonstrated.

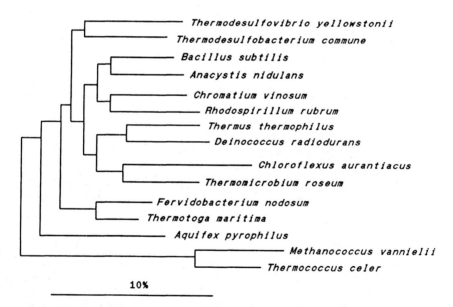

**Figure 1.** Phylogenetic tree with gram-negative thermophilic sulfate-reducing bacteria and other bacterial and archael groups. The tree is based on transversion distances between 16S rRNA. Scale bar represents 10 nucleotide changes per 100 sequence positions. (After Henry *et al.*)

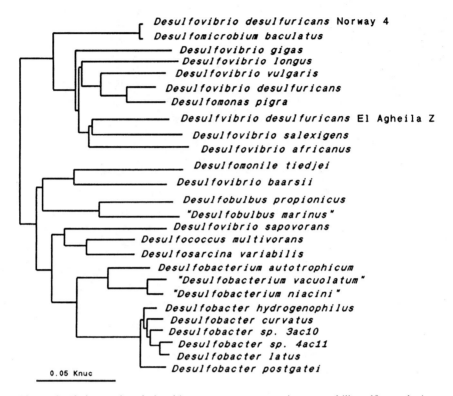

**Figure 2.** Phylogenetic relationships among gram-negative mesophilic sulfate-reducing bacteria based on 16S rRNA sequence comparisons. Scale bar is in units of fixed nucleotide substitutions per sequence position.

Comparative 16S ribosomal RNA sequence analyses have placed the mesophilic gram-negative species of SRB within the delta subclass of the *Proteobacteria* (Fowler *et al.*, 1986; Oyaizu and Woese, 1985). The delta subclass contains the myxobacteria, bdellovibrios, SRB, and has recently been expanded to include the genus *Pelobacter* (Stackebrandt *et al.*, 1989; Schink, 1992; Evers *et al.*, 1993). Comparative sequence analysis has also placed the iron-reducing bacterium *Geobacter metallireducens* within the delta subclass (Lovley *et al.*, 1993) (see Fig. 4 and note added in proof).

Initial analysis of phylogenetic relationships among SRB used the 16S rRNA cataloging approach (Fowler *et al.*, 1986). The results gave an indication of the phylogenetic divergence among SRB that had been suggested by the nutritional variation between genera. The *D. desulfuricans* and *D. gigas* line of descent was shown to be well separated from the line containing the other gram-negative genera included in the study set. The genera of the latter group were *Desulfobacter, Desulfosarcina, Desulfonema, Desulfococcus,* and *Desulfobulbus.* Diversity among these five genera was suggested in that the only close relationship demonstrated was between *Desulfonema* and *Desulfosarcina.*

The availability of complete 16S rRNA sequences, initially facilitated by the development of sequencing techniques using the enzyme reverse transcriptase (Lane *et al.,* 1985), made possible more rapid and more precise studies of phylogenetic relationships. Thus, subsequent comparisons of near-complete sequences further defined phylogenetic relationships among the gram-negative mesophilic species (Devereux *et al.,* 1989). These groupings, in general, correlated well with the classification based on physiological and biochemical characteristics. Two recent reviews summarize taxonomic and phylogenetic relationships among the SRB (Widdel and Bak, 1992; Devereux and Stahl, 1993).

Examination of the phylogenetic relationships determined among SRB has lead to the proposal of two families among the gram-negative mesophilic species. These are *"Desulfovibrionaceae"* and *"Desulfobacteriaceae"* (Devereux *et al.,* 1990; Widdel and Bak, 1992). The phylogenetic structure of these two families is illustrated in Fig. 2.

## 3.1. The Family *"Desulfovibrionaceae"*

The comparison of near-complete 16S rRNA sequences revealed the genus *Desulfovibrio* to be a phylogenetically diverse group (Devereux *et al.,* 1989) and, as shown by the previous cataloging analysis (Fowler *et al.,* 1986), distinct form the line leading to the other gram-negative mesophilic genera. Sequences from two *Desulfovibrios, D. sapovorans* and *D. baarsii,* fell well outside the main *Desulfovibrio* group; it has subsequently been proposed they be renamed *"Desulfoarculus" baarsii* and *"Desulfobotulus" sapovorans,* respectively (Widdel and Bak, 1992). The diversity observed within the genus *Desulfovibrio* has prompted more extensive studies.

Sequence comparisons of additional species and strains of *Desulfovibrio* indicated the presence of at least five lines related to each other at the level that separates other genera of SRB (Devereux *et al.,* 1990). Relationships within the *"Desulfovibrionaceae"* are shown in Fig. 2. The recognized lines at genus depth were represented by: (1) *D. salexigens* and *D. desulfuricans* strain El Aghelia Z; (2) *D. africanus;* (3) *D. desulfuricans* ATCC 2774, *Desulfuromonas pigra,* and *D. vulgaris* Hildenborough; (4) *D. gigas;* and (5) *Desulfomicrobium baculatus* and *D. desulfuricans* strain Norway 4. A correlation coefficient developed between

16S rRNA sequence similarity and genomic DNA:DNA homology supports the interpretation of divergence among *Desulfovibrio* species at the genus level (Devereux *et al.*, 1990). Phylogenetic diversity among the *Desulfovibrio* species was equivalent to that within other recognized bacterial families (such as *Rhizobiaceae*, *Enterobacteriaceae*, and *Vibrionaceae*). Additionally, a signature of 16S rRNA nucleotides was defined that distinguished *Desulfovibrio* species from the other genera of SRB. These observations lead to the proposal of the family "*Desulfovibrionaceae*" (Devereux *et al.*, 1990) in recognition of both the diversity within the present genus and the monophylogenetic origin of the group.

As additional physiological information is obtained, the present genus may be divided into separate genera as has been previously done by establishment of the genus *Desulfomicrobium* (Rozanova *et al.*, 1988). Unfortunately, chemotaxonomic properties that have been shown in many cases to be excellent markers for the delineation of phylogenetically defined clusters are sparse, and if available, are not characterized for all species. The presence of isoprenoid quinones of the MK-6 type, cytochrome $c_3$ and the presence of desulfoviridin appear to characterize members of *Desulfovibrio*. The mole% G+C range from 48 (*D. simplex, D. salexigens*) to 67 mole% (*D. termitidis*) and this parameter can most likely not be used to exclude the low G+C members from the genus as judged from the branching point of *D. salexigens* within the radiation of *Desulfovibrio* (Fig. 2). It is not surprising that the dendrogram based on fatty acids does not match that of the 16S rDNA. *Desulfomicrobium* strains, for example, characterized by cytochromes *b* and *c* and the lack of desulfoviridin, cluster with *D. africanus, D. halophilus*, and *D. salexigens* rather than outside the *Desulfovibrio* proper, as revealed by the phylogenetic analysis.

### 3.2. The Family "*Desulfobacteriaceae*"

Similar to the phylogenetic diversity among *Desulfovibrio* species, relationships among the remaining genera of gram-negative mesophilic species are of the order that they too can be considered to comprise a separate family (Devereux *et al.*, 1989, 1990). In addition, there is a unique 16S rRNA sequence signature to unite the members. Thus, the family "*Desulfobacteriaceae*" has been proposed to encompass them (Widdel and Bak, 1992). Relationships among these genera (Fig. 2) are based on the 16S rRNA studies of Devereux *et al.* (1989) and DeWeerd *et al.* (1990). These relationships are in agreement with the classical taxonomy (Widdel and Bak, 1992).

*Desulfobacter* species comprise a closely related group with members sharing more than 95% sequence similarity. Likewise, *Desulfobacterium autotrophicum* shares at least 95% sequence similarity with "*Desulfobacterium niacini*" and "*Desulfobacterium vacuolatum*," which supports the assignment of the latter two organisms to the genus. However, this placement is not obvious from the phenotypic properties, except perhaps from the presence of menaquinones with seven isoprene units. The two genera are derived from a recent common ancestor and are related by as much as a 90% similarity in 16S rRNA sequence.

The two *Desulfobulbus* species analyzed, *Desulfobulbus propionicus* and "*Desulfobulbus marinus*," though phenotypically homogeneous, appear to be phylogenetically more diverse than either *Desulfobacter* or *Desulfobacterium*. The characterized *Desulfobulbus* species share 92% similarity in 16S rRNA sequence and no more than 86% similarity with any member of any other group.

Among the higher carbon chain fatty acid oxidizers, *Desulfococcus multivorans* and *Desulfosarcina variabilis* share 92% similarity in 16S rRNA sequence. "*Desulfoarculus*" *sapovorans* appears to be related to this group sharing 90% sequence similarity with *Desulfococcus multivorans*.

*Desulfomonile tiedjei*, initially described as strain DCB-1, is an unusual SRB with the capability of reductive dechlorination of 3-chlorobenzoate (DeWeerd *et al.*, 1990). Initial 16S rRNA sequence comparisons with *Desulfomonile tiedjei* were against those from a sulfur-reducer and a limited number of sequences from SRB. The closest relationship found was to the sulfur-reducer. As shown in Figure 2, *Desulfomonile tiedjei* can be related to "*Desulfoarculus baarsii*"; the 16S rRNA sequences of the two are 88% similar.

A more precise phylogenetic placement of *Desulfonema limicola* must await analyses using complete 16S rDNA sequence. The $S_{AB}$ value of 0.53 derived from 16S rRNA cataloguing studies between this species and *Desulfosarcina variabilis* corresponds roughly to 88% sequence similarity. This value indicates that the branching point of this species is below the *Desulfosarcina/Desulfococcus* bifurcation point and it is low enough to support the validity of the genus *Desulfonema*.

## 4. THE PHYLOGENY OF THE THERMOPHILIC GRAM-NEGATIVE BACTERIA

For some time it has been shown that the thermophilic species are distantly related to the mesophilic species of SRB (Widdel, 1992b). Comparison of the 16S rRNA sequences from *T. commune* and a newly described thermophilic isolate, strain YP87 from a Yellowstone Park hot spring, has recently been completed (Henry *et al.,* 1994). The results demonstrated the origins of the gram-negative thermophilic SRB deep within the bacterial domain (Fig. 1). This analysis made use of transversion distances in tree construction. Thermophilic bacteria often possess ribosomal RNAs of high G+C content (from 60% to 66%) in contrast to mesophilic species (around 55%) (Woese *et al.,* 1991).

Sequences of high G+C content tend to cluster together when analyzed with parsimony or evolutionary distance methods (Weisburg *et al.,* 1989). However, the purine (and hence pyrimidine) content of ribosomal RNA sequences is fairly constant to a first approximation, and a phylogenetic analysis based on transversion distances is less susceptible to artifacts produced by G+C content (Woese *et al.,* 1991). The gram-negative thermophilic line of SRB, as shown in Fig. 1, is clearly distinct from (1) Gram-negative mesophilic SRB more closely related to the *Proteo-bacteria* (Fowler *et al.,* 1986; Devereux *et al.,* 1989, 1990); (2) from Gram-positive species (Fowler *et al.,* 1986; Devereux *et al.,* 1989) and (3) from the hyperthermophilic archaeal *Archaeoglobus* species (Woese *et al.,* 1991).

Despite considerable physiological similarity, the transversion distance between *T. commune* and strain YP87 (and slight differences in G+C content between YP87, the other described *Thermodesulfobacterium* species) indicated significant phylogenetic divergence (Henry *et al.,* 1994). This observation is similar to the phylogenetic diversity observed for the physiologically similar *Desulfovibrio* species (see below). Hence, strain YP87 was considered to represent a species of a new genus and the name *Thermodesulfovibrio yellowstonii* was proposed (Henry *et al.,* 1994).

The branching point for *T. commune* and *T. yellowstonii* deep within the bacterial domain strengthens the case for the thermophilic origins of the Bacterial domain (Achenbach-Richter *et al.,* 1987). As was previously described (Burggraf *et al.,* 1992) the other most deeply branching bacterial lines, the *Aquifex–Hydrogenobacter* complex, the *Themotogales* and *Coprothermobacter proteolyticus* (Rainey and Stackebrandt, 1993), are composed entirely of thermophilic species.

## 5. THE PHYLOGENY OF THE GRAM-POSITIVE SULFATE REDUCERS

Using the 16S rRNA cataloging approach, *Dm. nigrificans* and *Dm. acetoxidans* grouped with gram-positive bacteria of the *Clostridium/Bacillus* subphylum, with which they share a DNA G+C content of less than about 55 mole% (see below). Both species clustered together, but showed only a distant relationship to each other (Fowler *et al.*, 1986) that corresponds to about 86% sequence similarity (Devereux and Stahl, 1993). This low degree of relationship was supported by distinct genomic and phenotypic differences (see Table II). The closest relatives of *Desulfotomaculum* species initially identified were certain thermophilic members of the genus *Clostridium*. The phylogeny of a different pair of species, *Dm. orientis* and *Dm. ruminis*, was investigated using near-complete 16S rRNA sequences determined by the reverse transcriptase sequencing method (Devereux *et al.*, 1989). These species were even more distantly related to each other (83% similarity) and could not be clearly assigned to the same genus. This was somewhat surprising because they showed greater genomic and phenetic similarities than observed between *Dm. nigrificans* and *Dm. acetoxidans*. Analysis of sequence data from *Dm. australicum* (Love *et al.*, 1993), obtained by direct sequencing of 16S rDNA PCR products, shows it to cluster with *Dm. ruminis* and *Dm. orientis,* but the length of internodes was so small that the order at which these species evolved from each other could not be determined with certainty. Studies of members of the genus *Clostridium* and related taxa (Rainey *et al.*, 1993) revealed that the order of higher taxa within the subphylum could not be determined unambiguously. The topology of the tree changed with the selection of reference organisms, the length and position of sequences, and depended to a certain extent on the differences in the DNA base composition of mesophilic and thermophilic species. Furthermore, differences in the quality of sequences and gaps caused by unsequenced regions prevented a thorough phylogenetic analysis.

Following a more recent analysis of eight species (Rainey, unpublished) the unresolved phylogenetic structure of the genus *Desulfotomaculum* and its position among clostridia becomes even more obvious (Fig. 3). Using a different selection of clostridium species as reference organisms, this study revealed four species clusters, containing *Dm. orientis*, *Dm. acetoxidans*, *Dm. australicum*, and the other *Desulfotomaculum* species. Yet other topologies could be generated by comparing different sequence regions and by omitting the partial sequences of *Dm. niger* and *Dm. thermobenzoicum*. The phylogenetic tree shown in Fig. 3 is based on 500 nucleotides and a selection of reference strains that results in the *Desulfotomaculum* species clustering together. Pairs of specific, tentative relationships are *Dm. thermobenzoicum* and *Dm. australicum, Dm. geothermicum* and *Dm. sapomandens,* and *Dm. nigrificans* and *Dm. ruminis.* The relationship between species of the latter two strains is supported by phenotypic characters, while those of the other two pairs are much less obvious. As already seen in the 16S rRNA cataloging analysis, the closest

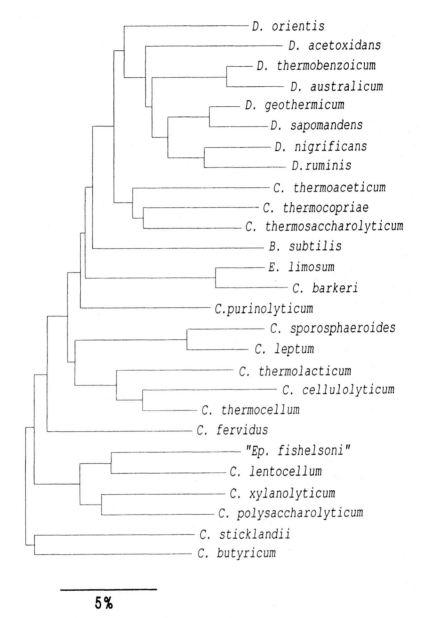

**Figure 3.** Phylogeny of members of the genus *Desulfotomaculum* within the radiation of gram-positive bacteria of the *Bacillus/Clostridium* subline of descent. The scale bar is in units of fixed nucleotide substitutions per 16S rRNA nucleotide position. The figure was kindly provided by Fred A. Rainey.

relatives of *Desulfotomaculum* are certain thermophilic clostridia. However, the relationships depicted in Fig. 3 need to be verified with complete 16S rDNA sequences some of which must be reanalyzed by more reliable methods. Note that *Dm. australicum,* as well as certain thermophilic clostridia, contains up to three rather large inserts (> 120 nucleotides) in their 16S rRNA genes that are not transcribed into mature rRNA species (Patel *et al.,* 1992; Rainey *et al.,* 1993). This feature has not been found in other prokaryotic rDNAs.

## 6. THE PHYLOGENY OF THE ARCHAEAL SULFUR-REDUCERS

With the isolation of *Archaeoglobus fulgidus* strain VC-16, a novel phenotype was included in the domain *Archaea.* This organism does not exclusively match any of the three major archaeal types in that it is not methanogenic, extremely halophilic, or extremely thermophilic, but it does reduce sulfate. It is the only archaeal organism recognized that uses this compound as a electron acceptor.

The observation that *A. fulgidus* also produces small amounts of methane made this organism a candidate for the missing link between the more ancient thermophilic sulfur reducers and the methanogenic archaeas (Achenbach-Richter *et al.,* 1987). Independent measurements of branching patterns by evolutionary distance and maximum parsimony and other strategies to determine the stability of tree topologies consistently placed the 16S rRNA between the two lines defined by *Thermococcus celer* (as a member of the *Euryarchaeota*) and *Methanococcus* species (Woese *et al.,* 1991). Interestingly, 16S rRNA cataloging of a highly related strain of VC-16 (Zellner *et al.,* 1989), namely strain Z, placed this isolate as a deep branching line within the radiation of methanogens (Stackebrandt and Zellner, unpublished). This discrepancy was resolved when Woese *et al.* (1991) observed the artificial clustering of lineages composed of thermophilic organisms. Their rRNA and rDNA may have a considerably higher mole% G+C base composition than mesophilic organisms. Elimination of the "composition-induced artifacts" by transversion analysis led to a change in the branching point of *A. fulgidus* strain VC-16, removing it from near the base of the euryarchaeota side of the archaeal tree and placing it within the phylogenetic cluster containing the *Methanomicrobiales* and the extreme halophiles (see also Brock and Madigan, 1991). The reason the cataloging approach provided a more accurate phylogeny than comparative analyses based on substantially more sequence information can be explained by the method used to generate 16S rRNA catalogs: the elimination of many of those positions with higher content of G-residues by RNase T1 introduces the bias of compositional disparities.

Today, members of *Archaeoglobus* are recognized to have evolved from methanogenic ancestors. Whether the enzymatic outfit for sulfur reduction is an ancient trait maintained in these taxa or whether it has been acquired by lateral gene transfer needs to be investigated and should be decided after comparative sequence comparison of orthologous genes. Structural studies on the adenylyl sulfate APS-reductase from *A. fulgidus* indicated the presence of [4Fe-4S] clusters, which are very similar to the corresponding enzyme from *Desulfovibrio gigas* (Lampreia *et al.*, 1991).

## 7. THE ANALYSIS OF GENETIC MARKERS OTHER THAN THE 16S rDNA

The rather specialized metabolism of SRB has raised considerable interest among microbial physiologists, biochemists, and geneticists. Most studies have been restricted to the *Desulfovibrio* species, some of which were investigated in detail in order to determine distribution, chemical composition, and primary structure of genes, amino acid composition of proteins, and the like. Only rarely were these studies initiated or interpreted from a taxonomic or phylogenetic perspective (although the electrophoretic motilities of ATP sulfurylases, sulfite reductases, and APS reductases have been considered of diagnostic value). Rather these earlier studies were mainly intended to provide information about the composition and function of genes and their products. Furthermore, most of these investigations were restricted to either defining the distribution of certain genes, for example, NifH gene (Kent *et al.*, 1989), and a gene tentatively identified as encoding a CO dehydrogenase (Stokkermans *et al.*, 1992), or to the genetic analysis of a few species only: genome size (*D. gigas, D. vulgaris* Hildenborough [Postgate, 1984b]), RNase P RNA (*D. desulfuricans* [Brown *et al.*, 1991]), succinate dehydrogenase gene (*Desulfobulbus elongatus*), protoporphyrinogen oxidase gene (*D. gigas*), orotidine-5'-phosphate decarboxylase gene (*D. vulgaris* [Li *et al.*, 1987]), chemoreceptor A gene (*D. vulgaris* Hildenborough [Dolla *et al.*, 1992]), rubredoxin (*D. vulgaris*, strains Miyazaki and Hildenborough [Shimizu *et al.*, 1989]), sulfite reductase (*D. vulgaris* Hildenborough [Pierik *et al.*, 1992]), desulforedoxin gene (*D. vulgaris* Hildenborough, *D. gigas* [Brumlik *et al.*, 1990]) and flavodoxin gene (*D. gigas, D. vulgaris, D. salexigens* [Helms and Swenson, 1992]). Only *D. gigas*,

*D. vulgaris* strains Hildenborough and Miyagawa, and *D. desulfuricans* were investigated in greater detail, but the data do not presently provide sufficient information that could be used to unravel events of the evolutionary history. The following is a short summary of those biochemical and genetic data that allow conclusions about the relationships of *Desulfovibrios* at the genetic level, other than those obtained from the analysis of 16S rDNA/DNA, to be drawn.

### 7.1. Ferredoxin

Members of *Desulfovibrio* have one to several ferredoxin types, (Fd I, II, III), distinguishable from each other by molecular weight, subunit structure, and Fe-S clusters per monomer (Cammack, 1983). Some species have a unique desulfoferredoxin (Moura *et al.*, 1990). Sequence comparison of *D. vulgaris* Miyazaki Fd II showed homology to Fd I of *D. desulfuricans* Norway, Fd I of *D. africanus*, and Fd of *D. gigas* (Okawara *et al.*, 1988a). Striking sequence similarities were found between Fd I of *D. vulgaris* Miyazaki and Fd III of *D. africanus* Benghazi (84%) as well as with Fd II of *D. desulfuricans* Norway (76%) (Okawara *et al.*, 1988b). In order to determine the phylogeny of ferredoxins, 31 prokaryotic type ferredoxins, including four types from three *Desulfovibrio* species, were compared (Fitch and Brutschi, 1987). The most parsimonious phylogenetic tree revealed that the *Desulfovibrios* appear as the most ancient group of the class Bacteria. *Thermus*, on the other hand, which is a line of rather ancient descent (based on comparative 16S rRNA sequencing), is closely related to other aerobic members of the genera *Bacillus* and *Mycobacterium* and members of the *Proteobacteria*. At least two duplications of the ferredoxin gene lead to the formation of paralogous genes. One of them was found in *D. desulfuricans* strain Norway. While the type I ferredoxin of this species clustered with the ferredoxins of *D. africanus* and *D. gigas*, the type II branched off later. Interesting, and not explainable, is the origin of the *Methanosarcina barkeri* ferredoxin within the radiation of the *Desulfovibrio* ferredoxins. Additional significant deviations are apparent between the branching patterns inferred from 16S rDNA and ferredoxin sequences (Fitch and Brutschi, 1987). This might in part be attributed to the poor confidence in the branching order of taxa below the upper quarter of the branching pattern inferred from ferredoxin sequence comparisons. Nevertheless, the three *Desulfovibrio* species group together and the large distances separating the lines is consistent with the 16S rDNA data indicating these three species are only remotely related.

## 7.2. Cytochrome *C*

As shown in Table I, the dissimilatory SRB exhibit different cytochromes. The principal cytochrome of *Desulfovibrio* is cytochrome $c3$ (Postgate, 1954), a small tetrahem cytochrome of very low redox potential that has so far only been found in members of this genus. Other cytochromes occur as well in *Desulfovibrios*, for example, C553 (Liu *et al.*, 1988; Nakagawa *et al.*, 1990; Bruschi *et al.* 1992), and it appears that the cytochrome system in strictly anaerobic sulfate reducers is complex (Ambler, 1991; Moura *et al.* 1991). The cytochrome systems of *Desulfotomaculum* species and other sulfate-reducers have been much less intensively investigated. Despite thorough genetic investigations on *Desulfovibrio* cytochromes, comprehensive phylogenetic comparisons of the primary structures of common genes and gene products have been restricted to only a few strains, including *D. desulfuricans* strains Berre-Eau (Moura *et al.*, 1987) and G 200 (Voordouw *et al.*, 1990a), *D. vulgaris* strains Hildenborough (Voordouw *et al.*, 1990a), Miyazaki (Tasaka *et al.*, 1991) and Norway (Loutfi *et al.*, 1989). Analyses of cytochrome $c_3$ for phylogenetic purposes are not greatly informative (Bruschi, 1981), probably because of the comparison of nonhomologous molecules.

## 7.3. Hydrogenases

Desulfovibrios contain three types of hydrogenases (hyd), which differ in their subunit and metal composition, amino acid sequence, gene structure, and catalytic and immunological properties (Fauque *et al.*, 1988). The complexity of the hydrogenase system, their nonubiquitous distribution in all species, and the unresolved questions about whether these proteins are of paralogous or orthologous origin, limits the use of hydrogenases as markers for evolution. All species investigated so far contain a [NiFe] hyd with two (4Fe-4S) centers and one (3Fe-xS) cluster in addition to nickel. Both periplasmic and membrane-bound hydrogenases have been identified. Conserved sequences in the genes coding for the periplasmic [NiFe] hyd have been found by hybridization between several species of *Desulfovibrio* (Li *et al.*, 1987). Amino acid sequence comparison revealed that the two subunits of the periplasmic [NiFe] hyd of *D. vulgaris* Miyazaki F share 80% homology with those of the [NiFe] hyd of *D. gigas* (Deckers *et al.*, 1990), while the periplasmic [NiFe] hyd from *D. fructosovorans* shared only 65% homology with that *D. gigas* (Hatchikian *et al.*, 1990; Rousset *et al.*, 1990). Conserved motifs have also been found in the large subunits of nickel-containing hydrogenases from *Dmi. baculatus* and *D. gigas* and in those of *Rhodobacter cap-*

*sulatus,* and even in the N-termini of the alpha subunits of the F420 hydrogenases from *Methanobacterium formicicum* and *M. thermoautotrophicum.* The [Fe] and the [NiFe-Se] hydrogenases are not present in all species of *Desulfovibrio.* While substantial sequence homology has been found between the [NiFe] hyd of *D. gigas* and the [NiFe-Se] hyd of *Dmi. baculatus* (Voordouw *et al.,* 1989), the periplasmic [NiFe] hydrogenase from *Dmi. baculatus* exhibits no homology to a periplasmic [Fe] hyd from *D. vulgaris* (Menon *et al.,* 1987). However, the latter enzyme from *D. vulgaris* (Hildenborough) shares some common properties (length, cysteine position, sequence) with the hydrogenase I from *Clostridium pasteurianum,* especially in the C-terminal half of the molecule (Meyer and Gagnon, 1991). A soluble [NiFe] hyd from *Thermodesulfobacterium mobile* has recently been isolated and characterized (Fauque *et al.,* 1992), but comparative analysis with other sulfate-reducers is still lacking.

## 8. DETERMINATIVE TOOLS AND NEW DEVELOPMENTS

Comparative molecular and biochemical studies are changing the character of determinative tools available for identification of SRB. These new tools have far greater application than simply the identification of laboratory isolates. Their direct application to environmental samples promises to serve as the basis for rapidly expanding our appreciation of environmental diversity. Measurement of sulfate-reduction rates in the environment can now be directly linked to community structure using molecular phylogenetic techniques. Thus sulfate-reduction that cannot be attributed to known SRB implies the presence of novel organisms.

### 8.1. Lipid Biomarkers

Lipid biomarker analysis served early as the basis for developing biochemical probes for determinative and environmental studies of SRB. Recently, a detailed comparison of the cellular fatty acid composition of *Desulfovibrio* species (Vainshtein *et al.,* 1992) demonstrated an excellent correspondence between phylogenetic relationships inferred by comparative 16S rRNA sequencing and relationships defined by numerical cluster analysis of cellular fatty acid profiles. A major division within the *Desulfovibrio* was defined by high abundance of either iso-17:1 or anteiso-15:0 fatty acids. Anteiso-15:0 defined a lineage containing *D. gigas* and related organisms that was distinct from other *Desulfovibrio* species. Specific relatives of *D. gigas,* as defined by fatty acid profiles, include *D. giganteus* (DSM 4123 and 4370), *D. fructosovorans* (DSM 3604), *D. carbinolicus* (DSM 3852), *D. sulfodismutans* (DSM 3696), *Desulfovibrio* sp. (DSM 6133), "*D. alcoholovorans*" (DSM 5433). Of note was the observation that the two strains of *D. gigas* examined (DSM 1382$^T$ and 496) were distinct in fatty acid profiles, suggesting the need for reclassification. A similar study examining a broader collection of SRB also demonstrated favorable correspondence (Kohring *et al.*, 1994).

The identification of distinctive lipid signatures among different natural assemblages of SRB has also served as a basis for direct inspection of environmental populations (White *et al.*, 1979; Parkes, 1987; Parkes *et al.*, 1993; Coleman *et al.*, 1993).

## 8.2. The Use of Nucleic Acid Hybridization for Determinative Studies

The use of nucleic acid hybridization in determinative and diagnostic microbiology is well established. Although most developmental efforts have been in medical applications, there is increasing application of these techniques to general microbiology. Given the recognized importance of SRB (corrosion, oil well souring, and biogeochemical cycling), they have received greater attention than most "nonmedical" microorganisms as target populations for nucleic acid probes. There are two general categories of nucleic acid hybridization for application to determinative studies. The first is the use of specific genes as targets for nucleic acid hybridization. The second category is whole genome hybridization.

Among examined specific gene targets, the ribosomal RNAs (either RNA or DNA) have received the greatest attention. This is largely due to the great flexibility afforded by the use of these genes (or transcription products) as hybridization targets. In particular, both species- and genus-specific hybridization probes can be, and have been, designed by targeting regions of the rRNAs that demonstrate variable sequence conservation (Stahl and Amann, 1991). These have been applied to both determinative and environmental studies (Devereux *et al.*, 1992; Kane *et al.*, 1993; Risatti *et al.*, 1995).

Although a variety of genes have been identified of potential utility for phylogenetic and determinative purposed (cytochrome $c_3$, bisulfite reductase, APS reductase) as detailed in section 7 of this chapter, such application has been essentially restricted to the hydrogenases. It is somewhat surprising that no genes in the pathway for dissimilatory sulfate-reduction have so far been used as probe targets.

## 8.3. Gene Probes

### 8.3.1. Ribosomal RNA Targeted Probes

A set of six oligonucleotide probes targeting conserved tracts of the 16S rRNAs from phylogenetically defined groups of SRB was recently described (Devereux *et al.*, 1992). The target groups for these probes are shown in Fig. 4. As for a variety of similar probes (ca. 20–30 nucleotides in length) that target the ribosomal RNAs, both radioactively labeled

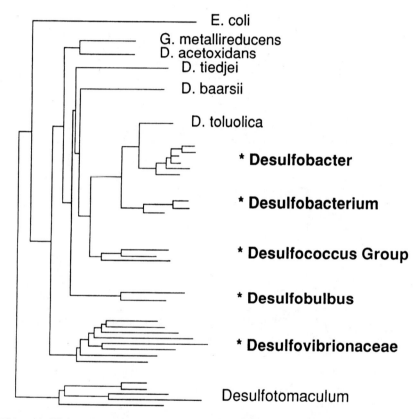

**Figure 4.** Oligonucleotide probe target groups in relationship to the phylogeny of sulfate-reducing bacteria. The five probe-target groups (bold with asterisk) are shown in relationship to different phylogenetic groups and species of sulfate-reducing bacteria. The *Desulfococcus*-group probe targets organisms from the genera *Desulfococcus, Desulfosarcina,* and *Desulfobotulus*. The distance tree was constructed as previously described (Devereux *et al.*, 1989). The probe sequences and demonstration of target group specificity were previously reported (Devereux *et al.*, 1992).

(for bulk nucleic acid hybridization) and fluorescent-dye labeled (for single-cell identification) probes have been used for determinative and environmental studies of SRB (Devereux *et al.*, 1992; Amann *et al.*, 1992). The specificity of the existing suite of probes was documented both empirically and within the context of environmental studies (Devereux *et al.*, 1992; Risatti *et al.*, submitted). The empirical demonstration of specificity tested probe hybridization against nontarget organisms. Because it is not realistic to experimentally test hybridization of these probes against all other known 16S rRNAs, this analysis used a reference collection applied to a single membrane ("phylogrid" membrane). The phylogrid membrane tests the hybridization of each probe to nucleic acids from 64 nontarget taxa, which together represent a diverse collection of eukarya, archaea, and bacteria. Each probe hybridized strongly to nucleic acids from target SRB but not to those of the other 64 representative taxa present on the phylogrid membranes. More recently, these probes have been used to study the abundance and distribution of SRB in a saline microbial mat community (Risatti *et al.*, 1995). This study demonstrated well-resolved stratification of SRB genera with depth in the mat. The observation of mostly nonoverlapping stratification of different SRB populations, in combination with direct measurement of associated process sulfate-reduction, supports the general utility of these probes in either determinative or environmental application.

### 8.3.2. Hydrogenases

The genes for three different hydrogenases ([Fe], [NiFe] and [NiFeSe]) have been cloned from *Desulfovibrio* species (see section 7.3., above). The DNA from all examined members of the genus *Desulfovibrio* hybridized to a [NiFe] hydrogenase gene probe from *D. vulgaris* Miyazaki F (Voordouw *et al.*, 1990b). Little or no hybridization was observed for other genera of SRB (for example, *D. baarsi*). Since different species carry this gene on differently sized restriction fragments, multiple species could be examined in a single sample (that is, within the same enrichment culture). However, only a subset of the inspected *Desulfovibrio* isolates hybridized with the [Fe] and [NiFeSe] hydrogenase gene probes. These same authors demonstrated the use of the [NiFe] hydrogenase to identify novel *Desulfovibrio* species, both in pure and enrichment culture (Voordouw *et al.*, 1990b).

### 8.3.3. Whole Genome Probing

A standard method to assess the genetic relationship between two organisms is DNA reassociation using total genomic DNA isolated from each. The extent of reassociation or hybridization between the DNAs provides an average measurement of the nucleotide sequence similarity between each organism. This general technique, used in a variety of experimental formats, is an established technique in molecular taxonomy (Johnson, 1991). A slight variation of this technique was recently described by Voordouw and colleagues (Voordouw *et al.*, 1992) to differentiate and identify SRB isolated from oil field samples. These authors used a collection of "standard" genome probes derived from appropriate reference organisms to screen a panel of DNA from uncharacterized SRB. The term "standard" is used to represent organisms that have little or no cross-reactivity in dot blot hybridization assays. In a variation of this standard approach, they reversed the procedure (reverse sample genome probing) by using the collection of reference organism DNA as the target for probes derived from environmental samples. The environmental samples were enrichment cultures (using alternative electron donors) derived from oil field samples. The later approach, using a reference panel of 35 sulfate-reducing standards, identified distinct communities of SRB in different oil fields (Voordouw *et al.*, 1991, 1992).

## 9. ORIGIN AND EVOLUTION OF SULFATE-REDUCING BACTERIA IN RELATIONSHIP TO THE BIOSPHERE

It has been proposed, based on geochemical data, that the first biologically derived oxygen (of presumed photosynthetic origin) was consumed in the oxidation of then abundant soluble iron [Fe(II)] (Walker, 1987). The oxidation of iron served to convert the oxidized end product of photosynthesis (oxygen) to insoluble iron hydroxides and oxides. As most oxidized forms of iron are very insoluble, the oxidized iron accumulated in sediments. Thus, it has been suggested that the Archaean biosphere was "upside down" in relation to the current biosphere: oxidized sediments and reduced atmosphere (Walker, 1987). The past activity of iron respiring microorganisms in the Archaean sediments is consistent with the isotopic fractionation record dating from that time (Walker, 1987).

As a contemporary group, the iron reducers are now recognized as important populations of both the subsurface and near-surface sediments (for example, of lake and marine systems) (Lovley, 1991; Perry *et al.*, 1993). Thus an outstanding question is the evolutionary relationships between sulfate- and iron-respiring anaerobes. It has been suggested, based on the above-noted hypothesis that abundant Fe(III) was available before abundant sulfur oxyanions on the early earth, that iron-based respiration predated sulfate-based respiration (Walker, 1987; Lovley, 1991). If so, this should be reflected in the phylogenetic relationships of organisms using these alternative electron accepting species. That is, if organisms using iron-based respiration emerged before those using sulfate, they should subsume (in phylogenetic depth and distribution) those capable of sulfate-reduction. This may explain the relatively limited phylogenetic distribution of organisms capable of using sulfate as an electron accepting species (Devereux, 1989). Limited phylogenetic distribution is consistent with their being relative latecomers in the evolution of anaerobic respiration.

Sulfur is also a major terminal electron accepting species among many hyperthermophilic archaea and could represent a more ancient form of respiration (Schauder and Kröger, 1993). The use of sulfur as a terminal electron acceptor is widely distributed among the Archaea and the Bacteria, including certain SRB. Thus, if the sequence of geological availability of different electron accepting species corresponds to the origin of their use in respiration, the likely sequence of evolutionary emergence is from sulfur, to Fe(III), to sulfate/sulfite. This evolutionary progression should be reflected in the phylogenetic relationships of contemporary organisms and/or the corresponding enzyme systems. These relationships are only now beginning to be explored by comparative sequence analyses. Of particular interest is the relationship, and overlap, of the enzyme systems participating in the reduction of iron and/or sulfate.

## 10. UNCULTURED NOVEL SULFATE-REDUCING BACTERIA

Recent studies using radiotracers or comparative 16S rRNA sequencing to directly interrogate the environmental populations, suggest yet greater diversity among SRB.

### 10.1. Greigite Magnetosomes

Magnetotactic bacteria collected from sulfidic, brackish-to-marine habitats have been shown to contain ferrimagnetic greigite ($Fe_3S_4$) and nonmagnetic pyrite ($FeS_2$) [Mann *et al.*, 1990; Farina *et al.*, 1990]. A recent comparison of 16S rRNA sequences obtained from enrichment cultures revealed that *Desulfosarcina variabilis* is among the closest relatives (DeLong *et al.*, 1993). However, since these organisms have yet to be brought into pure culture, it is unknown whether they grow via dissimilatory sulfate reduction. The observation of biotic pyrite formation is very intriguing with respect to the primitive metabolism proposed by Wächtershauser (1988). In the theory of the surface metabolist, the energy derived from pyrite formation was utilized by the earliest life forms.

### 10.2. Hyperthermophilic Sulfate-Reducing Bacteria

The indication that SRB are capable of growth at temperatures greater than 100 °C comes from recent radiotracer studies conducted at the Guaymas Basin tectonic spreading center in the Gulf of California (Jørgensen *et al.*, 1992). These studies demonstrated sulfate reduction at temperatures up to 106 °C, suggesting the existence of yet-to-be described hyperthermophilic SRB.

NOTE ADDED IN PROOF. 1. Liesack and Finster (1994) recently described a new genus, *Desulfuromonas*, which contains the species *D. kysingii*, *D. bakii*, and *D. succinoxidans*. This genus is related to the genera *Pelobacter* and *Desulfuromonas*. 2. *Desulfobacterium* has been validated by Galushko and Rozanova (1991) and the IJSB list (No. 49, IJSB 44, pp. 370–371). 3. Utkin *et al.* (1994) have recently designated a new genus, *Desulfitobacterium*, for a gram-positive bacterium, *D. dehalogenans*. This genus is related to *Desulfotomaculum orientis*.

### REFERENCES

Achenbach-Richter, L., Gupta, R., Stetter, K. O., and Woese, C. R., 1987, Were the original eubacteria thermophiles? *Syst Appl. Microbiol.,* **9:**34–38.

Adams, M. E., and Postgate, J. R., 1959, A new sulphate reducing vibrio. *J. Gen. Microbiol.,* **20:**252–257.

Amann, R. I., Stromley, J., Devereux, R., Key, R., and Stahl, D. A., 1992, Molecular and microscopic identification of sulfate-reducing bacteria in multispecies biofilms, *Appl. Environ. Microbiol.* **58:**614–623.

Ambler, R. P., 1991, Sequence variability in bacterial cytochromes c, *Biochim. Biophys. Acta* **1058:**42–47.

Baars, J. K., 1930, Over sulfaatreductie door bacteriën. Ph:D. Thesis, University of Delft, The Netherlands.

Beijerinck, W. M., 1895, Ueber *Spirillum desulfuricans* als Ursache von Sulfatreduktion, *Zentralbl. Bakteriol.* 2. Abt. **1:**1–9, 49–59, 104–114.

Brock, T. D., and Madigan, M. T. (eds.), 1991, *Biology of Microorganisms,* 6th. ed. Prentice-Hall International Editions, London.

Brown, J. W., Haas, E. S., James, B. D., Hunt, D. A., and Pace, N. R., 1991, Phylogenetic analysis and evolution of RNase P RNA in *Proteobacteria, J. Bacteriol.* **173:**3855–3863.

Brumlik, M. J., Leroy, G., Bruschi, M., and Voordouw, G., 1990, The nucleotide sequence of the *Desulfovibrio gigas* desulforedoxin gene indicates that the *Desulfovibrio vulgaris* rbo gene originated from a gene fusion event, *J. Bacteriol.* **172:**7289–7292.

Bruschi, M., 1981, The primary structure of the tetraheme cytochrome C3 from *Desulfovibrio desulfuricans* (strain Norway 4), *Biochim. Biophys. Acta* **671:**219–226.

Bruschi, M., Bertrand, P., More, C., Leroy, G., Bonicel, J., Haladjian, J., Chottard, G., Pollock, W. B., Voordouw, G., 1992, Biochemical and spectroscopic characterization of the high molecular weight cytochrome c from *Desulfovibrio vulgaris* Hildenborough expressed in *Desulfovibrio desulfuricans* G200, *Biochem*, **31:**3281–3288.

Burgraff, S., Olsen, G. J., Stetter, K. O., and Woese, C. R., 1992, A phylogenetic analysis of *Aquifex pyrophilus*, *Syst. Appl. Microbiol.* **15:**352–356.

Cammack, R., 1983, Evolution and diversity in the iron-sulphur proteins, *Chemica Scripta* **21:**87–95.

Campbell, L. L., and Postgate, J. R., 1965, Classification of the sporeforming sulfate-reducing bacteria, *Bacteriol. Rev.* **29:**359–363.

Campbell, L. L., and Postgate, J. R., 1969, Revision of the holotype strain of *Desulfotomaculum ruminis* (Coleman) Campbell and Postgate, *Int. J. Syst. Bacteriol.* **19:**139–140.

Campbell, L. L., and Singleton, R., 1986, Endospore-forming Gram-positive rods and cocci, in: *Bergey's Manual of Systematic Bacteriology* (P. H. A. Sneath, R. S. Mair, M. E. Sharpe, and J. G. Holt, eds.) Williams & Wilkins, Baltimore, pp. 1200–1202.

Campbell, L. L., Frank, H. A., and Hall, E. R., 1957, Studies on the thermophilic sulfate reducing bacteria. I. Identification of *Sporovibrio desulfuricans* as *Clostridium nigrificans*, *J. Bacteriol.* **73:**516–521.

Canfield, D. E., and Des Marais, D. J., 1991, Aerobic sulfate reduction in microbial mats, *Science* **251:**1471–1473.

Coleman, G. S., 1960, A sulphate-reducing bacterium from the sheep rumen, *J. Gen. Microbiol.* **22:**423–436.

Coleman, M. L., Hedrick, D. B., Lovley, D. R., White, D. C., Pye, K., 1993, Reduction of Fe (III) in sediments by sulphate-reducing bacteria, *Nature* **361:**436–438.

Collins, M. D., and Widdel, F., 1986, Respiratory quinones of sulfate-reducing and sulfur reducing bacteria: a systematic investigation, *Syst. Appl. Microbiol.* **8:**8–18.

Dahl, C., Koch, H.-G., Keuken, O., and Trüper, H. G., 1990, Purification and characterization of ATP sulfurylase from the extremely thermophilic archaebacterial sulfate-reducer, *Archaeoglobus fulgidus*, *FEMS Microbiol. Lett.* **67:**27–32.

Daumas, S. Cord-Ruwisch, R. and Garcia, J. L., 1988, *Desulfotomaculum geothermicum* sp. nov., a thermophilic, fatty acid-degrading, sulfate-reducing bacterium isolated with H$_2$ from geothermal ground water, *Antonie van Leewenhoek J. Microbiol. Ser.* **54:**165–178.

Deckers, H. M., Wilson, F. R. and Voordouw, G., 1990, Cloning and sequencing of a [NiFe] hydrogenase operon from *Desulfovibrio vulgaris* Miyazaki F, *J. Gen. Microbiol.* **136:**2021–2028.

DeLong, E. F., Frankel, R. B., and Bazylinski, D. A., 1993, Multiple evolutionary origins of magnetotaxis in bacteria, *Science* **259:**803–806.

Devereux, R., and Stahl, D. A., 1993, Phylogeny of sulfate-reducing bacteria and a perspective for analyzing their natural communities, in: *The Sulfate-Reducing Bacteria: Contemporary Perspectives* (J. M. Odom and R. Singleton, Jr., eds), Springer-Verlag, New York, pp. 131–160.

Devereux, R., Delaney, M., Widdel, F., and Stahl, D. A., 1989, Natural relationships among sulfate-reducing eubacteria, *J. Bacteriol.* **171:**6689–6695.

Devereux, R., He, S.-H., Doyle, C. L., Orkland, S., Stahl, D. A., LeGall, J., and Whitman, W. B., 1990, Diversity and origin of *Desulfovibrio* species: phylogenetic definition of a family, *J. Bacteriol.* **172:**3609–3619.

Devereux, R., Kane, M. D., Winfrey, J., and Stahl, D. A., 1992, Genus- and group-specific hybridization probes for determinative and environmental studies of sulfate-reducing bacteria, *Syst. Appl. Microbiol.* **15:**601–609.

DeWeerd, K., Mandelco, L., Tanner, R. S., Woese, C. R., and Suflita, J. M., 1990, De-

*sulfomonile tiedjei* gen. nov. and sp. nov., a novel anaerobic dehalogenating, sulfate-reducing bacterium, *Arch. Microbiol.* **154**:23–30.

Dolla, A., Fu, R., Brumlik, M. J. and Voordouw, G., 1992, Nucleotide sequence of DcrA, a *Desulfovibrio vulgaris* Hildenborough chemoreceptor gene, and its expression in *Escherichia coli, J. Bacteriol.* **174**:1726–1733.

Drzyzga, O., Kuever, J., and Blotevogel, K-H., 1993, Complete oxidation of benzoate and 4-hydroxybenzoate by a new sulfate-reducing bacterum resembling *Desulfoarculus*, *Arch. Microbiol.* **159**:109–113.

Elion, L., 1925, A thermophilic sulphate reducing bacterium, *Centralbl. Bakteriol (II. Abt.)* **63**:58–67.

Evers, S., Weizenegger, M., Ludwig, W., Schink, B., and Schleifer, K. H., 1993, The phylogenetic positions of *Pelobacter acetylenicus* and *Pelobacter propionicus*, *Syst. Appl. Microbiol.* **16**:216–218.

Farina, M., Esquivel, M. S., and Lins de Barros, H. G. P., 1990, Magnetic iron-sulphur crystals from a magnetotactic microorganism, *Nature* **343**:256–258.

Faquque, G., Czechowski, M., Berlier, Y. M., Lespinat, P. A., LeGall, J. and Moura, J. J., 1992, Partial purification and characterization of the first hydrogenase isolated from a thermophilic sulfate-reducing bacterium, *Biochem. Biophys. Res. Commum.* **184**:1256–1260.

Fauque, G., Lino, A. R., Czechowski, M., Kang, L., DerVartanian, D. V. Moura, J. J., LeGall, J., and Moura, I., 1990, Purification and characterization of bisulfite reductase (desulfofuscidin) from *Desulfovibrio thermophilus* and its complexes with exogenous ligands, *Biochim. Biophys. Acta* **1040**:112–118.

Fauque, G., Peck, H. D. Jr., Moura, J. J., Huynh, B. II., Berlier, Y., DerVartanian, D. V., Teixeira, M., Rzybyla, E. E., Lespinat, P. A., 1988, Three classes of hydrogenases from sulfate-reducing bacteria of the genus *Desulfovibrio, FEMS Microbiol. Rev.* **4**:299–344.

Fitch, W. M., and Bruschi, M., 1987, The evolution of prokaryotic ferredoxins—with a general method correcting for unobserved substitutions in less branched lineages, *Mol. Biol. Evol.* **4**:381–394.

Fowler, V. J., Widdel, F., Pfennig, F., Woese, C. R., and Stackebrandt, E., 1986, Phylogenetic relationships of sulfate- and sulfur-reducing bacteria, *Syst. Appl. Microbiol.* **8**:32–41.

Galushko, A. S., and Rozanova, E. P., 1991, *Desulfobacterium cetonicum* sp. nov.: a sulfate-reducing bacterium which oxidizes fatty acids and ketones, *Mikrobiologiya* **60**:102–107.

Gebhart, C. J., Barns, S. M., McOrist, S., Lin, G-F., and Lawson, G. H. K., 1993, Ileal symbiont intracellularis, an obligate intracellular bacterium of porcine intestines showing a relationship to *Desulfovibrio* species, *Int. J. Syst. Bacteriol.* **43**:533–538.

Gogotava, G. I., and Vainstein, M. B., 1989, Description of sulfate reducing bacterium *Desulfobacterium macestii* sp. nov. capable of autotrophic growth, *Microbiologiya* **58**:76–80.

Hatchikian, C. E., and Zeikus, J. G., 1983, Characterization of a new type of dissimilatory sulfite reductase present in *Thermodesulfobacterium commune, J. Bacteriol.* **153**:1211–1220.

Hatchikian, C. E., Traore, A. S., Fernandez, V. M. and Cammack, R., 1990, Characterization of the nickel-iron periplasmic hydrogenase from *Desulfovibrio fructosovorans, Eur. J. Biochem.* **187**:635–643.

Helms, L. R., and Swenson, R. P., 1992, The primary structure of the flavodoxins from two strains of *Desulfovibrio* gigas. Cloning and nucleotide sequence of the structural genes, *Biochim. Biophys. Acta* **1131**:325–328.

Henry, E. A., Devereux, R., Maki, J. S., Gilmour, C. C., Woese, C. R., Mandelco, L., Schauder, R., Remsen, C. C., and Mitchell, R., 1994, Isolation and characterization of *Thermodesulfovibrio yellowstonii* gen. nov. and sp. nov.: A New thermophilic sulfate-reducing bacterium from hydrothermal vent waters in Yellowstone Lake, Wyoming, USA, *Arch. Microbiol.* **161**:62–69.

Huber, R., Wilharm, T., Huber, D., Trincone, A., Burggraf, S., König, H., Rachel, R., Rockinger, I., Fricke, H., and Stetter, K. O., 1992, *Aquifex pyrophilus* gen. nov. sp. nov., represents a novel group of marine hyperthermophilic hydrogen-oxidizing bacteria, *System. Appl. Microbiol.* **15**:340–351.

Imhoff-Stuckle, D., and Pfennig, N., 1983, Isolation and characterization of a nicotinic acid-degrading sulfate-reducing bacterium, *Desulfococcus niacini* sp. nov, *Arch. Microbiol.* **136**:194–198.

Johnson, J. L., 1991, DNA Reassociation Experiments, in: *Sequencing and Hybridization Techniques in Bacterial Systematics* (E. Stackebrandt and M. Goodfellow, eds.), John Wiley & Sons, Chichester, England, pp. 21–44.

Jørgensen, B. B., Isaksen, M. F., and Jannasch, H. W., 1992, Bacterial sulfate reduction above 100 °C in deep sea hydrothermal vent sediments, *Science* **258**:1756–1757.

Kane, M. D., Poulsen, L. K., and Stahl, D. A., 1993, Monitoring the enrichment and isolation of sulfate-reducing bacteria by using oligonucleotide hybridization probes designed from environmentally derived 16S rRNA sequences, *Appl. Environ. Microbiol.* **59**:682–686.

Kent, H. M., Buck, M. and Evans, D. J., 1989, Cloning and sequencing of the NifH gene of *Desulfovibrio gigas*, *FEMS Microbiol. Lett.* **52**:73–78.

Kluyver, A. J., and Van Niel, C. B., 1936, Prospects for a natural system of classification of bacteria, *Zentralbl. Bakteriol. (2. Abt.)* **94**:369–403.

Kohring, L. L., Ringelberg, D. B., Devereux, R., Stahl, D. A., Mittelman, M. W., and White, D. C. (1994) Comparison of phylogenetic relationships based on phospholipid fatty acid profiles and ribosomal RNA sequence similarities among dissimilatory sulfate-reducing bacteria, *FEMS Microbiol. Lett.* **119**:303–308.

Kuever, J., Kulmer, J., Jannsen, S., Fischer, U., and Blotevogel, K-H., 1993, Isolation and characterization of a new spore-forming sulfate-reducing bacterium growing by complete oxidation of catechol, *Arch. Microbiol.* **159**:282–288.

Lampreia, J., Fauque, G., Speich, N., Dahl, C., Moura, I., Trüper, H. G., and Moura, J. J., 1991, Spectroscopic studies on APS reductase isolated from the hyperthermophilic sulfate-reducing archaebacterium *Archaeoglobus fulgidus*. *Biochem. Biophys. Res. Commun.* **181**:342–347.

Lane, D. J., Pace, B., Olsen, G. J., Stahl, D. A., Sogin, M. L., and Pace, N. R., 1985, Rapid determination of 16S ribosomal RNA sequences for phylogenetic analysis. *Proc. Natl. Acad. Sci. USA* **82**:6955–6959.

Langworthy, T. A., Holzer, G., Zeikus, J. G., and Tornabene, T. G., 1983, Iso- and anteiso-branched glycerol diethers of the thermophilic anaerobe *Thermodesulfobacterium commune*, *Syst. Appl. Microbiol.* **4**:1–7.

Li, C., Peck, H. D., LeGall, J., and Przybyla, A. E., 1987, Cloning, characterization, and sequencing of the genes encoding the large and small subunits of the periplasmatic [NiFe] hydrogenase of *Desulfovibrio gigas*, *DNA* **6**:539–551.

Liesack, W., and Finster, K., 1994, Phylogenetic analysis of five strains of gram-negative obligately anaerobic, sulfur-reducing bacteria and description of *Desulfuromonas* gen. nov., including *Desulfuromusa kysingii* sp. nov., *Desulfuromusa bakii* sp. nov., and *Desulfuromusa succinoxidans* sp. nov., *IJSB* **44**:753–758.

Liu, M. C., Costa, C., Coutinho, I. B., Moura, J. J., Moura, I., Xavier, A. V., and LeGall, J., 1988, Cytochrome components of nitrate -and sulfur-respiring *Desulfovibrio desulfuricans* ATCC 27774, *J. Bacteriol.* **170**:5545–5551.

Loutfi, M., Guerlesquin, F., Bianco, P., Haladjian, J., and Bruschi, M., 1989, Comparative studies of polyhemic cytochromes c isolated from *Desulfovibrio vulgaris* (Hildenborough) and *Desulfovibrio desulfuricans* (Norway), *Biochem. Biophys. Res. Commun.* **159**:670–676.

Love, C. A., Patel, B. K. C., Nichols, P. D. and Stackebrandt, E. 1993, *Desulfotomaculum australicum*, sp. nov., a thermophilic sulfate-reducing bacterium isolated from the great Artesian basis of Australia, *System. Appl. Microbiol.* **16**:244–251.

Lovley, D. R., 1991. Dissimilatory Fe(III) and Mn(IV) reduction, *Microbiol. Rev.* **55**:259–287.

Lovley, D. R., and Phillips, E. J. P., 1992, Reduction of uranium by *Desulfovibrio desulfuricans*, *Appl. Environ. Microbiol.* **58**:850–856.

Lovley, D. R., Giovannoni, S. J., White, D. C., Champine, J. E., Phillips, J. P., Gorby, Y. A., and Goodwin, S., 1993, *Geobacter metallireducens* gen. nov. sp. nov., a microorganism capable of coupling the complete oxidation of organic compounds to the reduction of iron and other metals, *Arch. Microbiol.* **159**:336–344.

Lovley, D. R., Roden, E. E., Phillips, E. J. P., and Woodward, J. C., 1993, Enzymatic iron and uranium reduction by sulfate-reducing bacteria, *Marine Geol.* **113**:41–53.

Mann, S., Sparks, N. H. C., Frankel, R. B., Bazylinski, D. A., and Jannasch, H., 1990, Biomineralization of ferrimagnetic greigite ($FeS_4$) and iron pyrite ($FeS_2$) in a magnetotactic bacterium, *Nature* **343**:258–261.

Marschall, C., Frenzel, P., and Cypionka, H., 1993, Influence of oxygen on sulfate reduction and growth of sulfate-reducing bacteria, *Arch. Microbiol.* **159**:168–173.

Menon, N. K., Peck, H. D., LeGall, J., and Przybyla, A. E., 1987, Cloning and sequencing of the genes encoding the large and small subunits of the periplasmatic (NiFeSe) hydrogenase of *Desulfovibrio baculatus.*, *J. Bacteriol.* **169**:5401–5407.

Meyer, J., and Gagnon, J., 1991, Primary structure of hydrogenase I from *Clostridium pasteurianum*, *Biochem.* **30**:9697–9704.

Migula, W., 1900, *System der Bakterien.* Gustav Fischer, Jena, Germany.

Min, H., and Zinder, S. H., 1990, Isolation and characterization of a thermophilic sulfate-reducing bacterium *Desulfotomaculum thermoacetooxidans*, *Arch. Microbiol.* **153**:399–404.

Möller-Zinkhan, D., Boerner, G., and Thauer, R. K., 1989, Function of methanofuran, tetrahydromethanopterin and coenzyme $F_{420}$ in *Archaeoglobus fulgidus*, **152**:362–368.

Moore, W. E. C., Johnson, J. L., and Holdeman, L. V., 1976, Emendation of the *Bacteroides* and *Butyrivibrio* and description of *Desulfomonas* gen. nov. and ten species of the genera *Desulfomonas*, *Butyrivibrio*, *Eubacterium*, *Clostridium* and *Ruminococcus*, *Int. J. Syst. Bacteriol.* **26**:238–252.

Moura, I., Fauque, G., LeGall, J., Xavier, A. V., and Moura, J. J., 1987, Characterization of the cytochrome c system of a nitrogen-fixing strain of a sulfate-reducing bacterium: *Desulfovibrio desulfuricans* strain Berre-Eau, *Eur. J. Biochem.* **162**:547–554.

Moura, I., Tavares, P., Moura, J. J., Ravi, N., Huynh, B. H., Liu, M. Y., and LeGall, 1990, Purification and characterization of desulfoferrodoxin. A novel protein from *Desulfovibrio desulfuricans* (ATCC 27774) and from *Desulfovibrio vulgaris* (strain Hildenborough) that contains a distorted rubredoxin center and a mononuclear ferrous center, *J. Biol. Chem.* **265**:21596–21602.

Moura, J. J., Costa, C., Liu, M. Y., Moura, I., and LeGall, J., 1991, Structural and functional approach toward a classification of the complex cytochrome c system found in sulfate-reducing bacteria, *Biochim. Biophys. Acta* **1058**:61–66.

Nakagawa, A., Higuchi, Y., Yasuoka, N., Katsube, Y., and Yagi, T., 1990, S-class cytochromes c have a variety of folding patterns: structure of cytochrome c-553 from *Desulfovibrio* vulgaris determined by the multi-wavelength anomalous dispersion method, *J. Biochem. Tokyo.* **108**:701–703.

Nazina, T. N., and Pivovarova, T. A., 1979, Submicroscopic organization and spore formation in *Desulfotomaculum nigrificans, Microbiologiya* (USSR) **57:**823–827.

Nazina, T. N., Ivanova, A. E., Kanchaveli, L. P., and Rozanova, E. P., 1988, *Desulfotomaculum kutznetsovii* sp. nov., a new spore-forming thermophilic methylotrophic sulfate-reducing bacterium. *Microbiologiya* (USSR) **57:**823–827.

Nazina, T. N., Poltaraus, A. B., and Rozanova, E. P., 1987, Estimation of genetic relationship between rod-shaped asporogenic sulfate-reducing bacteria, *Microbiologiya* (USSR) **56:**845–848.

Okawara, N., Ogata, M., Yagi, T., Wakabayashi, S., and Matsubara, H., 1988a, Characterization and complete amino acid sequence of ferredoxin II from *Desulfovibrio vulgaris* Miyazaki, *Biochim.* **70:**1815–1820.

Okawara, N., Ogata, M., Yagi, T., Wakabayashi, S., and Matsubara, H., 1988b, Amino acid sequence of ferredoxin I from *Desulfovibrio vulgaris* Miyazaki, *J. Biochem. Tokyo* **104:**196–199.

Ollivier, B., Hatchikian, C. E., Prensier, G., Guetennec, J., and Garcia, J.-L., 1991, *Desulfohalobium retbaense* gen. nov., sp. nov., a halophilic sulfate-reducing bacterium from sediments of a hypersaline lake in Senegal, *Int. J. Syst. Bacteriol.* **41:**74–81.

Oyaizu, H., and Woese, C. R., 1985, Phylogenetic relationships among the sulfate respiring bacteria, myxobacteria, and purple bacteria, *Syst. Appl. Microbiol.* **6:**257–263.

Pace, B., and Campbell, L. L., 1971, Homology of ribosomal ribonucleic acid of *Desulfovibrio* species with *Desulfovibrio vulgaris, J. Bacteriol.* **106:**717–719.

Parkes, R. J., 1987, Analysis of microbial communities within sediments using biomarkers, in: *Ecology of Microbial Communities,* Cambridge University Press, Cambridge, England, pp. 147–177.

Parkes, R. J., Dowling, N. J. E., White, D. C., Herbert, R. A., and Gibson, G. R., 1993, Characterization of sulphate-reducing bacterial populations within marine and estuarine sediments with different rates of sulphate reduction, *FEMS Microbiol. Ecol.* **102:**235–250.

Patel, B. K. C., Skeratt, J. H., and Nichols, P. D., 1991, The phospholipid ester-linked fatty acid composition of thermophilic bacteria, *Syst. Appl. Microbiol.* **14:**311–316.

Patel, B. K. C., Love, C. A., and Stackebrandt, E., 1992, Helix 6 of the 16S rDNA of the bacterium *Desulfotomaculum australicum* exhibits an unusual structural idiosyncrasy, *Nucl. Acids Res.* **20:**5483.

Perry, K. A., Kostka, J. E., Luther, G. W., and Nealson, K. H., 1993, Mediation of sulfur speciation by a black sea faculatative anaerobe, *Science* **259:**801–803.

Pfennig, N., and Biebl, H., 1981, The dissimilatory sulfur-reducing bacteria, in: *The Prokaryotes,* 2nd ed., (A. Balows, H. G. Trüper, M. Dworkin, W. Harder, and K.-H. Schleifer, eds.). Springer-Verlag, New York, pp. 941–947.

Pfennig, N., Widdel, F., and Trüper, H. G., 1981, The dissimilatory sulfate-reducing bacteria, in: *The Prokaryotes,* 2nd ed., (A. Balows, H. G. Trüper, M. Dworkin, W. Harder, and K.-H. Schleifer, eds.). Springer-Verlag, New York, pp. 926–940.

Pierik, A. J., Duyvis, M. G., van Helvoort, J. M., Wolbert, R. B., and Hagen, W. R., 1992, The third subunit of desulfoviridin-type dissimilatory sulfite reductases, *Eur. J. Biochem.* **205:**111–115.

Platen, H., Temmes, A., and Schink, B., 1990, Anaerobic degradation of acetone by *Desulfococcus baculatus* spec. nov., *Arch. Microbiol.* **154:**355–361.

Postgate, J. R., 1954, Presence of cytochrome in obligate anaerobe. *Biochem. J.* **56:**xi–xii.

Postgate, J. R., 1956, Cytochrome c3 and desulfoviridin; pigments of the anaerobe *Desulphovibrio desulphuricans, J. Gen. Microbiol.* **14:**545–572.

Postgate, J. R., 1984a, *The Sulfate-Reducing Bacteria.* Cambridge University Press, Cambridge.

Postgate, J. R., 1984b, Genus *Desulfovibrio*, in: *Bergey's Manual of Systematic Bacteriology*, Vol. 1 (N. R. Krieg and J. G. Holt, eds). Williams & Wilkins, Baltimore, pp. 666–672.

Postgate, J. R. and Campbell, L. L., 1966, Classification of *Desulfovibrio* species, the non-sporulating sulfate-reducing bacteria, *Bacteriol. Rev.* **30**:732–738.

Rabus, R., Nordhaus, R., Ludwig, W., and Widdel, F., 1993, Complete oxidation of toluene under strictly anoxic conditions by a new sulfate-reducing bacterium, *Appl. Environ. Microbiol.* **59**:1444–1451.

Rainey, F. A., and Stackebrandt, E., 1993, Transfer of the type species of the genus *Thermobacteroides* to the genus *Thermoanaerobacter* as *Thermoanaerobacter acetoethylicus* (Ben Basset and Zeikus 1981) comb. nov., description of *Coprothermobacter* gen. nov., and reclassification of *Thermobacteroides proteolyticus* as *Coprothermobacter proteolyticus* (Ollivier *et al.*, 1985) comb. nov., *Int. J. Syst. Bacteriol.* **43**:857–859.

Rainey, F. A., Ward, N. L., Morgan, H. W., and Stackebrandt, E., 1993, A phylogenetic analysis of anaerobic thermophilic bacteria: an aid for their reclassification, *J. Bacteriol.* **175**:4772–4779.

Risatti, J. B., Capman, W. and Stahl, D. A., (1995) Community structure of a microbial mat: the phylogenetic dimension, *Proc. Natl. Acad. Sci. USA*, in press.

Rousset, M., Dermoun, Z., Hatchikian, C. E., and Belaich, J. P., 1990, Cloning and sequencing of the locus encoding the large and small subunit genes of the periplasmatic [NiFe] hydrogenase from *Desulfovibrio fructosovorans*, *Gene* **94**:95–101.

Rozanova, E. P., and Khudyakova, A. I., 1974, A new non-sporeforming thermophilic sulfate reducing organism, *Desulfovibrio thermophilus* nov. spec, *Microbiologiya* (USSR) **43**:1069–1075.

Rozanova, E. P., and Nazina, T. N., 1976, A mesophilic, sulfate-reducing, rod-shaped nonspore-forming bacterium, *Microbiologiya* **45**:825–830.

Rozanova, E. P., and Pivovarova, T. A., 1988, Reclassification of *Desulfovibrio thermophilus*, *Microbiologiya* (Engl. Translation) **57**:85–89.

Rozanova, E. P., Nazina, T. N., and Galushko, A. S., 1988, A new genus of sulfate-reducing bacteria and the description of its new species *Desulfomicrobium apsheronum* gen. nov., sp. nov, *Mikrobiologiya* (USSR) **57**:634–641.

Schauder, R., and Kröger, A., 1993. Bacterial sulphur respiration, *Arch. Microbiol.* **159**:491–497.

Schink, B., 1992, The genus *Pelobacter*, in: *The Prokaryotes*, 2nd ed., (A. Balows, H. G. Trüper, M. Dworkin, W. Harder, and K.-H. Schleifer, eds.), Springer-Verlag, New York, pp. 3393–3399.

Schnell, S., and Schink, B., 1993, Anaerobic degradation of 3-aminobenzoate by a newly isolated sulfate reducer and a methanogenic enrichment culture, *Arch. Microbiol.* **158**: 328–334.

Schnell, S., Bak, F., and Pfennig, N., 1989, Anaerobic degradation of aniline and dihydroxybenzenes by newly isolated sulfate-reducing bacteria and description of *Desulfobacterium anilini*, *Arch. Microbiol.* **152**:556–563.

Shimizu, F., Ogata, M., Yagi, T., Wakabayashi, S., and Matsubara, H., 1989, Amino acid sequence and function of rubredoxin from *Desulfovibrio vulgaris* Miyazaki, *Biochimie* **71**:1171–1177.

Skerman, V. B. D., McGowan, V., and Sneath, P. H. A., 1980, Approved lists of bacterial names, *Int. J. Syst. Bacteriol.* **30**:225–420.

Sleytr, R., Adam, H., and Klaushofer, H., 1969, Die Feinstruktur der Zellwand und Cytoplasmamembran von *Clostridium nigrificans*, dargestellt mit Hilfe der Gefrierätz-und Ultradünnschnittechnik, *Arch. Mikrobiol.* **66**:40–58.

Speich, N., and Trüper, H. G., 1988, Adenylylsulphate reductase in a dissimilatory sulphate-reducing archaebacterium, *J. Gen. Microbiol.* **134**:1419–1425.

Stackebrandt, E., Wehmeyer, U., and Schink, B., 1989, The phylogenetic status of *Pelobacter acidigallici*, *Pelobacter venetianus*, and *Pelobacter carbinolicus*, *Syst. Appl. Microbiol.* **11**:257–260.

Stahl, D. A., and Amann, R., 1991, Development and application of nucleic acid probes in bacterial systematics, in: *Sequencing and Hybridization Techniques in Bacterial Systematics* (E. Stackebrandt and M. Goodfellow, eds). John Wiley & Sons, Chichester, England, pp. 205–248.

Starkey, R. L., 1933, Formation of sulfide by some sulfur bacteria, *J. Bacteriol.* **33**:545–571.

Stetter, K. O., 1988, *Archaeoglobus fulgidus* gen. nov., spec. nov.: a new taxon of extremely thermophilic archaebacteria, *Syst. Appl. Microbiol.* **10**:171–173.

Stokkermans, J. P., Pierik, A. J., Wolbert, R. B., Hagen, W. R., Van Dongen, W. M., and Veeger, C., 1992, The primary structure of a protein containing a putative [6Fe-6S] prismane cluster from *Desulfovibrio vulgaris* (Hildenborough), *Eur. J. Biochem.* **208**: 435–442.

Tasaka, C., Ogata, M., Yagi, T., and Tsugita, A., 1991, Partial sequences of high molecular-weight cytochrome c isolated from *Desulfovibrio vulgaris* Miyazaki, *Protein Seq. Data Anal.* **4**:25–27.

Ueki, A., and Suto, T., 1979, Cellular fatty acid composition of sulfate-reducing bacteria, *J. Gen. Appl. Microbiol.* **25**:185–196.

Utkin, I., Woese, C., and Wiegel, J., 1994, Isolation and characterization of *Desulfitobacterium dehalogenans* gen. nov., sp. nov., an anaerobic bacterium which reductively dechlorinates chlorophenolic compounds, *IJSB* **44**:612–619.

Vainshtein, M., Hippe, H., and Kroppenstedt, R. M., 1992, Cellular fatty acid composition of *Desulfovibrio* species and its use in classification of sulfate-reducing bacteria, *System. Appl. Microbiol.* **15**:554–566.

Vestal, J. R., and White, D. C., 1989, Lipid analysis in microbial ecology. Quantitative approaches to the study of microbial communities. *Bioscience* **39**:535–541.

Voordouw, G., Menon, N. K., LeGall, J., Choi, E. S., Peck, H. D., and Przybyla, A. E., 1989, Analysis and comparison of nucleotide sequences encoding the genes for [NiFe] and [NiFeSe] hydrogenases from *Desulfovibrio gigas* and *Desulfovibrio baculatus*, *J. Bacteriol.* **171**:2894–2899.

Voordouw, G., Pollock, W. B., Bruschi, M., Guerlesquin, F., Rapp-Giles, B. J. and Wall, J. D., 1990a, Functional expression of *Desulfovibrio vulgaris* Hildenborough cytochrome c3 in *Desulfovibrio desulfuricans* G200 after conjugational gene transfer from *Escherichia coli.*, *J. Bacteriol.* **172**:6122–6126.

Voordouw, G., Niviere, V., Ferris, F. G., Fedorak, P. M., and Westlake, D. W. S., 1990b, Distribution of hydrogenase genes in *Desulfovibrio* spp. and their use in identification of species from the oil field environment, *Appl. Environ. Microbiol.* **56**:3748–3754.

Voordouw, G., Voordouw, J. K., Karkhoff-Schweizer, R. R., Fedorak, P. M., and Westlake, D. W. S., 1991, Reverse sample genome probing, a new technique for identification of bacteria in environmental samples by DNA hybridization, and its application to the identification of sulfate-reducing bacteria in oil field samples, *Appl. Environ. Microbiol.* **57**:3070–3078.

Voordouw, G., Voordouw, J. K., Jack, T. R., Foght, J., Fedorak, P. M., and Westlake, D. W. S., 1992, Identification of distinct communities of sulfate-reducing bacteria in oil fields by reverse sample genome probing, *Appl. Environ. Microbiol.* **58**:3542–3552.

Wächtershauser, G., 1988, Before enzymes and templates: theory of surface metabolism, *Microbiol. Rev.* **52**:452–484.

Walker, J. C. G., 1987, Was the Archaean biosphere upside down? *Nature* **329:**710–712.

Weisburg, W. G., Giovannoni, S. J., and Woese, C. R., 1989, The *Deinococcus-Thermus* phylum and the effect of rRNA composition on the phylogenetic tree construction. *Syst. Appl. Microbiol.* **11:**128–134.

Widdel, F., 1980, Anaerober Abbau von Fettsäuren und Benzoesäure durch neu isolierte Arten Sulfat-reduzierender Bakterien. Ph.D. Thesis, Göttingen, Germany.

Widdel, F., 1988, Microbiology and ecology of sulfate- and sulfur- reducing bacteria, in: *Biology of Anaerobic Microorganisms* (A. J. B. Zehnder, ed.), John Wiley & Sons, New York, pp. 469–585.

Widdel, F., 1992a. The genus *Desulfotomaculum*, in: *The Prokaryotes*, 2nd ed., (A. Balows, H. G. Trüper, M. Dworkin, W. Harder, and K.-H. Schleifer, eds.), Springer-Verlag, New York, pp. 1792–1799.

Widdel, F., 1992b, The genus *Thermodesulfobacterium*, in: *The Prokaryotes*, 2nd ed., (A. Balows, H. G. Trüper, M. Dworkin, W. Harder, and K.-H. Schleifer, eds.), Springer-Verlag, New York, pp. 3390–3392.

Widdel, F. and Bak, F., 1992, Gram-negative mesophilic sulfate-reducing bacteria, in: *The Prokaryotes*, 2nd ed., (A. Balows, H. G. Trüper, M. Dworkin, W. Harder, and K.-H. Schleifer, eds.), Springer-Verlag, New York, pp. 3352–3378.

Widdel, F., and Hansen, T. A., 1992, The dissimilatory sulfate- and sulfur bacteria, in: *The Prokaryotes*, 2nd ed., (A. Balows, H. G. Trüper, M. Dworkin, W. Harder, and K.-H. Schleifer, eds.), Springer-Verlag, New York, pp. 583–624.

Widdel, F., and Pfennig, N., 1977, A new anaerobic, sporing, acetate-oxidizing, sulfate-reducing bacterium *Desulfotomaculum* (emend.) *acetoxidans*, *Arch. Microbiol.* 112:119–122.

Widdel, F., and Pfennig, N., 1981, Sporulation and further nutritional characteristics of *Desulfotomaculum acetoxidans*, *Arch. Microbiol.* **129:**401–402.

Widdel, F., and Pfennig, N., 1982, Studies on dissimilatory sulfate-reducing bacteria that decompose fatty acids. II. Incomplete oxidation of propionate by *Desulfobulbus propionicus* gen. nov., sp. nov., *Arch. Microbiol.* **131:**360–365.

Widdel, F., and Pfennig, N., 1984, Dissimilatory sulfate-or sulfur-reducing bacteria, in: *Bergey's Manual of Systematic Bacteriology*, Vol. 1 (N. R. Krieg and J. G. Holt, eds). Williams & Wilkins, Baltimore, pp. 663–679.

Woese, C. R., Achenbach, L., Rouviere, P., and Mandelco, L., 1991, Archaeal phylogeny: reexamination of the phylogenetic position of *Archaeoglobus fulgidus* in light of certain composition-induced artifacts, *Syst. Appl. Microbiol.* 14:364–371.

Zeikus, J. G., Dawson, M. A., Thompson, T. E., Ingvorsen, K., and Hatchikian, E. C., 1983, Microbial ecology of volcanic sulphidogenesis: isolation and characterization of *Thermodsulfobacterium commune* gen. nov. and sp. nov., *J. Gen. Microbiol.* **129:**1159–1169.

Zellner, G., Stackebrandt, E., Kneifel, H., Messner, P., Sleytr, U. B., Conway de Macario, E., Zabel, H. P., Stetter, K. O., and Winter, J., 1989, Isolation and characterization of a thermophilic, sulfate-reducing archaebacterium *Archaeoglobus fulgidus* strain Z. *Syst. Appl. Microbiol.* **11:**151–160.

# Respiratory Sulfate Reduction    4

## J. M. AKAGI

## 1. INTRODUCTION

In the microbial world the dissimilatory sulfate-reducing bacteria (SRB) are unique in that they have the ability to utilize inorganic sulfate as a terminal electron acceptor. This respiratory process, occurring under anaerobic conditions, is conducted by SRB for the purpose of generating high energy compounds for biosynthetic reactions involved in their growth and maintenance. Because relatively large amounts of sulfate are required for this mode of life, a serious consequence resulting from the growth of sulfate reducers is the dissemination of massive quantities of hydrogen sulfide in their immediate vicinity. Because of the deleterious consequences that can result from the growth of sulfate reducers, microbiologists have been studying the physiology, biochemistry, and ecology of these microorganisms to learn how to control them in nature.

Because most of our knowledge of the sulfate-reducing process comes from studies on species belonging to the genera *Desulfovibrio* and *Desulfotomaculum*, this chapter uses findings obtained by investigators who worked with these microorganisms.

## 2. ACTIVATION OF SULFATE AND ITS REDUCTION TO BISULFITE

### 2.1. Adenylyl Sulfate Formation

The SRB reduce sulfate by oxidizing various organic compounds and directing the electrons arising from the oxidations to the sulfate-

J. M. AKAGI • Department of Microbiology, University of Kansas, Lawrence, Kansas 66045.

*Sulfate-Reducing Bacteria*, edited by Larry L. Barton. Plenum Press, New York, 1995.

reducing system. A wide variety of organic compounds can be utilized by SRB, ranging from simple fatty acids to more complex aromatic hydrocarbons. The initial reaction in the reduction of sulfate is an activation step where ATP and sulfate form adenylyl sulfate (APS), and pyrophosphate (PP) thus: ATP + $SO_4^{2-}$ $====$ APS + PP. The reaction is catalyzed by ATP sulfurylase (EC 2.7.7.4; ATP:sulfate adenylyltransferase), which was purified from yeast cells by Lipmann and his coworkers during their pioneer studies on the activation of sulfate (Robbins and Lipmann, 1958a). This enzyme was also purified from extracts of the SRB, *Desulfovibrio vulgaris* (Baliga *et al.*, 1961; Akagi and Campbell, 1962) and *Desulfotomaculum* (*Dt.*) *nigrificans* (Akagi and Campbell, 1962), and was found to be similar to the yeast ATP sulfurylase in many of its properties. Peck (1959), and Ishimoto and Fujimoto (1959) independently discovered that APS was formed by *D. vulgaris* and established that this compound was the active intermediate during the dissimilatory reduction process. Although the equilibrium for this reaction lies in the direction of ATP and sulfate, $K_{eq} = 10^{-8}$ (Robbins and Lipmann, 1958b; Akagi and Campbell, 1962), the reaction is pulled toward APS formation by the action of inorganic pyrophosphatase (EC 3.6.1.1; pyrophosphate phosphohydrolase) (Akagi and Campbell, 1963; Ware and Postgate, 1970), which irreversibly hydrolyzes PP to 2Pi. *Dt. nigrificans*, a thermophilic, spore-forming sulfate reducer, has weak pyrophosphatase activity and it has been postulated that, for this organism, pyrophosphatase does not play a big role in pulling the above reaction in the direction of APS formation. Instead, Liu and Peck (1981) proposed that a pyrophosphate-dependent acetate kinase utilizes the pyrophosphate, forming acetyl phosphate, and pulls the reaction in favor of APS formation. This explanation has been questioned by Thebrath, Dilling, and Cypionka (1989). These workers did not detect any pyrophosphate-dependent acetate kinase activity in extracts from several *Desulfotomaculum* species, and they believe that pyrophosphatase, although weak, favors APS formation from ATP and sulfate. To date, this aspect of APS formation has not been resolved.

### 2.2. Reduction of APS to AMP and Bisulfite

The reduction of APS to AMP and bisulfite is catalyzed by APS reductase, discovered independently by Peck (1961) and Ishimoto and Fujimoto (1961). This enzyme contains one flavin adenine dinucleotide (FAD), 12 nonheme irons and 12 acid labile sulfides per molecule of enzyme based on a monomeric molecular weight of 220,000 (Bramlett and Peck, 1975). The mechanism of APS reduction has been postulated to occur by the formation of a sulfite adduct to FAD (Peck and Bramlett, 1982) at the N-5 position of the isoalloxazine ring (Massey *et al.,* 1969). In this mechanism, APS transfers its sulfite group to a reduced FAD moiety of APS reductase. The sulfite adduct subsequently dissociates into sulfite and oxidized APS reductase. This is shown below:

$$E\text{-}FAD + \text{electron carrier(red)} \rightleftharpoons E\text{-}FADH_2$$
$$+ \text{electron carrier(ox)} \tag{1}$$

$$E\text{-}FADH_2 + APS \rightleftharpoons E\text{-}FADH_2(SO_3) + AMP \tag{2}$$

$$E\text{-}FADH_2(SO_3) \rightleftharpoons E\text{-}FAD + SO_3{}^{2-} \tag{3}$$

Ishimoto and Fujimoto (1961) reported that the APS reductase from *D. vulgaris* also reduced the guanosine (GPS), cytidine (CPS), and uridine (UPS) analogue of APS to their respective monophosphates and sulfite. They further discovered that AMP inhibited the reduction of APS but ATP, ADP, GMP, CMP, and UMP had no effect on APS reduction.

## 3. BISULFITE REDUCTION

The bisulfite arising from the reduction of APS is subsequently reduced to the end product, sulfide, through a pathway that has not been resolved to date. Historically, this reductive process was thought to be caused by a "sulfite reductase system"; however, current evidence suggests that bisulfite is reduced to sulfide by one of two possible mechanisms. One hypothesis is that a direct six-electron reduction of bisulfite to sulfide occurs without the formation of any isolable intermediate compound(s). Another hypothesis proposes the formation of two intermediates, trithionate and thiosulfate, with the terminal step being the reduction of thiosulfate to sulfide and bisulfite.

### 3.1. Detection of Trithionate and Thiosulfate in Reaction Mixtures

In 1969, Kobayashi, Tachibana, and Ishimoto isolated two fractions from *D. vulgaris* (Miyazaki) extracts that reduced sulfite to sulfide sequentially through trithionate and thiosulfate. One fraction reduced sulfite to trithionate and thiosulfate, while the second fraction reduced these compounds to sulfide. The pathway they proposed for dissimilatory sulfite reduction was:

$$3SO_3^{2-} \underset{2e^-}{\rightleftharpoons} S_3O_6^{2-} \underset{2e^-}{\rightleftharpoons} \underset{+ SO_3^{2-}}{S_2O_3^{2-}} \underset{2e^-}{\rightleftharpoons} S^{2-} + SO_3^{2-} \qquad (4)$$

In the same year Suh and Akagi (1969) reported isolating two fractions from D. vulgaris 8303, that formed thiosulfate from sulfite. This activity was designated the "thiosulfate-forming system." One of the components of this system was desulfoviridin and the other fraction was designated FII (fraction II). Although the role of each of the components was not known, incubation of these fractions with sulfite, hydrogenase, and methyl viologen, under a hydrogen atmosphere, resulted in thiosulfate formation. These findings from the laboratories of Ishimoto and Akagi suggested that a trithionate pathway was operating in Desulfovibrio. In the same study, Suh and Akagi concluded that the ionic species of the substrate acted upon by the "thiosulfate-forming system" was bisulfite instead of sulfite. This was determined by pH optimum studies, absorption spectra, and substrate concentration studies.

The trithionate pathway, as proposed by Kobayashi, Tachibana, and Ishimoto (1969) involves a recycling mechanism of sulfite that is released during the reduction of trithionate to thiosulfate and the reduction of thiosulfate to sulfide. Evidence for this mechanism came from the studies of Findley and Akagi (1970) and Drake and Akagi (1978) when they incubated extracts from D. vulgaris with [35]S-sulfonate-1 labeled thiosulfate. When samples of the incubation mixture were removed and analyzed at various time intervals, the results showed that doubly-labeled thiosulfate formed with increasing incubation time.

## 3.2. Bisulfite Reductase

A green pigment absorbing at 630, 585, and 411 nm was isolated from extracts of D. vulgaris (Hildenborough) by Postgate (1956). He described this pigment as an acidic porphyroprotein that decomposes under alkaline conditions to yield a red fluorescent chromophore group when exposed to ultraviolet light at 365 nm. The pigment was named desulfoviridin. This property of fluorescing red under ultraviolet light was introduced by Postgate (1959) to be a diagnostic reaction for Desulfovibrio species. All species of Desulfovibrio except for one mutant, Desulfovibrio desulfuricans Norway 4, isolated by Miller and Saleh (1964), have desulfoviridin. Postgate (1956) found no known function for this pigment although it was present in extracts of D. vulgaris in large amounts. Oxidation and reduction did not result in any spectral change of the pigment and there was no reaction with carbon monoxide, cyanide, or sodium azide. Every investigator who fractionated extracts of *Desulfovibrio* species (other than the Norway strain) probably observed the emerald green pigment precipitating down at certain concentrations of ammonium sulfate, or adsorbing to ion-exchange columns, but no one was able to determine the function of desulfoviridin in these organisms.

The elusive function of the green pigment was finally resolved by Lee and Peck (1971) when they reported that this protein catalyzed the reduction of bisulfite to trithionate. They proposed the name bisulfite reductase (BR) for this pigment to differentiate it from the assimilatory sulfite reductase that reduced sulfite to sulfide. Subsequent work by other investigators established that desulfoviridin contained a tetra-hydroporphyrin prosthetic group common to assimilatory sulfite reductase (see 5.2 below) and, depending on the species of *Desulfovibrio* from which it was isolated, the molecular weight of the pigment ranged from 180,000 to 226,800 (Lee and Peck, 1971; Lee, LeGall, and Peck, 1973; Aketagawa, Kojo, and Ishimoto, 1985). Suh and Akagi (1969) reported that polyacrylamide gel electrophoresis of desulfoviridin showed two closely migrating bands that were later designated as the major and minor bands (Jones and Skyring, 1974; Seki, Kobayashi, and Ishimoto, 1979). Jones and Skyring (1974) separated the major and minor bands from each other by gel electrophoresis and tested the ability of each fraction to reduce sulfite. The ratio of methyl viologen (MVH) oxidized to sulfide formed was 6:1 with the minor band while the major band reduced sulfite to sulfide with a ratio of 12:1. An explanation for this was not given.

Seki, Kobayashi and Ishimoto (1979) separated the two forms of desulfoviridin and compared several of their physical and biochemical properties. Essentially no difference was observed in the MVH-linked sulfite reduction in that both bands formed trithionate, thiosulfate, and sulfide in similar ratios to each other. Adsorption spectra, molecular weight, subunit composition, labile sulfur and iron content, amino acid composition, and circular dichroism spectra were virtually the same for both bands. The similarity of the above properties in the major and minor bands of desulfoviridin indicated that whatever difference(s) there is between the two forms was relatively minor. The difference in affinity for ion-exchange resin isoelectric point and electrophoretic mobility suggested to the investigators that the two forms may be charge isomers or that the difference was due to conformational differences in the protein structures.

The subunit structure of desulfoviridin from *D. vulgaris* Hildenborough and Miyazaki, *D. gigas* and *D. africanus* was determined to be of the $\alpha_2\beta_2$ type (Lee, LeGall and Peck, 1973; Aketagawa, Kojo and Ishimoto, 1985; Seki, Nagai, and Ishimoto, 1985; Hall and Prince, 1981). The molecular weights of the $\alpha$ and $\beta$ subunits ranged from 50–61,000 for the $\alpha$ and 39 to 42,000 for the $\beta$ subunit. More recently a third subunit was detected in desulfoviridin from *D. vulgaris* Hildenborough (Pierik *et al.*, 1992). The third subunit, $\gamma$, an 11 kDa polypeptide, did not dissociate from the native desulfoviridin complex during purification procedures, native electrophoresis, gel filtration and isoelectric focusing. Antibodies directed toward the three subunits were prepared and found to be specific for the respective antigens. No cross reactions were noted indicating that the $\gamma$ subunit was not a proteolytic fragment of the $\alpha$ or $\beta$ subunit. Since antibody directed toward the $\gamma$ subunit reacted with desulfoviridin from *D. vulgaris* oxamicus Monticello, *D. gigas* and *D. desulfuricans* 27774, Pierik *et al.* (1992) proposed that all bisulfite reductases of the desulfoviridin type contain the subunit composition, $\alpha_2\beta_2\gamma_2$.

In 1970, Trudinger succeeded in isolating a carbon monoxide-binding pigment, P582, from *Dt. nigrificans*, that catalyzed the reduction of sulfite to sulfide. He suggested that P582 was responsible for reducing sulfite directly to sulfide in a manner analogous to the assimilatory sulfite reductases. This view was questioned by Akagi and Adams (1972) when they found that P582 formed trithionate as the major product of bisulfite reduction, while thiosulfate and sulfide were formed in lesser quantities. They proposed that P582 functioned in *Dt. nigrificans* in a manner analogous to desulfoviridin in *D. vulgaris*. Another protein that was found to form trithionate and sulfide from bisulfite reduction was a reddish pigment, desulforubidin, isolated from extracts of *D. desulfuricans* (Norway) by Lee *et al.* (1973). The three pigments above, desulfoviridin, P582, and desulforubidin all appear to be bisulfite reductase, corresponding to the organisms *D. vulgaris*, *Dt. nigrificans*, and *D. desulfuricans* Norway 4, respectively. A fourth type of bisulfite reductase was isolated from a nonspore-forming thermophilic sulfate-reducing bacterium, *Thermodesulfobacterium commune*. This bisulfite reductase, named desulfofuscidin, also formed trithionate as the major product of bisulfite reduction, with thiosulfate and sulfide being formed in lesser amounts (Hatchikian and Zeikus, 1983).

## 4. THE TRITHIONATE PATHWAY

The product(s) of bisulfite reductase activity has not been established to the satisfaction of all investigators. Under the conditions of their assay, Lee and Peck (1971) found that desulfoviridin catalyzed the reduction of bisulfite to trithionate as the only product. Subsequent studies by Kobayashi, Takahashi, and Ishimoto (1972); Kobayashi, Seki, and Ishimoto (1974); and Jones and Skyring (1975) on bisulfite reductase catalysis, showed that in addition to trithionate, thiosulfate and sulfide were also formed. The reduction of trithionate to thiosulfate or thiosulfate to sulfide was not caused by bisulfite reductase, indicating that the three compounds were products of bisulfite reduction. The amount of each of the three products was dependent upon the assay conditions. Most investigators used the hydrogenase–methyl viologen (MV) assay which consisted of these two components plus bisulfite and BR under a hydrogen atmosphere as shown below:

$$H_2 + MV_{(oxidized)} + H_2ase \rightleftharpoons MV_{(reduced)} + 2H^+ \qquad (5)$$

$$MV_{(reduced)} + nHSO_3^- + BR \rightleftharpoons S_3O_6^{2-} + S_2O_3^{2-} + S^{2-} \qquad (6)$$

A relatively high concentration of hydrogenase or methyl viologen with low concentrations of bisulfite generally resulted in lower trithionate and higher sulfide levels, whereas under opposite conditions, the reverse pattern of products formation was observed. Table I shows the pattern of product formation under different assay conditions. Table II shows the rate of formation of each of the products with increasing time. Instead of hydrogen and hydrogenase, Drake and Akagi (1977a) used pyruvate and the pyruvate phosphoroclastic system of *D. vulgaris* as the source of electrons for bisulfite reduction. Under these conditions trithionate, thiosulfate, and sulfide were formed and the pattern of their

**Table I. Effect of Assay Conditions on Product Formation during Bisulfite Reduction by Bisulfite Reductase (P582)**[a]

| | | | | $\mu$moles | | |
|---|---|---|---|---|---|---|
| pH | P582(mg) | Methyl viologen | $HSO_3^-$ | $S_3O_6^{2-}$ | $S_2O_3^{2-}$ | $S^{2-}$ |
| 6.0 | 0.4 | 1.0 | 10 | 2.09 | 0.9 | 0.35 |
| 7.0 | 0.4 | 1.0 | 10 | 0.9 | 0.13 | 0.33 |
| 8.0 | 0.4 | 1.0 | 10 | 0.04 | 0.09 | 0.15 |
| 6.0 | 0.5 | 1.0 | 2 | 0.8 | 0.13 | 0.33 |
| 6.0 | 0.5 | 5.0 | 2 | 0.5 | 0.13 | 0.64 |
| 6.0 | 0.5 | 1.0 | 10 | 2.8 | 0.35 | 0.53 |

[a] Standard manometric techniques used. Hydrogenase, 1 mg; K phosphate buffer; total volume, 1.2 ml; temperature, 37°C; time, 60 minutes; gas phase, $H_2$.

**Table II. Effect of Time on Products Formation during Bisulfite Reduction by Bisulfite Reductase (P582)**[a]

| Product | μmoles | | | |
| --- | --- | --- | --- | --- |
| | 30 min | 60 min | 90 min | 120 min |
| $S_3O_6^{2-}$ | 0.75 | 1.62 | 2.12 | 3.42 |
| $S_2O_3^{2-}$ | 0.38 | 0.75 | 1.13 | 1.31 |
| $S^{2-}$ | 0.13 | 0.30 | 0.56 | 0.58 |

[a] Standard manometric techniques. Bisulfite reductase, 0.65 mg. In μmoles: Methyl viologen, 1.0; K phosphate buffer, pH 7.0, 100; $HSO_3^-$, 20; $H_2$ase, 1 mg. Temp., 37 °C.

formation depended upon the concentrations of pyruvate and bisulfite. These results were interpreted to be similar to those obtained by the hydrogenase assay. The conclusion was that a low bisulfite concentration with a strong reducing system resulted in more sulfide formation, while higher bisulfite concentrations with weak reductant led to trithionate and thiosulfate accumulation and little sulfide formation. A model for the products formation was presented by Drake and Akagi (1977a) as shown in Fig. 1. In Fig. 1 the active site for bisulfite reductase contains the adjacent sites A, B, and C. Site C must be bound before sites A and B are in the proper conformation for catalysis. Once a bisulfite ion binds to site C, it is reduced to the level of sulfoxylate by two electrons. If another bisulfite ion is present, it occupies site A forming a two-sulfur intermediate. A third bisulfite ion may bind to site B and react with the two-sulfur intermediate, forming trithionate. If the electron concentration (pressure) is high, the sulfoxylate ion may be reduced to sulfide or, the two sulfur intermediate may be reduced to form thiosulfate. If the reducing pressure is relatively less, trithionate is the initial product. As the reaction proceeds and bisulfite concentration decreases, a critical point is

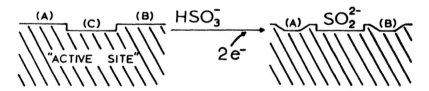

**Figure 1.** Model for bisulfite reductase active site (Drake and Akagi, 1977a).

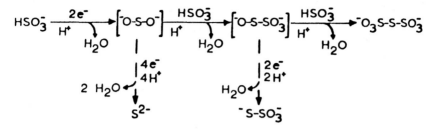

**Figure 2.** Proposed mechanism for product formation during bisulfite reduction (Drake and Akagi, 1977a).

reached and site B is not occupied, leaving the two-sulfur intermediate available for reduction to thiosulfate. As bisulfite ions become depleted to the point where sites A and B no longer are occupied, the sulfoxylate is reduced to sulfide. Figure 2 shows the pathway proposed for bisulfite reduction to thiosulfate and trithionate.

A slightly different scheme for sulfite reduction to sulfide, involving trithionate and thiosulfate formation, was proposed earlier by Kobayashi, Seki, and Ishimoto (1974). They visualized sulfite being reduced to an intermediate X ($SO_2{}^{2-}$?), which combines with two sulfites to form trithionate, or is reduced to another intermediate Y ($S°$?). This intermediate can then combine with a sulfite molecule to form thiosulfate or be reduced to sulfide (Fig. 3). According to this scheme, if trithionate is formed, it is reduced to thiosulfate and it, in turn, is reduced to sulfide.

The possibility that sulfide, once formed, reacted with bisulfite to form thiosulfate at pH 6.0 was considered as unlikely by Kobayashi, Seki, and Ishimoto (1974). They found that addition of cadmium carbonate to

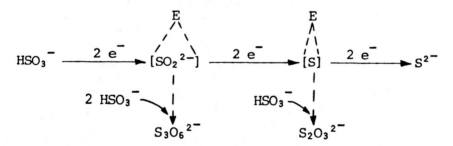

**Figure 3.** Proposed pathway for bisulfite reduction (Kobayashi, Seki, and Ishimoto, 1974).

reaction mixtures, to trap any sulfide formed, did not affect the formation of trithionate or thiosulfate, or on hydrogen untilization during bisulfite reduction studies.

### 4.1. Trithionate Reductase(s)

The reduction of trithionate to thiosulfate by a purified protein was first reported by Drake and Akagi (1977b) when they reinvestigated the "thiosulfate-forming system" reported previously (Suh and Akagi, 1969). Further purification of the FII moiety to homogeneity, as judged by polyacrylamide gel electrophoresis, resulted in a fraction that formed thiosulfate from bisulfite and trithionate. The enzyme, designated TF, required the presence of both bisulfite and trithionate. By $^{35}$S-labeling studies, the mechanism of this reaction was elucidated. The bisulfite ion was found to react with the inner sulfane atom of the trithionate molecule, forming thiosulfate, and releasing the two sulfonate groups as bisulfite ions (see below). This reaction required a reductive step and it was assumed that this occurred concomitantly with thiosulfate formation. The bisulfite molecules that were released from trithionate recycled to participate as free bisulfite in subsequent reactions. Although this enzyme was not a characteristically typical trithionate reductase, it was the first purified enzyme isolated that formed thiosulfate reductively from trithionate.

$$HS^*O_3^- + O_3S - S^* - SO_3^{2-} + 2e^- \rightleftharpoons S^* - S^*O_3^{2-}$$
$$+ HSO_3^- + SO_3^{2-} \tag{7}$$

The presence of this enzyme in D. vulgaris was confirmed in Peck's laboratory (Peck and LeGall, 1982); however, since they could not detect this enzyme in Desulfovibrio gigas, they suggested that it was not uniformly distributed in the Desulfovibrio species. Another trithionate-reducing system was isolated from extracts of D. vulgaris (Kim and Akagi, 1985). This system consisted of bisulfite reductase and another fraction that was designated TR-1. This activity has also been called bisulfite reductase-dependent trithionate reductase (Akagi, 1981). A TR-1 was also purified from Dt. nigrificans extracts and functioned with P582 as a trithionate reductase system. The TR-1 from Dt. nigrificans was capable of using desulfoviridin in reducing trithionate, indicating that the TR-1 and bisulfite reductase from D. vulgaris and Dt. nigrificans were interchangeable with each other (unpublished results). A typical trithionate reductase activity was discovered in whole cells of D. desulfuricans, strain Essex 6, by Fitz and Cypionka (1990). This activity was assayed by incubating bisulfite and thiosulfate with whole cells or cell extracts with ferricyanide. Concomitant to ferricyanide reduction was the formation of trithionate. The principle of this assay is the same as that developed for APS reductase by Peck (1961).

## 4.2. Thiosulfate Reductase

According to the trithionate pathway, the terminal step in the reduction of bisulfite to sulfide involves the enzyme, thiosulfate reductase. Ishimoto, Koyama and Nagai (1955) studied the reduction of thiosulfate by cell extracts from *D. vulgaris*. They observed that thiosulfate reduction occurred in two steps; the first phase was rapid, corresponding to the reduction of the sulfane sulfur atom to sulfide followed by a second phase, which was a slower reaction corresponding to the reduction of sulfite to sulfide. From reaction mixtures reducing thiosulfate, sulfite was isolated, indicating that the reduction of thiosulfate proceeded through the sequence shown below:

$$S\text{-}SO_3^{2-} \overset{2e^-}{\rightleftharpoons} S^{2-} + SO_3^{2-} \tag{8}$$

The formation of sulfite as one of the end products of thiosulfate reductase activity was confirmed by studies on partially purified thiosulfate reductase from *Dt. nigrificans* and *D. vulgaris* (Nakatsukasa and Akagi, 1969; Haschke and Campbell, 1971). Reduction of inner- and outer-labeled $^{35}$S-thiosulfate showed that the outer-labeled sulfane sulfur was reduced to sulfide with the inner sulfonate sulfur atom remaining as sulfite. If thiosulfate were reduced by cell extracts or whole cells, both sulfur atoms were reduced to sulfide at approximately equal rates (Nakatsukasa and Akagi, 1969; Chambers and Trudinger, 1975). The reason for this is that sulfite reductase(s) present in the extracts and whole cells reduce the sulfite as quickly as it is formed during thiosulfate reduction.

Thiosulfate reductase has been purified from *D. vulgaris 8303* (Haschke and Campbell, 1971), *D. vulgaris* Miyazaki F (Aketagawa, Kobayashi, and Ishimoto, 1985), *D. gigas* (Hatchikian, 1975) and *Dt. nigrificans* (Nakatsukasa and Akagi, 1969). The enzyme reduced thiosulfate to sulfide and sulfite and, in all cases, utilized methyl viologen as an electron donor. Cytochrome $c_3$ participated as an electron carrier between hydrogenase and the enzyme from *D. vulgaris* Miyazaki F and *D. gigas*. Reagents inhibiting the enzyme were sulfhydryl group reagents (Aketagawa, Kobayashi, and Ishimoto, 1985; Hatchikian, 1975; Nakatsukasa and Akagi, 1969). Sulfite was found to inhibit thiosulfate reductase activity (Hatchikian, 1975; Nakatsukasa and Akagi, 1969) and ferrous ions stimulated the activity of the enzyme from *D. gigas*. Thiosulfate reductase from *Dt. nigrificans* contained FAD as a coenzyme; removal of this moiety resulted in loss of activity. Riboflavin and FMN did not substitute for FAD in this case (Nakatsukasa and Akagi, unpublished results).

### 4.3. Bisulfite Reduction by Cell Extracts and Whole Cells

Studies on the overall reaction of bisulfite reduction to sulfide, by cell extracts and whole cells, have contributed some insight into the possible mechanism of this reductive process. Akagi (1983) observed that cell extracts from *Dt. nigrificans* rapidly reduced bisulfite to sulfide. When trithionate was added to this system, no sulfide was formed and thiosulfate accumulated in the reaction mixture. Thiosulfate alone was reduced rapidly to sulfide by these extracts, but if trithionate was added together with thiosulfate, no sulfide was formed. From these studies, it was concluded that trithionate, at the concentrations used, was an inhibitor of thiosulfate reductase. Further studies showed that (1) trithionate inhibited bisulfite reduction to sulfide with the accumulation of thiosulfate; (2) thiosulfate reduction to sulfide was inhibited by trithionate; (3) trithionate did not inhibit P582 activity, that is, bisulfite was reduced to trithionate (in the presence of trithionate); and (4) trithionate itself was reduced to thiosulfate by cell extracts. The inhibition of sulfide formation from sulfite and thiosulfate by trithionate was also observed to occur in washed cells of *D. desulfuricans* (Fitz and Cypionka, 1990). They found that these cells reduced micromolar concentrations (0.5 to 4 mM) resulted in thiosulfate accumulation. Furthermore, higher concentrations of trithionate inhibited the reduction of sulfate, sulfite, and thiosulfate to sulfide. The inhibition was reversed if the concentration of the electron donor system (formate) was increased.

If sulfide is the true product of bisulfite reductase catalysis, one would expect that a purified bisulfite reductase would have a significantly higher specific activity in sulfide formation than the crude extracts from which it was purified. This was not the case when purified P582 was compared to *Dt. nigrificans* crude extracts. The extracts reduced bisulfite to sulfide faster and to a greater extent than did the purified enzyme (Akagi, 1983). It is possible that, in this case, the presence of an assimilatory bisulfite reductase in the extracts may have influenced the rate and extent of sulfide formation.

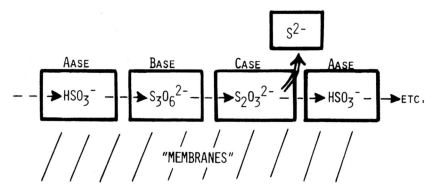

**Figure 4.** Role of membranes in bisulfite reduction (Drake and Akagi, 1978).

## 4.4. Role of Membranes in Bisulfite Reduction

Membranes were implicated in the bisulfite-reducing process by Drake and Akagi (1978). They found that cell extracts reduced bisulfite to sulfide without forming any isolable intermediates such as trithionate or thiosulfate. When the particulate fraction was removed by high-speed centrifugation, the soluble fraction reduced bisulfite to sulfide sequentially through trithionate and thiosulfate. Addition of the particulate fraction (membranes) the soluble fraction restored the original activity, that is, sulfide was formed without any intermediates. These workers proposed that membranes form a matrix in which the bisulfite, trithionate, and thiosulfate reductases operate to carry out the bisulfite reduction process (Fig. 4). Fitz and Cypionka (1990) reported that the reduction of sulfite or thiosulfate was sensitive to cell breakage and concluded that membrane-associated steps were involved in the reductive process.

## 5. DIRECT SIX-ELECTRON REDUCTION MECHANISM

Evidence against the formation of trithionate and thiosulfate during dissimilatory bisulfite reduction has been published by several investigators. Chambers and Trudinger (1975) carried out some isotope studies using $^{35}$S-labeled substrates and determined the fate of these labels during their metabolism by resting and growing cells of *D. desulfuricans*. They found that both the sulfane and sulfonate groups of thiosulfate were reduced to sulfide at approximately equal rates. The assumption made by these workers was that if thiosulfate were an intermediate during bisulfite reduction, the sulfane group should be reduced faster than the sulfonate group. Also, if thiosulfate were an intermediate, ". . . this would imply that there was no exchange between intracellular and extracellular thiosulfate." An experiment using $^{35}SO_4$ and unlabeled thiosulfate with resting cells of D. desulfuricans showed that intact cells incorporated $^{35}S$ into extracellular thiosulfate. Over 90% of the incorporated $^{35}S$ was present in the sulfonate group, suggesting that an exchange occurred between the sulfonate group and sulfite or some intermediate at the equivalent oxidation state. Theoretically, if sulfate is reduced through thiosulfate as an intermediate, the radioactivity should be distributed in the sulfane and sulfonate atoms equally. Because thiosulfate results from trithionate reduction, they concluded that trithionate and thiosulfate are not intermediates during bisulfite reduction. This conclusion supported the earlier claim by Jones and Skyring (1974) who found that desulfoviridin bands on polyacrylamide gels reduced sulfite to sulfide. They stated that sulfide "is at least a partial end-product" of sulfite reduction by desulfoviridin. They advocated the view that trithionate and thiosulfate were not intermediates during bisulfite reduction to sulfide. Chambers and Trudinger (1975) supported this view since they did not detect any radioactive product, other than sulfide, after incubating washed cells of D. desulfuricans anaerobically with $^{35}S$-sulfate. Conflicting results were reported by Fitz and Cypionka (1990); found that deenergized cells of D. desulfuricans formed thiosulfate and trithionate from sulfite when molecular hydrogen or formate was used as the electron source. Furthermore, they observed that up to five micromolar thiosulfate was produced by these cells growing with sulfate and hydrogen in a chemostat. Subsequent work by Sass et al. (1992) showed that washed cells of species belonging to the genera Desulfovibrio, Desulfobulbus, Desulfococcus, Desulfobacter and Desulfobacterium formed thiosulfate and/or trithionate when they were incubated with sulfite and a limiting supply of $H_2$. They also observed that growing cultures of D. desulfuricans formed up to 400 μM thiosulfate and 40 μM trithonate during growth in a chemostat under limited electron donor ($H_2$) conditions. Despite these findings, Sass et al. (1992) stated that the formation of these compounds is not proof that the trithionate pathway is operating in these organisms. These compounds were considered to be "by-products" of sulfite reduction and not necessarily intermediates in the reduction of sulfite to sulfide. Sass et al. did not make clear whether or not these "by-products" were one of the end products in addition to sulfide. They did state, however, that thiosulfate and trithionate were formed as physiological products of sulfite reduction. The accumulation of thiosulfate in the growth medium was made by another group previous to the work of Cypionka's group. Vainstein et al. (1979) reported that a growing culture of SRB produced thiosulfate that accumulated in the medium at the beginning of the growth phase.

### 5.1. Proton Translocation Studies

Kobayashi *et al.* (1982) demonstrated that proton translocation across biomembranes occurred in *D. vulgaris* in the presence of sulfite ions. Other compounds such as metabisulfite and dithionite also caused proton translocation but this was explained by the presence of bisulfite in these solutions. Trithionate, thiosulfate, and sulfate did not cause any pH change, indicating no proton translocation with these sulfur species.

In another study, a rapid proton production during sulfite reduction with hydrogen was observed (C-L. Liu and H. D. Peck, unpublished). No proton production was observed when trithionate or thiosulfate were used as electron acceptors during this study. Another study from Peck's laboratory showed that desulforubidin, reduced electrochemically, was oxidized by sulfite and produced 0.8 μmole of sulfide per μmole of enzyme, indicating a six-electron reduction (Peck and LeGall, 1982). The cumulative results of the above studies led investigators to advocate the hypothesis that bisulfite is not reduced to sulfide by the trithionate pathway, but rather, that bisulfite is reduced to sulfide by a direct six-electron reduction mechanism without forming any isolable intermediate compounds.

In contrast, Fitz and Cypionka (1989) studied proton translocation by washed cells of *D. desulfuricans* Essex 6 with sulfide electrode and pH measurements. Using electron acceptors such as sulfate, sulfite, and thiosulfate, proton translocation was observed with sulfide production, and in several experiments, proton translocation was observed without any sulfide formation, especially with thiosulfate. Fitz and Cypionka (1990) extended their studies to deenergized cells of *D. desulfuricans*. They found that cells treated with carbonyl-cyanide *m*-chloro phenylhydrazone (CCCP), an uncoupling agent, reduced bisulfite to sulfide with trithionate and thiosulfate accumulating in reaction mixtures. Sulfate did not replace bisulfite as the deenergized cells were not able to activate this compound. Thiosulfate was formed by cells growing in a chemostat with excess sulfate and molecular hydrogen. They also observed proton translocation during trithionate reduction with hydrogen. The $H+/H_2$ ratio of $1.5 \pm 0.6$ was less than with sulfite as electron acceptor, which they observed in an earlier study (Fitz and Cypionka, 1989). Nevertheless, these workers attributed trithionate reduction to be coupled to a respiration-driven proton translocation process. From their studies, they concluded that all of their data were consistent with the occurrence of a trithionate pathway for bisulfite reduction.

## 5.2. Siroheme and Tetrahydroporphyrin

The possibility that bisulfite reductase reduced bisulfite by a direct six-electron route, similar to assimilatory sulfite reductase, was further supported by spectral studies on the chromophore groups of *Escherichia coli* sulfite reductase and bisulfite reductase from *D. vulgaris* and *Dt. nigrificans*. Murphey *et al.* (1973) demonstrated that the assimilatory NADPH-sulfite reductase from *E. coli* contained a prosthetic group characterized as an iron-tetrahydroporphyrin with eight carboxylate side chains. Subsequently, Murphy and Siegel (1973) identified the prosthetic group of the bisulfite reductase (P582) from *Dt. nigrificans* as having identical spectral properties to the *E. coli* enzyme. They also found that the bisulfite reductase (desulfoviridin) from *D. gigas* yielded a chromophore group that exhibited spectral and fluorescence spectra indistinguishable from those found with demineralized heme tetrahydroporphyrins from *E. coli* sulfite reductase and *Dt. nigrificans* P582. When iron was inserted into the methyl ester of *D. gigas* desulfoviridin chromophore group, the absortion spectral properties of this compound was similar to the *E. coli* and *Dt. nigrificans* enzymes under the same conditions. They concluded that both the assimilatory and dissimilatory reductases contain a common tetrahydroporphyrin prosthetic group. This class of tetrahydroporphyrin compound was given the trivial name "sirotetrahydrochlorin." Because the *E. coli* and *Dt. nigrificans* enzymes were iron-chelated, they were given the name "siroheme." The structural correlation between the assimilatory sulfite reductase and dissimilatory bisulfite reductase chromophore groups was supportive of the direct reduction route of bisulfite reduction to sulfide.

## 5.3. Dithionite as a Possible Intermediate

Another compound that has been proposed as a possible intermediate during bisulfite reduction is dithionite (Skyring and Jones, 1977; Seki and Ishimoto, 1979). These investigators used the sirohydrochlorin chromophore from desulfoviridin to reduce sulfite and dithionite to sulfide. They acknowledged the earlier proposed scheme of Kobayashi, Seki, and Ishimoto (1974), which suggested that a reactive intermediate at the level of the sulfoxylate ion ($SO_2^{2-}$), reacted with bisulfite to form trithionate. In this light, the reduction of dithionite was thought to be significant since the sulfoxylate radical ($SO_2^{-}\cdot$) is in equilibrium with dithionite. It is possible that the sulfoxylate radical might be an intermediate during bisulfite reduction. Skyring and Trudinger (1972) electrophoresed crude extracts of *D. gigas* and *Dt. nigrificans* on polyacrylamide gels. They assayed the bands for several reductase activities and found that dithionite was rapidly reduced to sulfide by bands corresponding to sulfite reductase. No explanation for this phenomenon was given. The possible occurrence of dithionite as an intermediate during sulfite reduction was proposed earlier by Woolfolk (1962) during his studies on the reduction of inorganic sulfur compounds by extracts from *Micrococcus lactilyticus* (*Veillonella alcalescens*). He proposed that bisulfite, in equilibrium with metabisulfite, was reduced to dithionite, which was the intermediate between metabisulfite and thiosulfate. At that time, he also suggested that metabisulfite or bisulfite ion might be the actual species utilized by *Desulfovibrio* for reduction to sulfide.

During their studies on bisulfite reduction, Mayhew *et al.* (1977) noticed that reduced methyl viologen (MV·) was oxidized by bisulfite in the absence of any enzyme. They studied the nonenzymatic reaction and concluded that MV· plus bisulfite formed dithionite and MV. The reaction sequence for this reaction is presented below:

$$2MV\cdot\ +\ 2HSO_3^-\ \rightleftharpoons\ 2MV\ +\ 2SO_2\cdot^-$$
$$2SO_2\cdot^-\ \rightleftharpoons\ S_2O_4^{2-} \tag{9}$$

Dithionite was also formed when MV and bisulfite were incubated with hydrogenase under a hydrogen atmosphere. The same phenomenon was observed by Seki, Watanabe, and Seki (1989) during their studies on reconstituting the sulfite reductase system using hydrogenase, cytochrome $c_3$ and sulfite. These workers noticed that cytochrome $c_3$ reacted with sulfite in the same way as did methyl viologen in forming dithionite. Since dithionite and sulfoxylate ions can form thiosulfate and, possibly other compounds, and if such nonenzymatic reactions occur in reaction mixtures involving bisulfite reductase, it should serve as a caveat to investigators studying the bisulfite reduction process from interpreting results that may be due to nonenzymatic reactions. It should be noted that in the case of cytochrome $c_3$ (Seki, Watanabe, and Seki, 1989), no thiosulfate, trithionate or tetrathionate was detected in their reaction mixtures.

### 5.4. Is Assimilatory Bisulfite Reductase Involved?

The controversy over the mechanism of bisulfite reduction to sulfide has been going on since the introduction of the trithionate pathway by Kobayashi, Tachibana, and Ishimoto (1969) and the elucidation of the enzymatic nature of desulfoviridin in reducing bisulfite to trithionate (Lee and Peck, 1971). Investigators have debated the true product of bisulfite reductase (desulfoviridin, P582, desulforubidin, desulfofuscidin) activity: either trithionate or sulfide. So far, no convincing set of experiments has been performed to the satisfaction of all investigators that unequivocally establishes the true pathway of bisulfite reduction to sulfide. It should not be overlooked that another possibility might exist for the bisulfite reduction process occurring in SRBs. Chambers and Trudinger (1975) stated that the results they reported in their paper could be accounted for if the sulfite reduction process carried out by resting and growing cells of *Desulfovibrio* was due to assimilatory sulfite reductase. This enzyme has been isolated and purified from *D. vulgaris* (Lee *et al.*, 1973) and shown to reduce bisulfite to sufide without the formation of any other compound, such as trithionate or thiosulfate. A similar type of enzyme was isolated from *D. vulgaris* by Drake and Akagi (1976) but this enzyme was reported to form thiosulfate and sulfide from bisulfite. It is not known whether or not this enzyme is identical to the assimilatory sufite reductase described by Lee and Peck (1973). Nevertheless, if the assimilatory reductase is responsible for reducing bisulfite to sulfide, then the function of dissimilatory bisulfite reductase, present in cell extracts to the extent of approximately ten to twenty per cent of the total protein, becomes questionable if its role is to reduce bisulfite to sulfide by the direct route.

## 6. CONCLUDING REMARKS

The dissimilatory reduction of sulfate to sulfide is a necessary process for SRB to obtain energy for their survival. Although some of the energy is obtained through substrate-level phosphorylation reactions, the major source of ATP synthesis is probably due to the oxidative phosphorylation process occurring during bisulfite reduction to sulfide. The mechanism for this important process is not known since the site(s) of electron transport coupled to phosphate esterification has not been clearly elucidated.

The controversy over the pathway of bisulfite reduction to sulfide has persisted for more than two decades. Central to this controversy is the enzyme bisulfite reductase. After Lee and Peck (1971) reported that this enzyme reduced bisulfite to trithionate as the sole product, other investigators showed that thiosulfate and sulfide were formed in addition to trithionate; the pattern of products formed varied depending on assay conditions (Kobayashi, Takahashi, and Ishimoto, 1972; Kobayashi, Seki, and Ishimoto, 1974; Jones and Skyring, 1975). Since then, investigators advocated either the trithionate pathway or the direct six-electron reduction pathway for bisulfite reduction depending on the interpretation of their data. Certainly the conditions of assay for bisulfite reductase are important; however, because the isolated *in vitro* enzyme reaction may not reflect the actual conditions occurring within the organisms, it is not certain which condition actually exists *in vivo* where bisulfite reduction is taking place. Another concern is the conflicting results among investigators obtained for the same phenomenon. Examples of this are (1) Chambers and Trudinger (1975) reporting that they did not detect thiosulfate accumulating in growing cultures of *D. desulfuricans,* while Vainstein *et al.* (1980) reported thiosulfate formation by growing cells of a culture of SRB, and Fitz and Cypionka (1990) detected thiosulfate formation by washed cells of *D. desulfuricans* during reduction of sulfite. More recently, Sass *et al.* (1992) reported the formation of thiosulfate and trithionate by washed and growing cells of SRB. (2) Kobayashi *et al.* (1982) and Liu and Peck (unpublished) observed rapid proton translocation coupled to sulfite reduction with hydrogen, while Fitz and Cypionka (1989) observed proton translocation during trithionate reduction with hydrogen. (3) Lee, LeGall, and Peck (1973) isolated an assimilatory bisulfite reductase from *D. vulgaris* that reduced bisulfite to sulfide, while Drake and Akagi (1976) isolated a bisulfite reductase from *D. vulgaris* that reduced bisulfite to sulfide and thiosulfate. Presumably, both enzymes are identical but the reason for the difference in end products is not known. (4) Jones and Skyring (1974) reported that the minor band of desulfoviridin reduced sulfite to sulfide stoichiometrically, while the major band did not. Seki, Kobayashi, and Ishimoto (1979) found that there were no differences between the major and minor bands of desulfoviridin in enzyme catalysis, molecular weights, absorbance spectra, subunit composition, and a few other properties.

The differences in the results cited above obtained by different laboratories may suggest that different conditions were used for observing the same phenomenon. Although the results between different laboratories are relatively minor, they do exist and some of the data and interpretations may reflect the differences among investigators advocating the trithionate or the direct reduction pathway for bisulfite reduction. In all cases, it could probably be said that the data are correct but the interpretations are debatable.

So far, the pathway for bisulfite reduction to sulfide has not been established by biochemical means. Perhaps another approach to supplement current knowledge of bisulfite reduction may be through genetic analysis. If a mutant of one of the SRB were isolated (or created) that lacked an enzyme, such as trithionate or thiosulfate reductase, the pathway for bisulfite reduction could be resolved unequivocally. The author once heard a statement made by a microbial geneticist to the effect that any metabolic pathway is ultimately confirmed through genetic analysis. If this is true, then a genetic approach to solving the problem of the mechanism of bisulfite reduction may turn out to be fruitful.

## REFERENCES

Akagi, J. M., 1981, Dissimilatory sulfate reduction, mechanistic aspects, in: *Biology of Inorganic Nitrogen and Sulfur*, (H. Boethe and A. Trebst, eds.), Springer-Verlag, New York, pp. 178–187.

Akagi, J. M., 1983, Reduction of bisulfite by the trithionate pathway by cell extracts from *Desulfotomaculum nigrificans*, *Biochem. Biophys. Res. Commun.* **117:**530–535.

Akagi, J. M. and Adams, V., 1972, Isolation of a bisulfite reductase activity from *Desulfotomaculum nigrificans* and its identification as the carbon monxide-binding pigment P582, *J. Bacteriol.* **116:**392–396.

Akagi, J. M. and Campbell, L. L., 1962, Studies on thermophilic sulfate-reducing bacteria. III. Adenosine triphosphate-sulfurylase of *Clostridium nigrificans* and *Desulfovibrio desulfuricans*, *J. Bacteriol.* **84:**1194–1201.

Akagi, J. M., and Campbell, L. L., 1963, Inorganic pyrophosphatase of *Desulfovibrio desulfuricans*, *J. Bacteriol.* **86:**563–568.

Aketagawa, J., Kobayashi, K., and Ishimoto, M., 1985, Purification and properties of thiosulfate reductase from *Desulfovibrio vulgaris*, Miyazaki F, *J. Biochem.* **97:**1025–1032.

Aketagawa, J., Kojo, K., and Ishimoto, M., 1985, Purification and properties of sulfite reductase from *Desulfovibrio vulgaris*, Miyazaki F, *Agri. Biol. Chem.* **49:**2359–2365.

Baliga, B. S., Vartak, H. G. and Jagannathan, V., 1961, Purification and properties of sulfurylase from *Desulfovibrio desulfuricans*, *J. Scientific and Industrial Res.* **20C:**33–40.

Bramlett, R. N., and Peck, H. D., 1975, Some physical and kinetic properties of adenylyl sulfate reductase from *Desulfovibrio vulgaris*, *J. Biol. Chem.* **250:**2979–2986.

Chambers, L. A., and Trudinger, P. A., 1975, Are thiosulfate and trithionate intermediates in dissimilatory sulfate reduction?, *J. Bacteriol.* **123:**36–40.

Drake, H. L. and Akagi, J. M., 1976, Purification of a unique bisulfite-reducing enzyme from *Desulfovibrio vulgaris*, *Biochem. Biophys. Res. Commun.* **71:**1214–1219.

Drake, H. L. and Akagi, J. M., 1977a, Bisulfite reductase of *Desulfovibrio vulgaris:* Explanation for product formation, *J. Bacteriol.* **132:**139–143.

Drake, H. L. and Akagi, J. M., 1977b, Characterization of a novel thiosulfate-forming enzyme isolated from *Desulfovibrio vulgaris*, *J. Bacteriol.* **132:**132–138.

Drake, H. L., and Akagi, J. M., 1978, Dissimilatory reduction of bisulfite by *Desulfovibrio vulgaris*, *J. Bacteriol.* **136:**916–923.

Findley, J. E., and Akagi, J. M., 1970, Role of thiosulfate in bisulfite reduction as catalyzed by *Desulfovibrio vulgaris*, *J. Bacteriol.* **103:**741–744.

Fitz, R. M., and Cypionka, H., 1989, A study on electron transport-driven proton translocation in *Desulfovibrio desulfuricans*, *Arch. Microbiol.* **152:**369–376.

Fitz, R. M., and Cypionka, H., 1990, Formation of thiosulfate and trithionate during sulfite reduction by washed cells of *Desulfovibrio desulfuricans, Arch. Microbiol.* **154**:400–406.

Hall, H. M., and Prince, R. H., 1981, Iron-containing reductases: Investigation of sulphur oxyanions in desulphoviridin from *Desulphovibrio gigas, J. Inorg. Nucl. Chem.* **43**:815–823.

Haschke, R. H., and Campbell, L. L., 1971, Thiosulfate reductase of *Desulfovibrio vulgaris, J. Bacteriol.* **106**:603–607.

Hatchikian, E. C., 1975, Purification and properties of thiosulfate reductase of *Desulfovibrio gigas, Arch. Microbiol.* **105**:249–256.

Hatchikian, E. C., and Zeikus, J. G., 1983, Characterization of a new type of dissimilatory sulfite reductase present in *Thermodesulfobacterium commune, J. Bacteriol.* **153**:1211–1220.

Ishimoto, M., Koyama, J., and Nagai, Y., 1955, Biochemical studies on sulfate-reducing bacteria. IV. Reduction of thiosulfate by cell-free extract, *J. Biochem.* **42**:41–53.

Ishimoto, M., and Fujimoto, D., 1959, Adenosine-5'-phosphosulfate as an intermediate in the reduction of sulfate by a sulfate-reducing bacterium, *Proc. Japan Acad.* **35**:243–245.

Ishimoto, M. and Fujimoto, D., 1961, Biochemical studies of sulfate-reducing bacteria. X. Adenosine-5'-phosphosulfate reductase, *J. Biochem.* **50**:299–304.

Jones, H. E., and Skyring, G. W., 1974, Reduction of sulphite to sulphide catalyzed by desulfoviridin from *Desulfovibrio gigas, Aust. J. Biol. Sci.* **27**:7–14.

Jones, H. E., and Skyring, G. W., 1975, Effect of enzymic assay conditions on sulfite reduction catalyzed by desulfoviridin from *Desulfovibrio gigas, Biochem. Biophys. Acta* **377**:52–60.

Kim, J. H., and Akagi, J. M., 1985, Characterization of a trithionate reductase system from *Desulfovibrio vulgaris, J. Bacteriol.* **163**:472–475.

Kobayashi, K., Hasegawa, H., Takagi, M., and Ishimoto, M., 1982, Proton translocation associated with sulfite reduction in a sulfate-reducing bacterium, *Desulfovibrio vulgaris, FEBS Lett.* **142**:235–237.

Kobayashi, K., Seki, Y., and Ishimoto, M., 1974, Biochemical studies on sulfate-reducing bacteria. XII. Sulfite reductase from *Desulfovibrio vulgaris*—mechanism of trithionate, thiosulfate and sulfide formation and enzymatic properties, *J. Biochem.* **75**:519–529.

Kobayashi, K., Tachibana, S. and Ishimoto, M., 1969, Intermediary formation of trithionate in sulfite reduction by a sulfate-reducing bacterium, *J. Biochem.* **65**:155–157.

Kobayashi, K., Takahashi, E., and Ishimoto, M., 1972, Biochemical studies on sulfate-reducing bacteria. XI. Purification and properties of sulfite reductase, desulfoviridin, *J. Biochem.* **72**:879–887.

Lee, J-P., and Peck, H. D., 1971, Purification of the enzyme reducing bisulfite to trithionate from *Desulfovibrio gigas* and its identification as desulfoviridin, *Biochem. Biophys. Res. Commun.* **45**:583–589.

Lee, J-P., LeGall, J., and Peck, H. D., 1973, Isolation of assimilatory- and dissimilatory-type sulfite reductases from *Desulfoviribrio vulgaris, J. Bacteriol.* **115**:529–542.

Lee, J-P., Yi, C-S, LeGall, J., and Peck, H. D., 1973, Isolation of a new pigment, desulforubidin, from *Desulfovibrio desulfuricans* (Norway strain) and its role in sulfite reduction, *J. Bacteriol.* **115**:453–455.

Liu, C-L., and Peck, H. D., 1981, Comparative bioenergetics of sulfate reduction in *Desulfovibrio* and *Desulfotomaculum* spp., *J. Bacteriol.* **145**:966–973.

Massey, V., Muller, F., Feldberg, R., Schuman, M., Sullivan, P. A., Howell, L. G., Mayhew, S. G., Matthews, R. G., and Foust, G. P., 1969, The reactivity of flavoproteins with sulfite, *J. Biol. Chem.* **244**:3999–4006.

Mayhew, S. G., Abels, R., and Platenkamp, R., 1977, The production of dithionite and $SO_2^-$ by chemical reaction of (bi)sulphite with methyl viologen semiquinone, *Biochem. Biophys. Res. Commun.* **77**:1397–1403.

Miller, J. D. A., and Saleh, A. M., 1964, A sulphate-reducing bacterium containing cytochrome $c_3$ but lacking desulphoviridin, *J. Gen. Microbiol.* **37**:419–423.

Murphy, M. J. and Siegel, L. M., 1973, Siroheme and sirohydrochlorin, *J. Biol. Chem.* **248**:6911–6919.

Murphey, M. J., Siegel, L. M., Kamin, H. and Rosenthal, D., 1973, Reduced nicotinamide adenine dinucleotide phosphate-sulfite reductase of enterobacteria. II. Identification of a new class of heme prosthetic groups: an iron-tetrahydroporphyrin (isobacteriochlorin type) with eight carboxylic acid groups, *J. Biol. Chem.* **248**:2801–2814.

Nakatsukasa, W., and Akagi, J. M., 1969, Thiosulfate reductase isolated from *Desulfotomaculum nigrificans*, *J. Bacteriol.* **98**:429–433.

Peck, H. D., 1959, The ATP-dependent reduction of sulfite with hydrogen in extracts of *Desulfovibrio desulfuricans*, *Proc. Natl. Acad. Sci. U.S.A.* **45**:701–708.

Peck, H. D., 1961, Evidence for the reversibility of the reaction catalyzed by adenosine-5'-phosphosulfate reductase, *Biochim. Biophys. Acta* **49**:621–624.

Peck, H. D., and Bramlett, R. N., 1982, Flavoproteins in sulfur metabolism, in: *7th International Symposium on Flavin and Flavoproteins* (V. Massey and G. Williams, eds.), University Park Press, Tokyo, pp. 851–858.

Peck, H. D., and LeGall, J., 1982, Biochemistry of dissimilatory sulphate reduction, *Phil. Trans. R. Soc. Lond.* **B298**:443–466.

Pierik, A. J., Duyvis, M. G., Helvoort, J. M. L. M., Wolbert, B. G. and Hagen, W. R., 1992, The third subunit of desulfoviridintype dissimilatory sulfite reductases, *Eur. J. Biochem.* **86**:273–276.

Postgate, J. R., 1956, Cytochrome $c_3$ and desulphoviridin; pigments of the anaerobe *Desulphovibrio desulphuricans*, *J. Gen. Microbiol.* **37**:545–572.

Postgate, J. R., 1959, A diagnostic reaction of *Desulphovibrio desulphuricans*, *Nature* **37**:419–423.

Robbins, P. W., and Lipmann, F., 1958a, Separation of two enzymatic phases in active sulfate synthesis, *J. Biol. Chem.* **233**:681–685.

Robbins, P. W., and Lipmann, F., 1958b, Enzymatic synthesis of adenosine-5'-phosphosulfate, *J. Biol. Chem.* **233**:686–690.

Sass, H., Steuber, J., Kroder, M., Kronek, P. M. H., and Cypionka, H., 1992, Formation of thionates by freshwater and marine strains of sulfate-reducing bacteria, *Arch. Microbiol.* **158**:418–421.

Seki, Y., and Ishimoto, M., 1979, Catalytic activity of the chromophore of desulfoviridin, sirohydrochlorin, in sulfite reduction in the presence of iron, *J. Biochem.* **86**:273–276.

Seki, Y., Kobayashi, K., and Ishimoto, M., 1979, Biochemical studies on sulfate-reducing bacteria. XV. Separation and comparison of two forms of desulfoviridin, *J. Biochem.* **85**:705–711.

Seki, Y., Nagai, Y. and Ishimoto, M., 1985, Characterization of a dissimilatory-type sulfite reductase, desulfoviridin, from *Desulfovibrio africanus* Benghazi, *J. Biochem.* **98**:1535–1543.

Seki, Y., Watanabe, S., and Seki, S., 1989, Formation of dithionite by sulfite reduction with *Desulfovibrio vulgaris* hydrogenase and cytochrome $c_3$, *Chem. Pharm. Bull.* **37**:2573–2575.

Skyring, G. W. and Jones, H. E., 1977, Dithionite reduction in the presence of a tetrapyrrole-containing fraction from the desulfoviridin of *Desulfovibrio gigas*, *Aust. J. Biol. Sci.* **30**:21–31.

Skyring, G. W., and Trudinger, P. A., 1972, A method for the electrophoretic characterization of sulfite reductases in crude preparations from sulfate-reducing bacteria using polyacrylamide gels, *Can. J. Biochem.* **50:**1145–1148.

Suh, B. and Akagi, J. M., 1969, Formation of thosulfate from sulfite by *Desulfovibrio desulfuricans, J. Bacteriol.* **103:**741–744.

Thebrath, B., Dilling, W., and Cypionka, H., 1989, Sulfate activation in *Desulfotomaculum, Arch. Microbiol.* **152:**296–301.

Trudinger, P. A., 1970, Carbon monoxide-reacting pigment from *Desulfotomaculum nigrificans* and its possible relevance to sulfite reduction, *J. Bacteriol.* **104:**158–170.

Vainstein, M. B., Matrosoov, A. G., Baskunov, B. P., Zyakun, A. M., and Ivanov, M. V., 1980, Thiosulfate as an intermediate product of bacterial sulfate reduction, *Mikrobiologi* **49:**855–858.

Ware, D. and Postgate, J. R., 1970, Reductant-activation of inorganic pyrophosphatase: an ATP-conserving mechanism in anaerobic bacteria, *Nature* **226:**1250–1251.

Woolfolk, C. A., 1962, Reduction of inorganic compounds with molecular hydrogen by *Micrococcus lactilyticus.* II. Stoichiometry with inorganic sulfur compounds, *J. Bacteriol.* **84:**659–668.

# Characterization of Electron Transfer Proteins

# 5

## LIANG CHEN, MING-Y. LIU, and JEAN LE GALL

## 1. INTRODUCTION

The end-product of their respiration, hydrogen sulfide, is making sulfate-reducing bacteria (SRB) one of the main microbial agent that interacts with metals through formation of sulfide. As a consequence, most of the metals which are found to interact with living organisms have been found in these bacteria, with the exception of copper metal probably because the redox potentials of the $Cu^+/Cu^{2+}$ complexes are too high to be compatible with any of the physiological reactions common to SRB.

Several electron transfer proteins isolated from SRB have been utilized as models and also as tools for metal replacement or substitution due to their relative abundance, small molecular weight, stability, and easiness of purification. In this chapter, but for some recent advances, we shall limit the discussion of the physiological roles of SRB to a minimum because several recent reviews where this has been developed are available (Odom and Singleton, 1993; Fauque *et al.*, 1991) and, rather, concentrate on their properties as "tools" for new studies in the field of biochemistry, in general, and bioinorganic chemistry, in particular.

LIANG CHEN, MING-Y. LIU, and JEAN LE GALL • Department of Biochemistry, The University of Georgia, Athens, Georgia 30605

*Sulfate-Reducing Bacteria*, edited by Larry L. Barton. Plenum Press, New York, 1995.

## 2. METHODS OF CHARACTERIZATION

### 2.1. Prosthetic Groups

Electron carrier proteins, which have been so far well characterized in SRB, can be easily classified into groups, namely: nonheme iron proteins, flavoproteins and hemoproteins, and other metals-containing proteins.

#### 2.1.1. Flavoproteins

For many years the FMN-containing flavodoxin has been the only well-characterized electron carrier protein found to be of importance in sulfate-reducing bacteria metabolism (Faugue et al, 1991). Another protein, flavoredoxin, has been characterized recently; it couples the reduction of $H_2$ to the oxidation of sulfite (Chen et al, 1993c). It is also an FMN-containing protein. FAD has been found sometimes associated with FMN as in Desulfovibrio (D.) gigas NADH-rubredoxin oxidoreductase (Chen et al, 1993a), or in another important enzyme, APS reductase (Peck et al, 1965). Because of the increasing number of flavin-containing redox proteins found in sulfate-reducing bacteria, it is important to have a rapid method for distinguishing between FMN and FAD. A routine method to determine flavin is the use of paper chromatography with TCA extraction of flavoprotein (Dubourdieu and Le Gall, 1970; Chen et al., 1993a). However, fluorescent spectrophotometry allows to distinguish FMN from FAD without destroying the protein (Koziol, 1971).

#### 2.1.2. Cytochromes

With the exception of D. africanus (Jones, 1971), where the presence of a heme d-containing protein was reported, the cytochromes found in SRB are either of the b of the c-type. Another exception, rubredoxin-oxygen oxidoreductase, appears to contain a new type of heme, which remains to be identified (Chen et al, 1993b). Whenever a new cytochrome is described, its heme should be identified according to the recommendation of the I.U.B. "Enzyme Nomenclature" (1992). However, the authors have found difficult to follow the recommendation that "The position of the α band of the pyridine Fe(II) hemochrome should be determined after the acidified acetone-cleaved hemin has been extracted into ether and then extracted from the ether by dilute sodium hydroxide". This is because protoheme IX, the prosthetic group of cytochrome b is fairly insoluble in water; this treatment is not compatible with small amounts of material which are normally available after the first purification. We think that a spectrum of the pyridine hemochrome of the heme-preceding ether extraction, with its spectrum in ether, is sufficient to identify the chemical nature of the heme.

So far, the isobacteriochlorin "siro"-heme or porphyrin has been found only in sulfite reductase, but may be also be present in other proteins of clear "redox," beside enzymatic properties. For example, a small cobalt-siroporphyrin-containing protein (Hatchikian, 1981; Moura *et al.*, 1980b; Battersby and Sheng, 1982) has been found in some species of *Desulfovibrio;* its physiological role is still unknown.

### 2.1.3. Nonheme Iron Proteins

In contrast to cytochromes, which have very distinctive, sharp, spectral features, nonheme iron proteins are not that obvious and have a tendency to be "lost" among the hemoproteins which are so characteristic of *Desulfovibrio* species. The physiological role of nonheme iron proteins is often unknown so that their color remains a sure guide: reddish for 1 Fe-no inorganic sulfur-containing proteins, and also for [2Fe–2S], brownish for [3Fe–4S], [4Fe–4S] and [6Fe–6S] cluster-containing proteins.

Only a combination of spectroscopic properties will allow the characterization of these centers in addition to careful metal determination. For example, the optical spectrum of the highly disordered [6Fe–6S] cluster-containing protein is indistinguishable from a [4Fe–4S] one (Moura *et al.*, 1992b).

The characterization of the redox center of a nonheme iron protein is based on: 1) the number of metal per protein; 2) the characterization of its ligand; 3) its special characteristics using a variety of techniques from uv/visible spectroscopy to EPR, NMR, MCD, Raman spectroscopy, XAFS, and others that can lead not only to a precise description of the particular type of center but also to a lot of controversy. X-ray crystallography has often been the best method for resolving a lot of disputes concerning the structure of these centers. Because X-ray crystallography can provide only a static view of a particular redox center, other methods become essential to describe the redox properties.

### 2.2. Apo-Proteins

Redox proteins are usually small molecules; therefore, an analysis of the apo-protein can be informative. However, to be able to measure protein concentration with accuracy is a problem. Unfortunately, all the known colorimetric methods can lead to gross mistakes concerning the determination of prosthetic groups per molecule because with small proteins, the chromophores are not randomly present, as in BSA, for example. Also, prosthetic groups can influence the readings. A multiple-heme cytochrome *c* found in *D. vulgaris* has been described as a new triheme protein (Tan and Cowan, 1990) although all its characteristics are identical to the hexadecaheme protein found in the same organism (Higuchi *et al.,* 1987). This perfectly illustrates the care that should be taken to avoid confusion concerning sulfate-reducing bacteria redox proteins.

Accurate molecular weight determinations, generally with partial amino-acid sequences from the N-terminal readily obtained with modern automatic sequenators, are very informative and are often preferred to overall amino acid compositions. The latter results are very difficult to obtain with good accuracy and are very seldom sufficiently informative to characterize a newly discovered protein. Table I lists N-termini of a few families of redox proteins commonly found in sulfate-reducing bacteria; it shows that N-termini are very often the signature of a given family of proteins.

### 2.3. Native, Reconstituted, and Modified Proteins

Several sulfate-reducing bacteria redox proteins are quite stable and so are their apo-proteins. This has been at the origin of several modification such as metal or flavin replacement. Metal replacement can be either a change of the transition metal or the use of a different isotope of the same metal ($^{56}Fe \rightarrow {}^{57}Fe$, which allows the exploration of the redox center by Mössbauer spectrometry). Examples are found in Table II; the formation of mixed-metal clusters will be discussed later.

## 3. INDIVIDUAL ELECTRON TRANSFER PROTEINS

### 3.1. Non-Heme Iron Proteins

### 3.1.1. Rubredoxins

Rubredoxins are the simplest and smallest (6,000 Da) redox proteins found in sulfate-reducing bacteria. They contain one iron atom bound in a tetrahedral coordination by the sulfur atoms of four cysteinyl residues. Rubredoxins were found in the cytoplasm of several sulfate reducers including *D. gigas* (Le Gall and Dragoni, 1966), *D. vulgaris* Hildenborough (Bruschi and Le Gall, 1972), *D. desulfuricans* (Bruschi *et*

**Table I.  N-terminal Amino Acid Sequences of the Main Families of Redox Proteins Found in *Desulfovibrio*[a]**

| | 1 | 5 | 10 | 15 | 20 | 25 | 30 | 35 |
|---|---|---|---|---|---|---|---|---|
| Monoheme cytochromes $c_{553}$[b] [2]: | ADGAALYKSCVGCH | | | | | | | |
| Tetraheme cytochromes $c_3$[c] [6]: | --P-D-----------K--K-----------V-F-H--H | | | | | | | |
| Hexadecaheme cytochromes $c$[d] [2]: | -LP---GE-RADL-EIG-M--F--L-LPKV | | | | | | | |
| Diheme "Split Soret"[e] [1]: | GEAQPASGRFDQVG( )AFG( )KPH | | | | | | | |
| Rubredoxins[f] [4]: | f-M----C--CGY-YD | | | | | | | |
| Desulforedoxin[g] [1]: | ANEGDVYKCELCGQVVKVLEEGGGTLVCCGEDMVKQ | | | | | | | |
| Flavodoxins[h] [7] | M-K-L-V-GS-TGNTE | | | | | | | |
| Flavoredoxin[i] [1] | MKRSLGAKPLLFPTP | | | | | | | |

[a] References to individual proteins are to be found in the text. The number in brackets are the total number of sequences taken into consideration for each family.

[b] Cytochrome $c_{553}$: the only two references available are from two related strains of *D. vulgaris*, therefore, other members of the same family could have less identical N-termini. However, the proximity of the the the --CABCH-- heme attachment to the N-terminus is a very good signature of this type of cytochrome.

[c] Tetraheme cytochrome $c_3$: as seen, these N-termini are very variable except for a few residues. Histidine 36 serves as a ligand to the heme iron, with the exception of *D. desulfuricans* G200 cytochrome $c_3$ where it is replaced by a lysine residue (Voordouw *et al.*, 1990); clearly, this cytochrome deserves further characterization. The amino-acid sequence of the *Dm. baculatum* cytochrome $c_3$ has been used as reference for residue numeration.

[d] Hexadecaheme cytochrome $c$: as for cytochrome $c_{553}$, it is well possible that other cytochromes belonging to the same family do not show such a strong similarity since the only two sequences are from the same two *D. vulgaris* strains.

[e] Diheme "Split Soret": as stated in the text, this cytochrome has so far only been formed in *D. desulfuricans* strain 27774. Residues 15 and 19 appear as blanks since this sequence is from an automatic amino-acid analyzer. It would be interesting to determine if they are cysteinyl residues; in that case, lysine 20 could be a ligand for one of the hemes.

[f] Rubredoxins: their N-termini are very informative since the sequence --CABCGY-- is part of the Fe binding site.

[g] Desulforedoxin: this protein has been formed only in one strain of *D. gigas*, therefore, it is difficult to estimate if its sequence is really unique. However, note that the N-terminal part of desulfoferrodoxin is very closely related to its sequence (see text).

[h] Flavodoxins: their N-termini are very characteristic. In particular, note that residues 10, 12, 14, and 15 are involved in the binding of FMN.

[i] Flavoredoxin: is only found so far in *D. gigas*. Its N-terminus is very different from the flavodoxin one. Although it also binds FMN, the binding sequence --S-TGNTE-- is missing. Ferredoxins are not found in this table, because their N-termini are not very informative. In particular, they do not allow the distinction between one or two cluster-containing ferredoxin.

*al.*, 1977; Sieker *et al.*, 1983; Newman and Postgate, 1968; Zubieta *et al.*, 1973; Fauque *et al.*, 1987) and *D. salexigens* (Moura *et al.*, 1980a). The amino acid sequences of rubredoxins were determined from *Desulfovibrio* species (Vogel *et al.*, 1977; Hormel *et al.*, 1986; Shimizu *et al.*, 1989), and the gene sequence of rubredoxin is known form *D. vulgaris* strain Hildenborough (Voordouw, 1988). The structures of rubredoxins have been well studied and determined to atomic resolution from *D.*

**Table II. Isotope, Metal, and Prosthetic Group Substitutions in *Desulfovibrio* Redox Proteins**

| Type of substitution | Protein | Goal | Reference |
|---|---|---|---|
| $Fe \longrightarrow Co$ or $Ni^a$ | Rubredoxins Desulforedoxin | Exploration of new metal-protein complex | Moura et al., 1987b; 1991a; Huang et al., 1993 |
| $Fe \longrightarrow Ni^a$ | Rubredoxins | Study of new enzymatic activity ($H_2$ activation) | Saint Martin et al., 1988 |
| $^{56}Fe \longrightarrow ^{54}Fe^a$ | Rubredoxin Desulfoferrodoxin | Exploration of the Fe center using Resonance Raman | Czernuszewicz et al., 1986; Tavares et al., 1994 |
| $^{56}Fe \longrightarrow ^{57}Fe^b$ | 6 Fe-cluster protein | Determination of magnetic properties of a new Fe-S cluster using Mössbauer spectrometry | Moura et al., 1992b |
| $^{56}Fe \longrightarrow ^{57}Fe^{a,c}$ | Ferredoxin | Exploration of the Fe-S center sub-sites by Mössbauer spectrometry | Kent et al., 1982 |
| $Fe \longrightarrow Co^c$ $Fe \longrightarrow Zn^c$ $Fe \longrightarrow Ni^c$ $Fe \longrightarrow Cd^c$ | D. gigas ferredoxin II (3Fe-4S) cluster | Magnetic and electrochemical properties of new heterometal clusters | Moura et al., 1986 Surerus et al., 1987 Moreno et al., 1994 |
| $^{56}Fe \longrightarrow ^{57}Fe^b$ | Tetraheme cyctochrome $c_3$ | Exploration of heme magnetic properties by Mössbauer spectrometry | Moura et al., 1992a |
| $^{12}C \longrightarrow ^{13}C^a$ $^{14}N \longrightarrow ^{15}N^a$ | Flavodoxin | Study of the apoprotein/FMN binding by NMR | Vervoort et al., 1985 |
| $FMN \longrightarrow$ riboflavin 3′, 5′ biphosphate$^a$ | Flavodoxin | Properties of the new complex artificial prosthetic group/apoprotein | Vervoort et al., 1986 |

[a] Substitution by chemical methods (removal of the natural metal or prosthetic group following by replacement by new isotope, metal or prosthetic group).
[b] Substitution by growing the bacterial cells in a medium enriched with the isotope.
[c] Addition to the redox center.

*vulgaris* Hildenborough (Adman *et al.*, 1977; Dauter *et al.*, 1992; Adman *et al.*, 1991), *D. desulfuricans* (ATCC 27774) (Sieker *et al.*, 1986b; Stenkamp *et al.*, 1990), *D. gigas* (Frey *et al.*, 1987), *Clostridium* (*C.*) *pasteurianum* (Herriot *et al.*, 1970), and *Paracoccus furiosus* (Blake *et al.*, 1991). A large number of charged residues differ among rubredoxins (Vogel *et al.*, 1977) and the deletion of a seven-amino-acid-containing loop from *D. desulfuricans* might explain the different reactivity with a NADH-rubredoxin oxidoreductase isolated form *D. gigas* (Le Gall, 1968; Odom *et al.*, 1976; Chen *et al.*, 1993a). The redox potential of rubredoxin is between $-50$ and $0$ mV (Moura *et al.*, 1979). Such high redox potential makes this redox protein difficult to be fitted into the physiological reactions in sulfate reducers because dissimilatory sulfate reduction by *Desulfovibrio* species requires electrons from much lower reducing power ($-400$ to $-200$ mV) (Le Gall *et al.*, 1982; Le Gall *et al.*, 1979). Although rubredoxins from *C. thermoaceticum* and *Acetobacterium woodii* were found as electron acceptors for CO dehydrogenase (Ragsdale *et al.*, 1983), and a rubredoxin from *C. perfringens* is involved in the reduction of nitrate to ammonia with NAD(P)H, prompting to oxidation of organic substrates (Seki *et al.*, 1988; Seki *et al.*, 1989), the question of the physiological role of rubredoxin in sulfate-reducing bacteria has been raised since its discovery many years ago. Recently, the characterization of an NADH-rubredoxin oxidoreductase (Chen *et al.*, 1993a) and the finding of a rubredoxin-oxygen oxidoreductase (Chen *et al.*, 1993b) from *D. gigas* revealed a possible function because rubredoxin from *D. gigas* appears to be involved into an aerobic respiration chain that reduces oxygen to water from NADH. This new electron transfer chain is linked to ATP formation during the degradation of polyglucose, a carbon reservoirs in *D. gigas*. The proposed pathway is shown as following for the linkage between the degradation of polyglucose and oxygen reduction (Santos *et al.*, 1993):

$$\text{polyglucose} \\ \downarrow \qquad\qquad \Big(\ \text{NAD}^+ \\ \text{(glycolysis)} \qquad\qquad\qquad\qquad\qquad\qquad \Big(\ \text{O}_2 \\ \downarrow \qquad\qquad \text{NADH} \to \text{NRO} \to \text{Rd} \to \text{ROO} \\ \text{acetate} \qquad\qquad\qquad\qquad\qquad\qquad\qquad \text{H}_2\text{O}$$

(NRO: NADH-rubredoxin oxidoreductase; Rd: Rubredoxin; ROO: Rubredoxin–oxygen oxidoreductase)

In addition, rubredoxin from *D. vulgaris* Miyazaki (unspecified strain) has been suggested to function as an electron carrier for a cytoplasmic lactate dehydrogenase (Shimizu *et al.*, 1989). With the application of molecular modeling (section 4.3) and NMR studies, a hypothetic complex with tetraheme cytochrome $c_3$ was established (Stewart *et al.*, 1989; Stewart and Wampler, 1991). Tetraheme cytochrome $c_3$ is located in periplasm and rubredoxins are found in the cytoplasm, the interaction between these two proteins thus seems to be nonphysiological. However, molecular modeling shows a specific interaction of one of hemes in cytochrome $c_3$ with the [Fe-S] center of rubredoxin.

Spectroscopic studies show that the iron of rubredoxin can be substituted by cobalt and nickel (Moura *et al.*, 1991a; Huang *et al.*, 1993). The nickel-substituted rubredoxin was found to have hydrogenase-like activity (Saint-Martin *et al.*, 1988).

Rubrerythrin, a nonheme iron protein isolated from *D. vulgaris*, was initially described by Le Gall *et al.* (1988). Following the establishment of its primary structure, this protein was described as a homodimer containing four irons arranged into two rubredoxin center and one intersubunit dinuclear cluster (VanBeeumen *et al.*, 1991). The gene for rubrerythrin has been cloned and sequenced (Prickril *et al.*, 1991). There is no evidence for the presence of a leader sequence from the gene sequence of the protein. Recently, another nonheme iron protein, nigerythrin, with undetermined function was reported as having an analogy of rubrerythin (Pierik *et al.*, 1993). Nigerythrin is also a homodimer containing six irons per monomer. The molecular weight, pI, N-terminal sequence, antibody cross-reactivity, optical absorption, EPR spectroscopy, and redox potentials of rubrerythrin and nigerythrin show that they are two distinct proteins.

Another type of nonheme iron protein, desulfoferrodoxin, was isolated and characterized from *D. desulfuricans* (ATCC 27774) and *D. vulgaris* (Hildenborough) (Moura *et al.*, 1990). This protein is a monomer (16 kDa) with two irons per molecule. Mössbauer studies indicate the presence of two types of irons in the protein: an iron-sulfur center similar to that of desulforedoxin from *D. gigas* and an octahedral coordinated high-spin ferrous site probably with nitrogen/oxygen-containing ligand. The N-terminal sequence of the protein shows that it is the product of the *rbo* gene from *D. vulgaris* that is proposed to be a rubredoxin oxidoreductase (Brumlik and Voordouw, 1989). NAD(P)H are unable to reduce this new protein (Moura *et al.*, 1990).

### 3.1.2. Desulforedoxin

A new type of nonheme iron protein, desulforedoxin, was isolated and characterized from *D. gigas* (Moura *et al.*, 1977, 1978). Similar to rubredoxin, this protein also contains one iron atom that can be replaced by cobalt and nickel (Moura *et al.*, 1991a; Huang *et al.*, 1993). The gene for desulforedoxin (dsr) from *D. gigas* has been cloned and sequenced (Brumlik *et al.*, 1990). It indicates that the *rbo* gene product from *D. vulgaris*, which has a 4 kDa desulforedoxin domain near its N-terminus, may have arisen by gene fusion (Brumlik *et al.*, 1990).

### 3.1.3. Ferredoxins

#### 3.1.3.1. Classification of Ferredoxins.
Ferredoxins, the redox proteins containing iron and sulfur atoms, play a role in the electron transfer processes related to the phosphoroclastic reaction and the reduction of sulfite by sulfate-reducing bacteria. These small molecular weight (6 kDa) proteins have low redox potentials, characteristic electronic spectra, and typical EPR signals. Seven ferredoxins have been isolated from sulfate-reducing bacteria containing four types of cluster arrangement: [3Fe-4S], [4Fe-4S], [3Fe-4S], and [4Fe-4S], and 2x[4Fe-4S] clusters. These ferredoxins show a common structure because each iron atom is tetrahedrally coordinated with bridging inorganic sulfur atoms. Cysteinyl sulfur atoms are generally the terminal ligand for the clusters, however, other O- and N-containing ligands such as aspartic acid may also involved (Fukuyama *et al.*, 1989; Stout, 1988, 1989; Stout *et al.*, 1988; Kissinger *et al.*, 1989; Fukuyama *et al.*, 1980).

*D. gigas* has two forms of ferredoxin, which have been termed ferredoxin I (Fd I) and ferredoxin II (Fd II) (Bruschi *et al.*, 1976). Fd I, reported as trimer (Bruschi *et al.*, 1976) and a dimer, contains a single [4Fe-4S] cluster (Moura, J. J. G., personal communication). Fd II is a tetramer containing a [3Fe-4S] cluster per subunit. Both proteins are composed of the same polypeptide chain (Bruschi, 1979). Fd I has a redox potential of 450 mV and Fd II exhibits a $E_0'$ of $-130$ mV (Bruschi *et al.*, 1976; Cammack *et al.*, 1977).

A single [4Fe–4S] containing ferredoxin I and a 2x[4Fe–4S] containing ferredoxin II have been isolated from *Desulfomicrobium* (*Dm.*) *baculatum* Norway 4 (former names: *Desulfovibrio desulfuricans* Norway 4, also *Desulfovibrio baculatum* Norway 4) (Bruschi *et al.*, 1977; Guerlesquin *et al.*, 1980). Fd I and FD II of *Dm baculatum* have redox potentials of $-374$ mV and $-500$ mV, respectively (Bruschi *et al.*, 1985; Guerlesquin*et al.*, 1982) and are composed of different polypeptide chains.

There are three ferredoxins in *D. africanus*. These ferredoxins are dimers with 6 kDa per monomer. Fd I contains a single [4Fe–4S] cluster while Fd II is a minor component, not well characterized and seems to contain a [4Fe–4S] center as well (Hatchikian and Bruschi, 1981; Bruschi and Hatchikian, 1982). Fd III contains a [3Fe–4S] ($E_0' = -140$ mV) and a [4Fe–4S] ($E_0' = -410$ mV) (Hatchikian and Bruschi, 1981;

Bovier-Lapierre *et al.*, 1987; Armstrong *et al.*, 1989a,b); only seven cysteine residues are present in its polypeptide chain which are ligands of the two clusters.

D. *vulgaris* Miyazaki contains two types of ferredoxins. Fd I, containing two redox centers with distinct behavior (Ogata *et al.*, 1988), is a high amino acid sequence homology to *D. africanus* Fd III. This dimeric protein contains a [3Fe–4S] cluster and a [4Fe–4S] cluster. Fd II, also a dimer, contains only one [4Fe–4S] center per monomer with a redox potential of −405 mV (Ogata *et al.*, 1988; Okawara *et al.*, 1988a, 1988b).

Ferredoxins are also present, but not well characterized in the other sulfate-reducing bacteria strains including *D. vulgaris* Hildenborough, *D. salexigens*, *D. desulfuricans* ATCC 27774, *Dm. baculatum* strain 9974, and *Desulfotomaculum* (Postgate, 1984; Le Gall and Postgate, 1973).

**3.1.3.2. Structures of Ferredoxins.**   [3Fe-4S] cluster-containing *D. gigas* Fd II is the only ferredoxin from sulfate-reducing bacteria whose structure is well studied (Kissinger *et al.*, 1988, 1989; Sieker *et al.*, 1984). So far, the structure of [4Fe-4S] clusters-containing ferredoxins from sulfate-reducing bacteria have not been solved. The [3Fe-4S] center is a unique iron–sulfur core whose structure was studied by spectroscopic methods (Antonio *et al.*, 1982), similar to the one found in other proteins including aconitase (Beinert and Kennedy, 1989; Robbins and Stout, 1989) and *Azotobacter vinelandii* ferredoxin (Stout, 1988; Stout *et al.*, 1988).

The X-ray structure analysis of *D. gigas* ferredoxin II (Kissinger *et al.*, 1989, 1991) shows that the [3Fe-4S] cluster is bound to the polypeptide chain by three cysteinyl residues: Cys 8, Cys 14, and Cys 50. The residue Cys 11, a potential ligand for the fourth site of a [4Fe-4S] cluster, is twisted away from the cluster. A disulfide bridge between Cys 18 and Cys 42 is located where the second iron–sulfur cluster is found in 2x[4Fe-4S] ferredoxins. Recently, it was observed that this disulfide bridge can be open during the reduction of the Fd II (Macedo *et al.*, 1994). The potential role of this disulfide bridge in the physiological behavior of the ferredoxin remains to be studied. Preliminary crystallographic data have been reported for the ferredoxin I from *Dm. baculatum* Norway 4 (Guerlesquin *et al.*, 1983).

The [3Fe-4S] center-containing *D. gigas* Fd II have proven to be a useful tool for the detailed spectroscopic properties of this type of iron–sulfur clusters. Many techniques have been used for the characterization of its properties including EPR (Cammack *et al.*, 1977), Mössbauer (Huynh *et al.*, 1980; Papaefthymiou *et al.*, 1987), Resonance Raman (Johnson *et al.*, 1981), MCD (Thomson *et al.*, 1981), EXAFS (Antonio *et al.*, 1982), saturation magnetization (Day *et al.*, 1988), electrochemistry (Moreno *et al.*, 1993), and NMR (Macedo *et al.*, 1993). Recent reports also focus on the characterization of ferredoxins using different techniques from different strains, such as *D. vulgaris* Miyazaki ferredoxin I (Park *et al.*, 1991), *D. africanus* ferredoxin III (Thomson *et al.*, 1992; George *et al.*, 1989; Armstrong, 1989a,b), *Dm. baculatum* Norway (Marion and Guerlesquin, 1989).

Amino acid sequences of ferredoxins have been determined from several sulfate reducers including *D. gigas* Fd I and Fd II (Bruschi, 1979), *Dm. baculatum* Norway 4 Fd I and Fd II (Bruschi *et al.*, 1985; Guerlesquin *et al.*, 1983), *D. vulgaris* Miyazaki Fd I and Fd II (Okawara *et al.*, 1988a,b), *D. africanus* Fd I and Fd II (Bruschi and Hatchikian, 1982; Bovier-Lapierre *et al.*, 1987).

### 3.1.3.3. Physiological Roles.

Most ferredoxins have biological activity in the stimulation of hydrogen consumption with sulfite as a terminal electron acceptor or in hydrogen production from pyruvate (Odom and Peck, 1984; Bruschi and Guerlesquin, 1988; Le Gall and Fauque, 1988; Moura *et al.*, 1984; Hatchikian and Le Gall, 1970b; Le Gall and Dragoni, 1966). It has been demonstrated that the tetraheme cytochrome $c_3$ is an intermediate between hydrogenase and ferredoxin (Akagi, 1967; Suh and Akagi, 1969). This electron transfer chain is also active in the coupling of hydrogenase with sulfite reductase (Suh and Akagi, 1969) or during the autotrophical growth under an $H_2$ atmosphere (Le Gall and Postgate, 1973).

In *D. gigas,* Fd I and Fd II function respectively in different metabolic pathways. Fd I is required in the phosphoroclastic reaction in which hydrogen is evolved form the oxidation of pyruvate (Moura *et al.*, 1978). Fd II also stimulates this phosphoroclastic reaction after a long lag phase when it is added to ferredoxin-depleted *D. gigas* crude extracts. This observation is consistent with the appearance of an EPR signal of g = 1.94, suggesting the conversion of [3Fe-4S] cluster to [4Fe-4S] cluster in *D. gigas* crude extracts (Moura *et al.*, 1984). The evolution of hydrogen from pyruvate in *D. vulgaris* Miyazaki can be reconstituted in a system containing hydrogenase, cytochrome $c_3$, Fd I, partially purified pyruvate dehydrogenase, and CoA, while its Fd II is only 40% as effective as Fd I in the reaction (Ogata *et al.*, 1988).

*D. gigas* Fd II or flavodoxin is required between hydrogenase and sulfite reductase (Bruschi *et al.*, 1976). Recently, it has been reported that sulfite reduction coupling with hydrogen oxidation in *D. gigas* can be *in vitro* reconstituted by an electron transfer chain containing cytochrome $c_3$ or a membrane-bound cytochrome c, Fd II, and flavoredoxin, and stimulated by a $Ca^{2+}$-binding protein in the presence of calcium ions (Chen *et al.*, 1991; Chen *et al.*, 1993c).

### 3.1.4. [3Fe-4S]/[4Fe-4S] Cluster Interconversions and Heterometal Clusters

As we have just seen, pyruvate can induce the conversion of the [3Fe-4S]-cluster of *D. gigas* Fd II into the [4Fe-4S] cluster-containing Fd I in the presence of *D. gigas* crude extracts (Moura *et al.*, 1984). Purified *D. gigas* Fd II can also be converted to Fd I following incubation with excess amount of $Fe^{2+}$ in the presence of dithiothretol (Moura *et al.*, 1982; Kent *et al.*, 1982), demonstrating that the polypeptide chain of *D. gigas* ferredoxin can accommodate both [3Fe-4S] and [4Fe-4S] clusters.

The [3Fe-4S] cluster of *D. africanus* Fd III can be interconverted to [4Fe-4S] center in the presence of $Fe^{2+}$ (Armstrong *et al.*, 1989a,b; George *et al.*, 1989). The carboxylate side chain of Asp 14 is proposed as the most likely fourth ligand of the converted [4Fe-4S] because it is occupying a cysteine position of typical 8Fe ferredoxins. The [3Fe-4S]/[4Fe-4S] interconversion previously found with *D. gigas* ferredoxins was an indication that transition metals, other than iron, could be incorporated into the [3Fe-4S] center. Indeed cubanlike centers of the type [M, 3Fe-4S] can be formed with $Co^{2+}$, $Zn^{2+}$, and $Ni^{2+}$. *D. gigas* Fd II [Co, 3Fe-4S] was the first heterometal center ever reported (Moura *et al.*, 1986). Similar results were obtained from *D. africanus* Fd III (Butt *et al.*, 1991a,b) and *Pyrococcus furiosus* Fd (Conover *et al.*, 1990a,b). [Zn, 3Fe-4S] center and [Ni, 3Fe-4S] center were also produced from *D. gigas* Fd II (Surerus *et al.*, 1987; Moreno *et al.*, 1993).

### 3.2. Flavoproteins

#### 3.2.1. Flavodoxins

Flavodoxins, low molecular weight electron carrier flavoproteins (15 kDa) that contain FMN as the prosthetic group, are found in a variety of microorganisms (Meyer and Cusanovich, 1989). Flavodoxins have been identified in several members of the *Desulfovibrio* species including *D. gigas* (Le Gall and Hatchikian, 1967), *D. vulgaris* (Dubourdieu *et al.*, 1973), *D. desulfuricans* (Palma *et al.*, 1994), and *D. salexigens* (Moura *et al.*, 1980a). The flavodoxin of *D. vulgaris* has been extensively studied (Dubourdieu and Le Gall, 1970; Dubourdieu *et al.*, 1975; Favaudon *et al.*, 1976; Visser *et al.*, 1980; Leenders *et al.*, 1990). Its primary (Dubourdieu *et al.*, 1973) and tertiary structures (Watenpaugh *et al.*, 1972; Watt *et al.*, 1991) have been established. A hypothetical model of the flavodoxin-tetraheme cytochrome $c_3$ electron-transfer complex has been constructed from *D. vulgaris* by using interaction computer graphics based on electrostatic potential field calculations and NMR experiments (Stewart *et al.*, 1988). Although the complex between these two proteins may not be of direct physiological significance due to their compartmentalization, it is an excellent model for the study of the mechanism of electron transport between hemes and flavin groups (section 4.3). Flavodoxin genes were cloned and characterized from *D. gigas* (Helms and Swenson, 1992), *D. vulgaris* (Hildenborough) (Krey *et al.*, 1988; Curley and Voordouw, 1988; Curley *et al.*, 1991), *D. salexigens* (Helms *et al.*, 1990), and *D. desulfuricans* Essex 6 (Helms and Swenson, 1991). Two of these genes, *D. salexigens* and *D. vulgaris*, have been expressed in *E. coli*.

Flavodoxin can replace ferredoxin in hydrogen-producing reactions and in hydrogen-utilizing reactions (Hatchikian and Le Gall, 1970b; Le Gall and Hatchikian, 1967). The biochemical similarity between flavodoxin and ferredoxin is that flavodoxin has two stable oxidation-reduction states: the semiquinone-hydroquinone form ($E_0' = -440$ mV) and the semiquinone-quinone form ($E_0' = -150$ mV) (Dubourdieu *et al.*, 1975). These two states of flavodoxin are compatible with the redox potentials of ferredoxin I ($E_0' = -440$ mV) and ferredoxin II ($E_0' = -130$ mV) from *D. gigas* (Moura *et al.*, 1978). This oxidation-reduction behavior of the two electron-transfer proteins is consistent with the observed biological interchangeability. Similar results were obtained with *C. tyrobutyricum* where flavodoxin can also be substituted for ferredoxin in the oxidation of NADH by NADH-ferredoxin oxidoreductase (Petitdemange *et al.*, 1979). Electron transfer between hydrogenase and sulfite reductase of *D. gigas* was reported to be restored by the addition of either ferredoxin or flavodoxin, although flavodoxin was less effective in the coupling reactions (Barton *et al.*, 1972). Electron-transfer-linked phosphorylation was only observed by the addition of ferredoxin, but not flavodoxin (Barton and Peck, 1970; Peck and Le Gall, 1982).

In *C. pasteurianum*, flavodoxin is only synthesized under iron-deficiency growth conditions and substitutes for a 2x[4Fe-4S] ferredoxin (Knight and Hardy, 1966). In contrast, in *D. vulgaris* Hildenborough, flavodoxin appears in large amounts (Dubourdieu and Le Gall, 1970; Mayhew *et al.*, 1978; Le Gall and Peck, 1987) even in the presence of excess iron in the growth medium. *Dm. baculatum* Norway 4 and DSM 1743 do not contain flavodoxin (Bruschi et al., 1977; Fauque *et al.*, 1991), but have a one [4Fe-4S]-cluster ferredoxin analogous to *D. gigas* ferredoxin and a 2x[4Fe-4S]-cluster ferredoxin similar to clostridial ferredoxins (Guerlesquin *et al.*, 1980). This could indicate that in some metabolic pathways of these bacteria, flavodoxin and ferredoxin cannot substitute for each other.

### 3.2.2. Flavoredoxin

Flavoredoxin is a new flavoprotein isolated from *D. gigas* (Chen *et al.*, 1993c). This protein is a homodimer (25 kDa per monomer) with 2 FMN per molecule. The redox potential of flavoredoxin is −348 mV at pH 7.5. During its reduction with sodium dithionite, no semiquinone form can be observed, which is a significant difference from flavodoxin isolated from the same organism. The comparison of the N-terminal sequence of flavoredoxin with those of flavodoxins from *D. gigas* (Helms and Swenson, 1992) and *D. vulgaris* (Dubourdieu *et al.*, 1973) indicates a low homology between flavoredoxin and flavodoxin, consistent with their different roles and specificities in electron transport, although their optical spectra exhibit relatedness (Chen *et al.*, 1993c). Flavoredoxin is required for the coupling of hydrogen oxidation with sulfite reduction (section 3.1.3.3).

### 3.3. Cytochromes

### 3.3.1. Cytochrome b

Cytochrome b appears to be the only type of cytochrome found within the genus *Desulfotomaculum* (Le Gall and Fauque, 1988), one of the numerous characters that indicates their close relationship with some clostridial species because this cytochrome has also been found in some species of these latter organisms (Gottwald *et al.*, 1975).

The cytochrome b found in *D. gigas* seems to be related to fumarate reductase since its synthesis is increased when the bacterium uses fumarate instead of sulfate as a terminal electron acceptor (Hatchikian and Le Gall, 1972). Cytochrome b, as in other such proteins, is also part of the succinate dehydrogenase which is one of the major proteins in *Desulfobulbus elongatus* (Samain *et al.*, 1987).

A very intriguing problem is the fact that every strain of *Desulfovibrio* tested so far does contain quinones including *D. vulgaris* Hildenborough. However, no cytochrome b could be detected in this organism (Hatchikian and Le Gall, 1970a; Jones, 1971). Thus, the classical relationship quinone/cytochrome b does not appear to exist in this strain of *Desulfovibrio*: clearly, a thorough investigation of the role of these essential redox components is still needed in sulfate-reducing bacteria. Recently, in an analysis of a cluster of genes following the gene cloning for the *D. vulgaris* hexadecaheme cytochrome c (section 3.3.2), an open reading frame corresponds gene could code for a protein having analogies with other cytochrome b (Rossi *et al.*, 1993). So, the presence of a physiologically active cytochrome b in *D. vulgaris* Hildenborough deserves to be reviewed.

### 3.3.2. Cytochrome c

The finding of cytochrome $c_3$ in sulfate-reducing bacteria (Postgate, 1954; Ishimoto *et al.*, 1954) was a very challenging discovery since at the time it was thought that an essential step in protoheme synthesis (protoporphyrinogen to porphyrin) was oxygen dependent. However, it has been clearly demonstrated that $O_2$ could be replaced by other terminal electron acceptors such as fumarate (Klemm and Barton, 1985).

Since then, several other c-type cytochromes have been discovered in these organisms. Although a complete description of these hemoproteins would be beyond the scope of this review, we will present their main characteristics; their properties are given in detail by Moura and colleagues (1991b).

**3.3.2.1. Monoheme cytochrome c (Heme Ligation: Methionine-Heme-Histidine).**    First isolated by Le Gall and Bruschi (1968) in *D. vulgaris* strain Hildenborough, this cytochrome has now been found in several other strains (Yagi, 1979; Moura *et al.*, 1987a; Fauque *et al.*, 1988). Since its unique heme is located near its N-terminus, this cytochrome belong to Ambler's class I. It is related to mitochondrial cytochrome c but differ by its more negative redox potentials (Bianco *et al.*, 1982; Koller *et al.*, 1987). Named cytochrome $c_{553}$ from the position of its α peak, it has been proposed to be the natural electron acceptor of *D. vulgaris* Miyazaki fumarate dehydrogenase (Yagi, 1979).

**3.3.2.2. Diheme Cytochrome c (Heme Ligation: Histidine-Heme-Histidine).**    This cytochrome, also termed *Split Soret* to account for its characteristic optical spectrum, has so far been only found in *D. desulfuricans* 27774 (Liu *et al.*, 1988). This dimeric protein contains two hemes per monomer of 26 kDa with distinct redox potentials of -168 mV and -330 mV (Moura *et al.*, 1991b).

No physiological function is known for this protein and the analysis of its N-terminus shows no relation with any other known cytochromes.

**3.3.2.3. Tetraheme Cytochrome $c_3$ (Heme Ligation: Histidine-Heme-Histidine).**    This is the original cytochrome $c_3$ first discovered by Postgate (1954) and Ishimoto (1954), but it was at the time thought to be a diheme protein until the first establishment of its primary structure by Ambler (1968) clearly demonstrated the presence of four hemes per molecule. This protein is not only found in *Desulfovibrio* species but also in other sulfate-reducing bacteria such as *Desulfomicrobium* (Bruschi *et al.*, 1977), *Thermodesulfobacterium* (Hatchikian *et al.*, 1984) and *Desulfobulbus* (Samain *et al.*, 1986). It's X-ray structure is well established from *D. vulgaris* Miyazaki (Higuchi *et al.*, 1984), *D. gigas* (Sieker *et al.*, 1986a; Kissinger, 1989), and *Dm. baculatum* (Haser *et al.*, 1979b; Pierrot *et al.*, 1982), it is to be noted that this latter structure has recently been found erroneous and corrected following $^1$H-NMR investigations (Coutinho *et al.*, 1992), It's four hemes, arranged as a cluster are in different protein environments giving rise to very low redox potentials (from -30 to -400 mV). This protein has been proposed as a cofactor for hydrogenase required for electron transfer to other electron carriers; however, slow reduction of ferredoxin, rubredoxin, and flavodoxin have been noted (Bell *et al*, 1978) in their absence and more recently direct electron transfer between *D. gigas* hydrogenase and ferredoxin II has been demonstrated by Moreno *et al.* (1993). Reported as "stimulating" electron transfer in various electron transfer chains linked to sulfur metabolism (Moura *et al.*, 1991b), it is also capable of acting as terminal electron donors in the reduction of sulfur in *Dm. baculatum* and *D. gigas* (Fauque *et al.*, 1979). Tetraheme cytochrome $c_3$ from *D. vulgaris* strain Hildenborough, a bacterium which cannot grow by using sulfur as terminal electron acceptor, is unable to perform this reaction.

**3.3.2.4. Octaheme Cytochrome $c_3$.**    This protein has been found in various species of *Desulfovibrio* such as *D. gigas* (Hatchikian *et al.*, 1969) and *Dm. baculatum* (Loutfi *et al.*, 1989). This cytochrome has been reported by these later authors to be composed of two identical subunits of 13.5 kDa with some of the hemes forming bridges between each monomer. This unique structure deserves more investigations. The *D. gigas* protein is a very efficient carrier from $H_2$ to thiosulfate (Hatchikian *et al.*, 1969). The results of a preliminary X-ray crystallography of this cytochrome has been published (Sieker *et al.*, 1986a).

**3.3.2.5. Hexadecaheme Cytochrome c.**    A cytochrome containing 16 hemes has first been fully characterized in *D. vulgaris* Hildenborough (Higuchi *et al.*, 1987). This has been confirmed by the cloning of the corresponding gene (Pollock *et al*, 1991). These two findings confirm that the report of Tan and Cowan (1990), that this cytochrome, which contains both high spin and low spin hemes, has only three hemes per molecule was erroneous (Moura *et al*, 1991b). The discovery of other genes adjacent to the gene coding to this very unusual hemoprotein will be discussed in section 4.4.

## 4. RECONSTITUTION OF ELECTRON TRANSFER CHAINS

Different kinds of difficulties hamper the solution of the problem of electron transport and energy conservation in sulfate-reducing bacteria: the diversity among the different species (some proteins found in large amounts in one species are absent in others); some discrete proteins in one species are found incorporated in a more complex one in another (i.e., desulforedoxin in *D. gigas* is found as a part of desulfoferrodoxin in *D. vulgaris* (Moura *et al.*, 1978; Moura *et al.*, 1990); the lack of specificity of certain electron carriers toward electron donors or acceptors. The multiheme cytochromes $c_3$ are a perfect example of this problem (Moura *et al.*, 1991b): cytochrome $c_7$, a triheme hemoprotein closely related to the cytochrome $c_3$ and found in *Desulforomonas acetoxidans* (Haser *et al.*, 1979a), an organism devoid of hydrogenase, is readily reduced by the latter enzyme. Other results are possible such as unspecific reductions through the production of $O_2^-$ from flavins if small amounts of $O_2$ are present. A typical artifact is the reduction of flavodoxin to the semiquinone level in light, this reduction being due to trace amount of free FMN. Any reduction of a redox protein should be tested in the presence of SOD whenever the presence of even small amounts of $O_2$ are suspected (the problem of compartmentalization still brings a lot of controversy).

### 4.1. Protein Localization and Reconstitution with Soluble and/or Solubilized Proteins

We shall concentrate on *D. gigas* because recent works have resulted in the *in vitro* reconstitution of complete electron transfer chain from the energy substrate to the terminal electron acceptor. (Barata *et al.*, 1993), It takes eleven discrete redox centers to produce $H_2$ from aldehydes; the question is, how close are they to the physiological reality? *D. gigas* has some advantages since it seems "simpler" than other strains in having only one hydrogenase whereas *D. vulgaris* has a total of three (Fauque *et al.*, 1988) and also one ferredoxin versus two in *D. desulfuricans* (section 3.1.3).

So far, the hydrogen cycling hypothesis proposed by Odom and Peck (1981) is the only model that explains energy conservation in sulfate-reducing bacteria and is the most readily adaptable to experi-

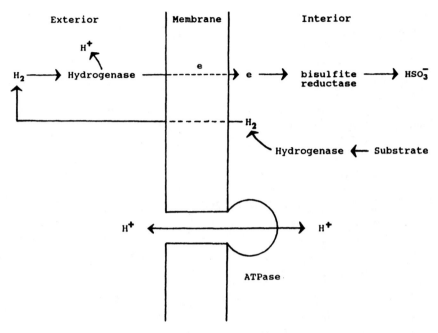

**Figure 1.** Hydrogen cycling model proposed by Odom and Peck (1981) for sulfate-reducing bacteria.

mental data. This model requires the presence of two hydrogenases, or the same hydrogenase located in different cell compartments (Fig. 1).

Recently, several reports have described electron transfer chains which are compatible with such a model (Fig. 2). Thus, two complete sets of electron transfer proteins have been discovered explain energy conservation in *D. gigas*. All proteins but one (the poorly identified membrane-bound cytochrome c) are soluble and their proposed localization is compatible with experimental data with one major exception: the [Ni-Fe] hydrogenase which localization is still raising controversies. Experimentally, *D. gigas* hydrogenase activity can be found in the "periplasmic" fraction (obtained by washing the whole cells in the buffer) and the "cytoplasmic" fraction (obtained after breaking the washed cells in a French pressure cell and removing the "membrane" fraction by high speed centrifugation).

This double localization of hydrogenase is compatible with other results showing that the same protein can be found in different cell compartments (Carlson and Bolstein, 1982; Natsoulis *et al.*, 1986; Wu and Tzagoloff, 1987). However, these results have been challenged by Nivière *et al.* (1991) who have shown electron micrograph pictures of *D.*

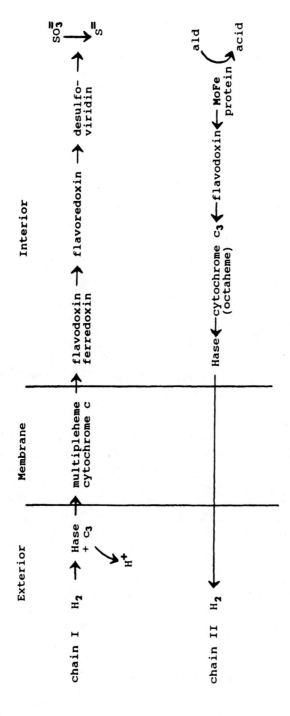

**Figure 2.** Two electron transfer chains involved in hydrogen evolution and utilization in *D. gigas*.

gigas that indicate an almost complete localization of the hydrogenase outside of the cytoplasm. This, in turn, can be criticized since the cells appear to have been damaged during the preparation showing large clear "dilated perisplasmic zones" (Nivière *et al.*, 1991) which are not apparent in another publication dealing with the localization of APS reductase and sulfite reductase in *D. gigas* (Kremer *et al.*, 1988). If all hydrogenase molecules were indeed physiologically present in the periplasm, then a model such as the one proposed by Le Gall and Fauque (1988) (Fauque *et al.*, 1988), and used by Nivière *et al.* (1991) to explain their results could be valid (Fig. 3). In such a scheme, electron transfer chain I (Fig. 2) remains unchanged; in contrast, electron transfer chain II now requires an additional transmembrane electron transfer carrier such as MK 6 present in *D. gigas* (Maroc *et al.*, 1970). Although such a reaction between quinone and multipleheme cytochromes c has not been demonstrated, it is significant that a thermodynamic model for important contributions such as the elaboration of the hydrogen cycling hypothesis (Odom and Peck, 1981) or the localization of the hydrogenases in *D. multispirans* (Czechowski *et al.*, 1984). Curiously, there are no systematic reports of the structure of these spheroplasts and of their relationships with soluble proteins in *Desulfovibrio* species.

A membrane preparation, obtained after breaking the cells with a French pressure cell and ultracentrifugation of the resulting extract, was used by Peck (1966) to demonstrate the existence of oxidative phosphorylations in *D. gigas*. Soluble proteins were necessary in these experiments to couple ATP formation with the reduction of sulfite from molecular hydrogen. The unusual results obtained with chemical uncouplers could be explained by some ATP being formed at the substrate level from polyglucose in the same experimental conditions (Santos *et al.*, 1993).

ATP formation during fumarate reduction from $H_2$ has also been demonstrated in *D. gigas* (Barton *et al.*, 1970). No soluble protein was necessary in this reaction; fumarate reductase (Hatchikian and Le Gall, 1970a), cytochrome b (Hatchikian and Le Gall, 1972), and menaquinone are involved in this electron transfer chain. However, the presence of the "soluble" hydrogenase in these preparations is not well explained (section 4.1) unless the enzyme was entrapped in inverted vesicles or that an unidentified membrane-bound hydrogenase was responsible for the activity. Like spheroplasts, a systematic study of "membrane" *D. gigas* tetraheme cytochrome $c_3$ has been proposed which demonstrates the existence of a proton-linked and redox-linked conformational switch ("redox-Bohr" effect) in this protein (Coletta *et al.*, 1991).

Could *D. gigas* contain a yet undetected hydrogenase? None of the genes coding for either *Desulfovibrio* [Fe]-only hydrogenase or [Ni-Fe-Se]

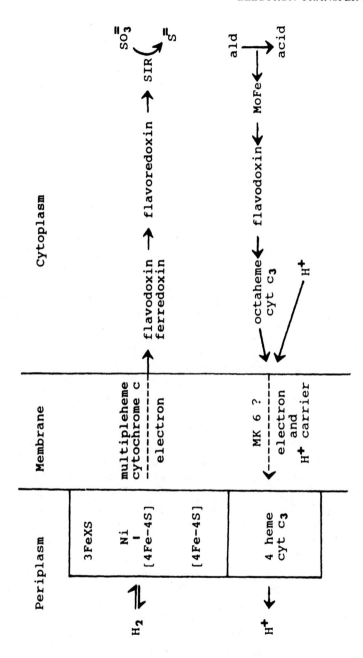

**Figure 3.** A modified hydrogen cycling model by Le Gall and Fauque (1988) suggested that transmembrane electron and proton carriers are needed.

hydrogenase appear to be present in *D. gigas* (Voordouw *et al.*, 1990b) that contains only the gene for the [Ni-Fe] hydrogenase, one of the three hydrogenases found so far in all *Desulfovibrio* and *Desulfomicrobium* (Prickril *et al.*, 1987; Fauque *et al.*, 1988). However, this negative result does not exclude the presence of another hydrogenase yet undetected (because of its high oxygen sensitivity). The closely related genus *Desulfobulbus* also has no gene corresponding to the three known *Desulfovibrio* hydrogenases (Voordouw *et al.*, 1990b), however, it has been shown to contain enzymatic activity (Samain *et al.*, 1986). Thus it is possible that a similar protein, or proteins could be found in *D. gigas;* their discovery would, of course, nullify all previously proposed schemes.

### 4.2. Spheroplasts, Membrane Preparations, with or without Addition of Soluble Proteins

The use of spheroplasts has been at the origin of some important contributions such as the elaboration of the hydrogen cycling hypothesis (Odom and Peck, 1981) or the localization of the hydrogenases in *D. multispirans* (Czechowski *et al.*, 1984). Curiously, there are no systematic reports of the structure of these spheroplasts and of their relationships with soluble proteins in *Desulfovibrio* species.

A membrane preparation, obtained after breaking the cells with a French pressure cell and ultracentrifugation of the resulting extract, was used by Peck (1966) to demonstrate the existence of oxidative phosphorylations in *D. gigas*. Soluble proteins were necessary in these experiments to couple ATP formation with the reduction of sulfite from molecular hydrogen. The unusual results obtained with chemical uncouplers could be explained by some ATP being formed at the substrate level from polyglucose in the same experimental conditions (Santos *et al.*, 1993).

ATP formation during fumarate reduction from $H_2$ has also been demonstrated in *D. gigas* (Barton *et al.*, 1970). No soluble protein was necessary in this reaction; fumarate reductase (Hatchikian and Le Gall, 1970a), cytochrome b (Hatchikian and Le Gall, 1972), and menaquinone are involved in this electron transfer chain. However, the presence of the "soluble" hydrogenase in these preparations is not well explained (section 4.1) unless the enzyme was entrapped in inverted vesicles or that an unidentified membrane-bound hydrogenase was responsible for the activity. Like spheroplasts, a systematic study of "membrane" preparations, including good electron microscope pictures, is badly needed.

### 4.3. Protein–Protein Interactions; Computer Modeling

X-crystallography has had very important successes in establishing some *Desulfovibrio* redox protein structures: (1) of *D. vulgaris* flavodoxin, the first known structure of flavoprotein (Watenpaugh, 1972); (2) of the tetraheme cytochrome $c_3$ (Haser *et al*, 1979b); (3) of the first ferredoxin containing a redox active disulfide bridge (Kissinger, 1989); and (4) of rubredoxins from *D. vulgaris* (Adman *et al*, 1977), *D. desulfuricans* (Sieker *et al*, 1986b), and *D. gigas* (Frey *et al*, 1987).

As already mentioned, the use of $^1$H-NMR in establishing three dimensional structure used for the correction of X-ray data on cytochrome $c_3$ (Coutinho *et al*, 1992) will undoubtedly be important in the future.

However, because of the complexity of *Desulfovibrio* electron transfer chains, these successful contributions are still relatively limited. As a consequence, these limited data have been used to interpret the particularities of protein-protein electron transfer mechanism even when the models are not entirely compatible with physiological data (referred as protein localization in section 4.1). Several hypothetical complexes have been studied between cytochrome $c_3$ and other small electron carriers: (1) ferredoxin I from *Dm. baculatum* strain Norwy 4 (Cambillau *et al.*, 1988); this interaction was also studied in detail using NMR and cross-linking, giving a binding constant of $6 \times 10^6$ M$^{-1}$ for the complex (Guerlesquin *et al*, 1985, 1987; Capeiltere-Blandin *et al*, 1986; Dolla and Bruschi, 1988); (2) flavodoxin from *D. vulgaris* with a binding constant of $1 \times 10^3$ M$^{-1}$ (Stewart *et al*, 1988); and (3) rubredoxin from *D. vulgaris*, in this case, the binding constant with cytochrome $c_3$ is $1 \times 10^4$ M$^{-1}$ (Stewart *et al*, 1989).

In all these cases, the profile of the NMR chemical shifts during protein-protein titration was interpreted with a 1:1 stoichiometry. However, in a more recent contribution (Park *et al.*, 1991) it was found that NMR data concerning the interaction between cytochrome $c_3$ from *D. vulgaris* Miyazaki F and ferredoxin I from the same organism did fit better assuming a binding of two cytochrome $c_3$ for one molecule of ferredoxin with a binding constant of $1 \times 10^8$ M$^{-1}$. Another recent study (Palma *et al.*, 1994) of the interaction between *D. salexigens* flavodoxin and three tetraheme cytochrome $c_3$ (chosen because of the large variations of their isoelectric points, namely *D. vulgaris* Hildenborough, pI = 9.5; *D. desulfuricans* 27774, pI = 7.0; and *D. gigas*, pI = 5.5) also indicates a 1:2 stoichiometry.

Stoichiometry reflects the fact that tetraheme cytochrome $c_3$, which seems to possess in every case a signal peptide (*vide infra*) and is experimentally found in the periplasmic space, and flavodoxin or ferredoxin, which are cytoplasmic proteins, cannot really physiologically interact but that the real partner is another, homologous, cytochrome c, which would be located in the cytoplasm. This hypothesis appears to have some experimental ground since the redox chain that links aldehydes oxidation to hydrogen evolution in *D. gigas* can be reconstituted using flavodoxin and either the tetraheme cytochrome $c_3$ or the octaheme cytochrome $c_3$ which is thought to be a cytoplasmic protein (Barata *et al.*, 1993). Of course, the interactive model between flavodoxin and octaheme cytochrome $c_3$ could be 1:1 since the latter protein has twice the number of hemes of the periplasmic protein. The confirmation of such a model must await more structural information; X-ray crystallographic studies are now in progress concerning *D. gigas* octaheme cytochrome $c_3$ (Sieker *et al.*, 1986a).

### 4.4. Molecular Biology

The use of the powerful tool represented by molecular biology has begun to bear fruits in the study of electron transfer chains in sulfate-reducing bacteria: (1) The tendency to name "operon" what is but a cluster of genes with no experimental proof that the genes involved are indeed part of a physiologically competent unit. (2) The attribution of a physiological activity deduced from the proximity of a given gene to another (for example, a gene upstream of the gene of flavodoxin does not make it automatically a flavodoxin-oxidoreductase even if the first gene has some resemblance with a redox protein)—this can only be established after isolation of the protein coded by the gene and experimental evidence that electrons are transferred between the two proteins. Conversely, there are cases where an oxidoreductase is to be found in a very different part of the genome other than the one in which the gene coding for the redox protein has been found. One is the case for *Pseudomonas oleovorans* rubredoxin and its NADH oxidoreductase (Eggink *et al.*, 1990). Another example, *D. vulgaris* Hildenborough Fe only-hydrogenase operon, shows that it is constituted by the two genes coding for the enzyme with the exclusion of any other redox proteins (van den Berg *et al.*, 1993). Thus, no conclusion can be drawn concerning the physiological partner of this enzyme. (3) The attribution of the localization of a given protein deduced only from the presence or absence of a "leader sequence" coded by its gene. As already mentioned, the presence of such a sequence is not a sure warrant of the final location of the mature protein (Le Gall and Fauque, 1988; Prickril *et al.*, 1986). Also, the absence of a leader sequence in the *D. vulgaris* rubrerythrin gene (Prickril *et al*, 1991) contradicts the experimental finding that the mature protein is exclusively found in the periplasm (Le Gall *et al.*, 1988). It could be possible that this latter protein is released through another protein and dissociate when its final location has been reached. This would be similar to *D. vulgaris* hydrogenase which is constituted of two polypeptide chains, one bearing the redox center and the other the leader sequence (Prickril *et al.*, 1987). However, in this case, the two chains do not dissociate in the periplasm.

The cloning of the tetraheme cytochrome $c_3$ showing that its gene is coding for a signal peptide is in agreement with several experimental evidence showing that this protein is periplasmic (Voordouw and Brenner, 1986). The same results have been obtained with *D. vulgaris* monoheme cytochrome $c_{553}$ (Van Rooijen *et al.*, 1989). The state of the art with cytochrome $c_3$ is such that site directed mutagenesis is possible showing that the replacement of a heme axial histidine by a methionine induces an increase of the redox potential of that particular heme (Mus-Veteau *et al.*, 1992). This result is expected because all mitochondrial cytochromes c have a histidine-heme-methionine type of ligation and a very positive redox potential.

As already reported, several flavodoxins genes (Helms and Swenson, 1991, 1992; Helms *et al.*, 1990a; 1990b; Krey *et al.*, 1988; Curley and Voordouw, 1988) have also been cloned; they have no region corresponding to a leader sequence, in agreement with the established cytoplasmic location of the protein (Mayhew *et al.*, 1978).

Finally, the cloning of the hexadecaheme cytochrome found by Higuchi *et al.* (1987) in *D. vulgaris* Hildenborough (Pollock *et al.*, 1991) has been followed by an analysis of a gene cluster following its gene. Eight open-reading frames have been found and a scheme presenting the corresponding proteins on a transmembrane operational unit has been presented (Rossi *et al.*, 1993). This scheme is highly speculative and needs to be verified experimentally.

## REFERENCES

Adman, E. T., Sieker, L. C., Jensen, L. H., Bruschi, M., and Le Gall, J., 1977, A structural model of rubredoxin from *D. vulgaris* at 2 Å resolution, *J. Mol. Biol.* **112:**113–120.

Adman, E. T., Sieker, L. C., Jensen, L. H., 1991, Structure of rubredoxin from *Desulfovibrio vulgaris* at 1.5 Å resolution, *J. Mol. Biol.* **217:**337–352.

Akagi, J. M., 1967, Electron carriers for the phosphoroclastic reaction of *Desulfovibrio desulfuricans*, *J. Biol. Chem.* **242:**2478–2483.

Ambler, R. P., 1968, The amino acid sequence of cytochrome $c_3$ from *Desulfovibrio vulgaris* (N.C.I.B. 8303), *Biochem. J.* **109:**47P.

Andrews, A. T., 1981, Electrophoresis: theory, techniques, and biochemical and clinical applications, in: *Monographs on Physical Biochemistry* (A. R. Peacocke and W. F. Harrington, eds). Clarendon Press, Oxford.

Antonio, M. R., Averill, B. A., Moura, I., Moura, J. J. G., Orme-Johnson, W. H., Teo, B. K., and Xavier, A. V., 1982, Core dimensions in the 3 Fe cluster of *Desulfovibrio gigas* ferredoxin II by extended X-ray absorption fine structure spectroscopy, *J. Biol. Chem.* **257:**6646–6649.

Armstrong, F. A., Butt, J. N., George, S. J., Hatchikian, E. C., and Thomson, A. J., 1989a, Evidence for reversible multiple redox transformations of [3Fe-4S] clusters, *FEBS Lett.* **259**:15–18.

Armstrong, F. A., George, S. J., Cammack, R., Hatchikian, E. C., and Thomson, A. J., 1989b, Electrochemical and spectroscopic characterization of the 7 Fe form of ferredoxin III from *Desulfovibrio africanus, Biochem. J.* **264**:265–273.

Barata, B. A. S., Le Gall, J., and Moura, J. J. G., 1993, Aldehyde oxidoreductase activity in *Desulfovibrio gigas: in vitro* reconstitution of an electron-transfer chain from aldehydes to the production of molecular hydrogen. *Biochemistry* **32**:11559–11568.

Barton, L. L., Le Gall, J., and Peck, H. D., Jr., 1970, Phosphorylation coupled to oxidation of hydrogen with fumarate in extracts of the sulfate reducing bacterium, *Biochim. Biophys. Acta* **243**:53–65.

Barton, L. L., Le Gall, J., and Peck, H. D., Jr., 1972, Oxidative phosphorylation in the obligate anaerobe, *Desulfovibrio gigas,* in: *Horizons of Bioenergetics,* (A. San Pietro, and H. Gest, eds.), Academic Press, New York, pp. 33–51.

Barton, L. L., and Peck, H. D., Jr., 1970, Role of ferredoxin and flavodoxin in the oxidative phosphorylation catalyzed by cell-free preparations of the anaerobe *Desulfovibrio gigas, Bacteriol. Proc.* p. 134, American Society for Microbiology, Washington, DC.

Battersby, A. R. and Sheng, Z. C., 1982, Preparation and spectroscopic properties of Co[III]-isobacteriochlorins: relationship to the cobalt-containing proteins from *Desulfovibrio gigas* and *D. desulphuricans, J. Chem. Soc. Chem. Commun.* 1393–1394.

Beinert, H., and Kennedy, M. C., 1989, Engineering of protein bound iron-sulfur clusters. A tool for the study of protein and cluster chemistry and mechanism of iron-sulfur enzymes, *Eur. J. Biochem.* **186**:5–15.

Bell, G. R., Lee, J. P., Peck, H. D., Jr., and Le Gall, J., 1978, Reactivity of *D. gigas* hydrogenase toward artificial and natural electron donors or acceptors, *Biochimie* **60**:315–320.

Bianco, P., Haladjian, J., Bruschi, M., and Gillard, R., 1982, Electrochemistry of c-type cytochromes electrode reactions of cytochrome $C_{553}$ from *Desulfovibrio vulgaris* Hildenborough, *J. Electroanal. Chem.* **136**:291–299.

Blake, P. R., Park, J-B., Bryant, F. O., Aono, S., Magnuson, J. K., Eccleston, E., Howard, J. B., Summers, M. F., and Admans, M. W. W., 1991, Determinants of protein hyperthermostability: purification and amino acid sequence of rubredoxin from the hyperthermophilic archaebacterium *Pyrococcus furiosus* and secondary structure of the zinc adduct by NMR, *Biochemistry* **30**:10885–10895.

Bovier-Lapierre, G., Bruschi, M., Bonicel, J., and Hatchikian, E. C., 1987, Amino acid sequence of *Desulfovibrio africanus* ferredoxin III: a unique structural feature for accommodating iron-sulfur cluster, *Biochim. Biophys. Acta* **913**:20–26.

Brumlik, M. J., Leroy, G., Bruschi, M., and Voordouw, G., 1990, The nucleotide sequence of the *Desulfovibrio gigas* desulforedoxin gene indicates that the *Desulfovibrio vulgaris rbo* gene originated from a gene fusion event, *J. Bacteriol.* **172**:7289–7292.

Brumlik, M. J., and Voordouw, G., 1989, Analysis of the transcriptional unit encoding the genes for rubredoxin (rub) and a putative rubredoxin oxidoreductase (rbo) in *Desulfovibrio vulgaris* Hildenborough, *J. Bacteriol.* **171**:4996–5004.

Bruschi, M., 1979, Amino acid sequence of *Desulfovibrio gigas* ferredoxin revisions, *Biochem. Biophys. Res. Commun.* **91**:623–628.

Bruschi, M., and Guerlesquin, F., 1988, Structure, function and evolution of bacterial ferredoxins. *FEMS Microbiol. Rev.* **54**:155–176.

Bruschi, M., Guerlesquin, F., Bovier-Lapierre, G., Bonicel, J. J.., and Couchond, P. M., 1985, Amino acid sequence of the (4Fe-4S) ferredoxin isolated from *D. desulfuricans* Norway, *J. Biol. Chem.* **260**:8292–8296.

Bruschi, M., and Hatchikian, E. C., 1982, Non heme iron proteins of *Desulfovibrio:* the primary structure of ferredoxin I from *D. africanus, Biochimie* **64:**503–507.

Bruschi, M., Hatchikian, E. C., Golovleva, L., and Le Gall, J., 1977, Purification and characterization of cytochrome $c_3$, ferredoxin and rubredoxin isolated from *D. desulfuricans* Norway, *J. Bacteriol.* **129:**30–38.

Bruschi, M., Hatchikian, E. C., Le Gall, J., Moura, J. J. G., and Xavier, A. V., 1976, Purification, characterization and biological activity of three forms of ferredoxin from the sulfate-reducing bacterium *D. gigas, Biochim. Biophys. Acta* **449:**275–284.

Bruschi, M., and Le Gall, J., 1972, Purification et propriétés d'une rubrédoxine isolée à partir de *D. vulgaris* (souche NCIB 8380). *Biochim. Biophys. Acta* **263:**279–282.

Butt, J. N., Armstrong, F. A., Breton, J., George, S. J., Thomson, A. J., and Hatchikian, E. C., 1991a, Investigation of metal ion uptake reactivities of [3Fe-4S] cluster in proteins: voltammetry of co-adsorbed ferredoxin–aminocyclitol films at graphite electrodes and spectroscopic identification of transformed clusters, *J. Am. Chem. Soc.* **113:**6663–6670.

Butt, J. N., Sucheta, A., Armstrong, F. A., Breton, J., Thomson, A. J., and Hatchikian, E. C., 1991b, Binding of thallium(I) to a [3Fe-4S] cluster: evidence for rapid and reversible formation of [Ti3Fe-4S]$^{2+}$ and [Ti3Fe-4S]$^{1+}$ centers in a ferredoxin, *J. Am. Chem. Soc.* **113:**8948–8950.

Cambillau, C., Frey, M., Mossé, J., Guerlesquin, F., and Bruschi, M., 1988, Model of a complex between the tetrahemic cytochrome $c_3$ and the ferredoxin I from *Desulfovibrio desulfuricans, Proteins: Struct. Funct. Genet.* **4:**63–70.

Cammack, R., Rao, K. K., Hall, D. O., Moura, J. J. G., Xavier, A. V., Bruschi, M., Le Gall, J., Deville, A., and Gayda, J. P., 1977, Spectroscopic studies of the oxidation reduction properties of three forms of ferredoxin from *D. gigas, Biochim. Biophys. Acta* **490:**311–321.

Capeillere-Blandin, C., Guerlesquin, F., and Bruschi, M., 1986, Rapid kinetics studies of the electron-exchange reaction between cytochrome $C_3$ and ferredoxin from *Desulfovibrio desulfuricans* Norway strain and their individual reactions with dithionite, *Biochim. Biophys. Acta* **848:**279–293.

Carlson, M., and Bolstein, D., 1982, Two differentially regulated mRNAs with different 5' ends encode secreted and intracellular forms of yeast invertase, *Cell* **28:**145–154.

Chen, L., Liu, M-Y., and Le Gall, J., 1991, Calcium is required for the reduction of sulfite from hydrogen in a reconstituted electron transfer chain from the sulfate reducing bacterium, *Desulfovibrio gigas, Biochem. Biophys. Res. Commun.* **180:**238–242.

Chen, L., Liu, M-Y., Le Gall, J., Fareleira, P., Santos, H., and Xavier, A. V., 1993a, Purification and characterization of an NADH-rubredoxin oxidoreductase involved in the utilization of oxygen by *Desulfovibrio gigas, Eur. J. Biochem.* **216:**443–448.

Chen, L., Liu, M-Y., Le Gall, J., Fareleira, P., Santos, H., and Xavier, A. V., 1993b, Rubredoxin oxidase, a new flavo-hemo-protein, is the site of oxygen reduction to water by the "strict anaerobe" *Desulfovibrio gigas, Biochem. Biophys. Res. Commun.* **193:**100–105.

Chen, L., Liu, M-Y., and Le Gall, J., 1993c, Isolation and characterization of flavoredoxin, a new flavoprotein that permits *in vitro* reconstitution of an electron transfer chain from molecular hydrogen to sulfite reduction in the bacterium *Desulfovibrio gigas, Arch. Biochem. Biophys.* **303:**44–50.

Coletta, M., Catarino, T., Le Gall, J., and Xavier, A. V., 1991, A thermodynamic model for the cooperative functional properties of the tetraheme cytochrome $c_3$ from *Desulfovibrio gigas, Eur. J. Biochem.* **202:**1101–1105.

Conover, R. C., Park, J.-B., Adams, M. W. W., and Johnson, M. K., 1990a, Formation and properties of a NiFe$_3$S$_4$ cluster in *Pyrococcus furiosus* ferredoxin, *J. Am. Chem. Soc.* **112:**4562–4564.

Conover, R. C., Kawal, A. T., Fu, W., Park, J.-B., Aono, S., Adams, M. W. W., and Jonhnson, M. K., 1990b, Spectroscopic characterization of the novel iron-sulfur cluster in *Pyroccus furiosus* ferredoxin, *J. Biol. Chem.* **265**:8533–8541.

Curley, G. P., Carr, M. C., Mayhew, S. G., and Voordouw, G., 1991, Redox and flavin-binding properties of recombinant flavodoxin from *Desulfovibrio vulgaris* (Hildenborough), *Eur. J. Biochem.* **202**:1091–1100.

Curley, G. P., and Voordouw, G., 1988, Cloning and sequencing of the gene encoding flavodoxin from *Desulfovibrio vulgaris* Hildenborough, *FEMS Microbiol. Lett.* **49**:295–299.

Coutinho, I. B., Turner, D. L., Le Gall, J., and Xavier, A. V., 1992, Revision of the haem-core architecture in the tetrahaem cytochrome c$_3$ from *Desulfovibrio baculatus* by two-dimensional $^1$H-NMR, *Eur. J. Biochem.* **209**:329–333.

Czechowski, M. H., He, S. H., Nacro, N., DerVartanian, D. V., Peck, H. D., Jr., and Le Gall, J., 1984, A cytoplasmic nickel-iron hydrogenase with high specific activity from *Desulfovibrio multispirans*, a new species of sulfate reducing bacterium, *Biochem. Biophys. Res. commun.* 125:1025–1032.

Czernuszewicz, R. S., Le Gall, J., Moura, I., and Spiro, T. G., 1986, Resonance raman spectra of rubredoxin: new assignments and vibrational coupling mechanism from iron 54/iron 56 isotope shifts and variable wavelength excitation, *Inorg. Chem* **25**:696–700.

Day, E. P., Peterson, J., Bonvoish, J. J., Moura, I., and Moura, J. J. G., 1988, Magnetization of the oxidized and reduced three-iron cluster of *Desulfovibrio gigas* ferredoxin II, *J. Biol. Chem.* **263**:3684–3689.

Dauter, Z., Sieker, L. C., Wilson, K. S., 1992, Refinement of rubredoxin from *Desulfovibrio vulgaris* at 1.0 Å with and without restraints, *Acta Crystallogr. B.* **48**:42–59.

Dolla, A., and Bruschi, M., 1988, The cytochrome c$_3$-ferredoxin electron transfer complex: crossing-linking studies, *Biochim. Biophys. Acta* **932**:26–32.

Dubourdieu, M., and Le Gall, J., 1970, Chemical study of two flavodoxins extracted from sulfate-reducing bacteria, *Biochem. Biophys. Res. Commun.* **38**:965–972.

Dubourdieu, M., Le Gall, J., and Favaudou, V., 1975, Physicochemical properties of flavodoxin from *D. vulgaris, Biochim. Biophys. Acta* **376**:519–532.

Dubourdieu, M., Le Gall, J., and Fox, J. L., 1973, The amino acid sequence of *Desulfovibrio vulgaris* flavodoxin, *Biochem. Biophys. Res. Commun.* **52**:1418–1425.

Eggink, G., Engel, H., Vriend, G., Terpstra, P., Witholt, B., 1990, Rubredoxin reductase of *Pseudomonas oleovorans*, structural relationship to other flavoprotein oxidoreductases based on one NAD and two FAD fingerprints, *J. Mol. Biol.* **212**:135–142.

*Enzyme Nomenclature*, 1992, published for International Union of Biochemistry and Molecular Biology by Academic Press, San Diego.

Fauque, G., Herve, O., and Le Gall, J., 1979, Structure function relationships in hemoproteins: The role of cytochrome c$_3$ in the reduction of colloidal sulfur by sulfate-reducing bacteria, *Arch. Microbiol.* **121**:261–264.

Fauque, G., Le Gall, J., and Barton, L. L., 1991, Sulfate-reducing and sulfur-reducing bacteria, in: *Variations in Autotrophic Life*, (J. M. Shively and L. L. Barton, eds), Academic Press, New York, pp. 271–337.

Fauque, G. D., Moura, I., Moura, J. J. G., Xavier, A. V., Galliano, N., and Le Gall, J., 1987, Isolation and characterization of a rubredoxin and a flavodoxin from *Desulfovibrio desulfuricans* Berre-Eau, *FEBS Lett.* **215**:63–67.

Fauque, G., Peck, H. D., Jr., Moura, J. J. G., Huynh, B. H., Berlier, Y., DerVartanian, D. V., Texeira, M., Przybyla, A. E., Lespinat, P. A., Moura, I., and Le Gall, J., 1988, The three classes of hydrogenases from sulfate-reducing bacteria of the genus *Desulfovibrio, FEMS Microbiol. Rev.* **54**:299–344.

Favaudon, V., Le Gall, J., and Lhoste, J. M., 1976, Proton magnetic resonance of *D. vulgaris* and *D. gigas* flavodoxins, in: *Flavins and Flavoproteins*, (T. P. Singer, ed.), Elsevier, New York, pp. 434–438.

Frey, M., Seiker, L., Payne, F., Haser, R., Bruschi, M., Pepe, G., and Le Gall, J., 1987, Rubredoxin from *Desulfovibrio gigas*. A molecular model of the oxidized form at 1.4 Å resolution. *J. Mol. Biol.* **197:**525–535.

Fukuyama, K., Tsukihara, T., and Katsube, Y., 1989, Structure of [4Fe-4S] ferredoxin from *Bacillus thermoproteolytrcus* refined at 2.3 Å resolution. Structural comparison of bacterial ferredoxins, *J. Mol. Biol.* **210:**383–398.

Fukuyama, K., Hase, T., Matsumoto, S., Tsukihara, T., Katsube, Y., Tanaka, N., Kakudo, M., Wada, K., and Matsubara, H., 1980, Structure of *S. platenisis* [2Fe-2S] ferredoxin and evolution of chloroplast-type ferredoxins, *Nature* **286:**522–524.

George, S. J., Armstrong, F. A., Hatchikian, E. C., Thomson, A. J., 1989, Electrochemical and spectroscopic characterization of the conversion of the 7 Fe into 8 Fe form of ferredoxin III from *Desulfovibrio africanus*. Identification of a [4Fe-4S] cluster with one non-cysteine ligand, *Biochem. J.* **264;**275–284.

Gottwald, M., Andreesen, J. R., Le Gall, J., and Ljungdahl, L. G., 1975, Presence of cytochrome and menaquinone in *Clostridium formicoaceticum* and *C. thermoaceticum*, *J. Bacteriol.* **122:**325–328.

Guerlesquin, F., Bruschi, M., Bovier-Lapierre, G., Bonicel, J., and Couchand, P., 1983, Primary structure of the two (4Fe-4S) clusters ferredoxin from *D. desulfuricans* (strain Norway 4), *Biochimie* **65:**43–47.

Guerlesquin, F., Bruschi, M., Bovier-Lapierre, G., and Fauque, G., 1980, Comparative study of two ferredoxins from *Desulfovibrio desulfuricans* Norway, *Biochim. Biophys. Acta* **626:**127–135.

Guerlesquin, F., Moura, J. J. G., and Cammack, R., 1982, Iron-sulfur cluster composition and redox properties of two ferredoxins from *Desulfovibrio desulfuricans* Norway strain, *Biochim. Biophys. Acta* **679:**422–427.

Guerlesquin, F., Noailly, M., and Bruschi, M., 1985, Preliminary $^1$H-NMR studies of the interaction between cytochrome $c_3$ and ferredoxin I from *Desulfovibrio desulfuricans* Norway, *Biochem. Biophys. Res. Commun.* **130:**1102–1108.

Guerlesquin, F., Sari, J. C., and Bruschi, M., 1987, Thermodynamic parameters of cytochrome $c_3$-ferredoxin complex formation, *Biochemistry* **26:**7438–7443.

Haser, R., Payan, F., Bache, R., Bruschi, M., and Le Gall, J., 1979a, Crystallization and preliminary crystallographic data for cytochrome $c_{551.5}$ ($c_7$) from *Desulfuromonas acetoxidans*, *J. Mol. Biol.* **130:**97–98.

Haser, R., Pierrot, M., Frey, M., Payan, F., Astier, J. P., Bruschi, M., and Le Gall, J., 1979b, Structure and sequence of cytochrome $c_3$, a multi-heme cytochrome, *Nature* **282:**806-810.

Hatchikian, E. C., 1981, A cobalt prophyrin containing protein reducible by hydrogenase isolated from *Desulfovibrio desulfuricans* (Norway), *Biochem. Biophys. Res. Commun.* **103:**521–530.

Hatchikian, E. C., and Bruschi, M., 1981, Characterization of a new type of ferredoxin form *Desulfovibrio africanus*, *Biochim. Biophys. Acta* **634:**41–51.

Hatchikian, E. C., Bruschi, M., Le Gall, J., and Dubourdieu, M., 1969, Cristallisation et proprietes d'un cytochrome intervenant dans la reduction du thiosulfate par *D. gigas*, *Bu.. Soc. Fr. Physiol. Veg.* **15:**381.

Hatchikian, E. C., Jones, H. E., and Bruschi, M., 1979, Isolation and characterization of a rubredoxin and two ferredoxins from *Desulfovibrio africanus*, *Biochim. Biophys. Acta* **548:**471–483.

Hatchikian, E. C., and Le Gall, J., 1970a, Etude du metabolisme des acides dicarboxyliques et du pyrubate chez les bacteries sulfato-reductrices. I. Etude enzymatique de l'oxydation du fumarate en acetate, *Ann. Inst. Pasteur* **118**:125–142.

Hatchikian, E. C., and Le Gall, J., 1970b, Etude du métabolisme des acides dicarboxyliques et du pyruvate chez les bactéries sulfato-réductrices. II. Transport des électrons-acdepteurs finaux, *Ann. Inst. Pasteur* **118**:288–301.

Hatchikian, E. C., and Le Gall, J., 1972, Evidence for the presence of a b-type cytochrome in the sulfate-reducing bacterium *Desulfovibrio gigas*, and its role in the reduction of fumarate by molecular hydrogen, *Biochim. Biophys. Acta* **267**:479–484.

Hatchikian, E., Papavassiliou, P., Bianco, P., and Haladjian, J., 1984, Characterization of cytochrome c₃ from the thermophilic sulfate reducer *Thermodesulfobacterium commune*, *J. Bacteriol.* **159**:1040–1046.

Helms, L. R., Krey, G. D., and Swenson, R. P., 1990, Identification, sequence determination, and expression of the flavodoxin gene from *Desulfovibrio salexigens*, *Biochem. Biophys, Res. Commun.* **168**:809–817.

Helms, L. R., and Swenson, R. P., 1991, Cloning and characterization of the flavodoxin gene from *Desulfovibrio desulfuricans*, *Biochim. Biophys. Acta* **1089**:417–419.

Helms, L. R., and Swenson, R. P., 1992, The primary structures of the flavodoxins from two strains of *Desulfovibrio gigas*. Cloning and nucleotide sequence of the structure genes, *Biochim. Biophys. Acta* **1131**:325–328.

Herriot, J. R., Sieker, L. C., Jensen, L. H., and Lovenberg, W., 1970, Structure of rubredoxin: an X-ray study to 2.5 Å resolution, *J. Mol. Biol.* **50**:391–406.

Higuchi, Y., Inaka, K., Yasuoka, N., and Yagi, T., 1987, Isolation and crystallization of high molecular weight cytochrome from *Desulfovibrio vulgaris* Hildenborough, *Biochim. Biophys. Acta* **911**:341–348.

Higuchi, Y., Kusunoki, M., Matsuura, Y., Yasuoka, N., and Kakudo, M., 1984, Refined structure of cytochrome c₃ at 1.8 Å resolution, *J. Mol. Biol.* **172**:109–139.

Hormel, S., Walsh, K. A., Prickril, B. C., Titani, K., Le Gall, J., and Sieker, L. C., 1986, Amino acid sequence of rubredoxin from *Desulfovibrio desulfuricans* strain 27774, *FEBS Lett.* **201**:147–150.

Huang, Y.-H., Moura, I., Moura, J. J. G., Le Gall, J., Park J.-B., Adams, M. W. W., and Johnson, M. K., 1993, Resonance raman of nickel tetrathiolates and nickel-substituted rubredoxins and desulforedoxin. *Inorgan. Chem.* **32**:406–412.

Huynh, B. H., Moura, J. J. G., Moura, I., Kent, T. A., Le Gall, J., Xavier, A. V., and Münck, E., 1980, Evidence for a three-iron center in a ferredoxin from *Desulfovibrio gigas, J. Biol. Chem.* **255**:3242–3244.

Ishimoto, M., Koyama, J., and Nagai, Y., 1954, Role of a cytochrome in thiosulfate reduction by a sulfate-reducing bacterium, *Seikagaku Zasshi* **26**:303.

Johnson, M. K., Ware, Y. W., Spiro, T. G., Moura, I., Xavier, A. V., and Le Gall, J., 1981, Resonance raman spectra of three-iron centers in ferredoxin from *Desulfovibrio gigas*, *J. Biol. Chem.* **256**:9806–9808.

Jones, H. E., 1971, A re-examination of *Desulfovibrio africanus*, *Arch. Mikrobiol.* **80**:78–86.

Kent, T. A., Moura, I., Moura, J. J. G., Lipscomb, J. D., Huynh, B. H., Le Gall, J., Xavier, A. V., and Münck, E., 1982, Conversion of [3Fe-4S] into [4Fe-4S] clusters in a *Desulfovibrio gigas*: ferredoxin and isotopic labeling of iron-sulfur cluster subsites, *FEBS Lett.* **138**:55–58.

Kissinger, C. R., 1989, Crystal structures of ferredoxin II and cytochrome c₃ from *Desulfovibrio gigas*, *Ph. D. thesis*, University of Washington, Seattle.

Kissinger, C. R., Adman, E. T., Sieker, L. C., Jensen, L. H., and Le Gall, J., 1989, The crystal structure of the three iron ferredoxin II from *Desulfovibrio gigas, FEBS Lett.* **244**:437–442.

Kissinger, C. R., Adman, E. T., Sieker, L. C., and Jensen, L. H., 1988, Structure of the 3Fe-4S cluster in *Desulfovibrio gigas* ferredoxin II, *J. Am. Chem. Soc.* **110**:8721–8723.

Kissinger, C. R., Sieker, L. C., Adman, E. T., and Jensen, L. H., 1991, Refined crystal structure of ferredoxin II from *Desulfovibrio gigas* at 1.7 Å, *J. Mol. Biol.* **219**:693–715.

Klemm, D. J., and Barton, L. L., 1985, Oxidation of protoporphyrinogen in the obligate anaerobe *Desulfovibrio gigas, J. Bacteriol.* **164**:316–320.

Knight, E., and Hardy, R. W. F., 1966, Isolation and characteristics of flavodoxin from nitrogen-fixing *Clostridium pasteurianum, J. Biol. Chem.* **241**:2752–2756.

Koller, K. B., Hawkridge, F. M., Fauque, G., and Le Gall, J., 1987, Direct electron transfer reactions of cytochrome $c_{553}$ from *Desulfovibrio vulgaris* Hildenborough at indium oxide electrodes, *Biochem. Biophys. Res. Commun.* **145**:619–624.

Koziol, J., 1971, Fluorometric analyses of riboflavin and its coenzymes, *Methods in Enzymology* **18**:253–285.

Kremer, D. R., Veenhuis, M., Fauque, G., Peck, H. D., Jr., Le Gall, J., Lampreia, J., Moura, J. J. G., and Hansen, T. A., 1988, Immunocytochemical localization of APS reductase and bisulfite reductase in three *Desulfovibrio* species, *Arch. Microbiol.* **150**:296–300.

Krey, G. D., Vanin, E. F., and Swenson, R. P., 1988, Cloning, nucleotide sequence and expression of the flavodoxin gene from *Desulfovibrio vulgaris* (Hildenborough), *J. Biochem. Chem.* **263**:15436–15443.

Leenders, H. R., Vervoort, J., van-Hoek, A., Visser, A. J., 1990, Time-resolved fluorescence studies of flavodoxin. Fluorescence decay and fluorescence anisotropy decay of tryptophan in *Desulfovibrio* flavodoxins, *Eur. Biophys. J.* **18**:43–55.

Le Gall, J., 1968, Purification partielle et étude de la NAD:rubrédoxine oxidoréductase de *D. gigas, Ann. Inst. Pasteur* **114**:109–115.

Le Gall, J., DerVartanian, D. V., and Peck, H. D., Jr., 1979, Flavoproteins, iron proteins, and hemoproteins as electron-transfer components at the sulfate-reducing bacteria, *Curr. Top. Bioenerg.* **9**:237–265.

Le Gall, J., and Bruschi, M., 1968, Purification and some properties of a new cytochrome from *D. deslfuricans*, in: *Structure and Function of Cytochrome* (K. Okunuki, M. D. Kamen, and I. Sekuzu, eds.), University of Tokyo Press and University Park Press, p. 467.

Le Gall, J., and Dragoni, N., 1966, Dependence of sulfite-reduction on a crystallized ferredoxin from *D. gigas, Biochem. Biophys. Res. Commun.* **23**:145–149.

Le Gall, J., and Fauque, G., 1988, Dissimilatory reduction of sulfur compounds, in: *Biology of Anaerobic Microorganisms* (A. J. B. Zehnder, ed.), J. Wiley and Sons, New York, pp. 587–639.

Le Gall, J., and Hatchikian, E. C., 1967, Purification et proprietes d'une flavodoxine intervenant dans la reduction du sulfite par *D. gigas, C. R. Acad. Sci.* **264**:2580–2583.

Le Gall, J., Moura, J. J. G., Peck, H. D., Jr., and Xavier, A. V., 1982, Hydrogenase and other iron-sulfur proteins from sulfate-reducing and methane-form bacteria, in: *Iron-Sulfur Proteins* (Spiro, T. G., ed), John Wiley & Sons, New York, pp. 177–248.

Le Gall, J., and Peck, H. D., Jr., 1987, $NH_2$-terminal amino-acid sequences of electron transfer proteins from Gram-negative bacteria as indicators of their cellular localization: the sulfate-reducing bacteria, *FEMS Microbiol. Rev.* **46**:35–40.

Le Gall, J., and Postgate, J. R., 1973, The physiology of the sulfate-reducing bacteria, *Adv. Microbiol. Physiol.* **20**:81–133.

Le Gall, J., Prickril, B. C., Moura, I., Xavier, A. V., Moura, J. J. G., and Huynh, B. H., 1988, Isolation and characterization of rubrerythrin, a non-heme iron protein from *De-*

*sulfovibrio vulgaris* that contains rubredoxin centers and a hemerythrin-like binuclear center, *Biochemistry* **27:**1636–1642.

Liu, M.-C., Costa, C., Coutinho, I. B., Moura, J. J. G., Moura, I., Xavier, A. V., and Le Gall, J., 1988, Studies on the cytochrome components at nitrate-respiring and sulfate-respiring *Desulfovibrio desulfuricans* ATCC 27774, *J. Bacteriol.* **170:**5545–5551.

Loutfi, M., Guerlesquin, F., Bianco, P., Haladjian, J., and Bruschi, M., 1989, Comparative studies of polyhemic cytochromes c isolated from *Desulfovibrio vulgars* (Hildenborough) and *Desulfovibrio desulfuricans* (Norway), *Biochem. Biophys. Res. Commun.* **159:**670–676.

Macedo, A. L., Moura, I., Moura, J. J. G., Le Gall, J., and Huynh, B. H., 1993, Temperature-dependent protein NMR investigation of the electronic structure of the trinuclear iron cluster of the oxidized *Desulfovibrio gigas* ferredoxin II, *Inorg. Chem.* **32:**1101–1105.

Macedo, A. L., Moura, I., Surerus, K. K., Papaefthymiou, V., Liu, M.-Y., Le Gall, J., Münck, E., and Moura, J. J. G., 1994, Thio/disulfide formation associated with the redox activity of the [Fe$_3$S$_4$] cluster of *Desulfovibrio gigas* ferredoxin II. *J. Biol. Chem.* **268:**8052–8058.

Marion, D., and Guerlesquin, F., 1989, Experimental evidence of an alpha helix in *Desulfovibrio desulfuricans* Norway ferredoxin I: a two-dimensional NMR study. *Biochem. Biophys. Res. Commun.*, **159:**592–598.

Maroc, J., Azerad, R., Kamen, M. D., and Le Gall, J., 1970, Menaquinone (MK-6) in the sulfate-reducing, obligate anaerobe *Desulfovibrio, Biochim. Biophys. Acta* **197:**87–89.

Mayhew, S. G., van Dijk, C., and van der Westen, H. M., 1978, Properties of hydrogenases from the anaerobic bacteria *Megasphaera elsdenii* and *Desulfovibrio vulgaris* (Hildenborough), in: *Hydrogenase: Their Catalytic Activity, Structure and Function* (H. G. Schlegel and K. Schneider, eds), Erich Göltze, KG, Göttingen, Germany. pp. 125–140.

Meyer, T. E., and Cusanovich, M. A., 1989, Structure, function and distribution of soluble bacterial redox proteins, *Biochim. Biophys. Acta* **975:**1–28.

Moreno, C., Franco, R., Le Gall, J., and Moura, J. J. G., 1993, Electrochemical studies of the electron transfer between the *Desulfovibrio gigas* hydrogenase and several small electron transfer carriers isolated from the same genus, *Eur. J. Biochem.* **271:**981–989.

Moreno, C., Macedo, A. L., Moura, I., Le Gall, J., and Moura, J. J. G., 1994, Redox properties of *Desulfovibrio gigas* [Fe$_3$S$_4$] and [Fe$_4$S$_4$] ferredoxins and heterometal cubane-type clusters formed within the [Fe$_3$S$_4$] core. Square wave voltammetric studies. *J. Inorg. Biochem.*, **53:**219–234.

Moura, I., Bruschi, M., Le Gall, J., Moura, J. J. G., and Xavier, A. V., 1977, Isolation and characterization of desulforedoxin, a new type of non-heme iron protein from *D. gigas, Biochem. Biophys. Res. Commun.* **75:**1037–1044.

Moura, I., Fauque, G., Le Gall, J., Xavier, A. V., and Moura, J. J. G., 1987a, Characterization of the cytochrome system of a nitrogen-fixing strain of a sulfate-reducing bacterium, *Desulfovibrio desulfuricans* strain Berre Eau, *Eur. J. Biochem.* **162:**547–559.

Moura, I., Teixeira, M., Xavier, A. V., Moura, J. J. G., Lespinat, P. A., Le Gall, J., Berlier, Y. M., Saint-Martin, P., and Fauque, G., 1987b, Spectroscopic studies of cobalt and nickel substituted rubredoxins, *Recueil. Trav. Chim. Pay-Bas* **106:**418.

Moura, I., Moura, J. J. G., Bruschi, M., and Le Gall, J., 1980a, Flavodoxin and rubredoxin from *Desulfovibrio salexigens, Biochem. Biophys. Acta* **613:**1–12.

Moura, J. J. G., Moura, I., Bruschi, M., Le Gall, J., and Xavier, A. V., 1980b, A cobalt-containing protein isolated from *D. gigas*, a sulfate reducer, *Biochem. Biophys. Res. Commun.* **92:**962–970.

Moura, I., Moura, J. J. G., Münck, E., Papaefthymiou, V., and Le Gall, J., 1986, Evidence for the formation of a CoFe$_3$S$_4$ cluster in *Desulfovibrio gigas* ferredoxin II, *J. Am. Chem. Soc.* **108**:349–351.

Moura, I., Moura, J. J. G., Santos, M. H., Xavier, A. V., and Le Gall, J., 1979, Redox studies of rubredoxins from sulphate and sulphur reducing bacteria, *FEBS Lett.* **107**:419–421.

Moura, I., Tavares, P., Moura, J. J. G., Ravi, N., Huynh, B. H., Liu, M.-Y., and Le Gall, J., 1990, Purification and characterization of desulfoferrodoxin. A novel protein from *Desulfovibrio desulfuricans* (ATCC 27774) and from *Desulfovibrio vulgaris* (strain Hildenborough) that contains a distorted rubredoxin center and a mononuclear ferrous center, *J. Biol. Chem.* **265**:21596–21602.

Moura, I., Ravi, N., Costa, C., Teixeira, M., Le Gall, J., Moura, J. J., and Huynh, B. H., 1992a, Mössbauer characterization of the tetraheme cytochrome c$_3$ from *Desulfovibrio baculatus* (DSM 1743). Spectral deconvolution of the heme components, *Eur. J. Biochem.* **204**:1131–1139.

Moura, I., Tavares, P., Moura, J. J. G., Ravi, N., Huynh, B. H., Liu, M.-Y., and Le Gall, J., 1992b, Direct spectroscopic evidence for the presence of a 6 Fe cluster in an iron-sulfur protein isolated from Desulfovibrio desulfuricans (ATCC 27774). *J. Biol. Chem.* **267**:4489–4496.

Moura, I., Teixeira, M., Le Gall, J., and Moura, J. J., 1991a, Spectroscopic studies of cobalt and nickel substituted rubredoxin and desulforedoxin. *J. Inorg. Biochem.* **44**:127–139.

Moura, J. J. G., Costa, C., Liu, M.-Y., Moura, I., and Le Gall, J., 1991b, Structural and functional approach toward a classification of the complex cytochrome c system found in sulfate-reducing bacteria. *Biochim. Biophys. Acta* **1058**:61–66.

Moura, I., Xavier, A. V., Cammack, R., Bruschi, M., and Le Gall, J., 1978, A comparative spectroscopic study of desulforedoxin and rubredoxin, two non-heme iron proteins from *D. gigas, Biochim. Biophys. Acta* **533**:156–162.

Moura, J. J. G., Le Gall, J., and Xavier, A. V., 1984, Interconversion from 3 Fe into 4 Fe clusters in the presence of *Desulfovibrio gigas* cell extracts, *Eur. J. Biochem.* **141**:319–322.

Moura, J. J. G., Moura, I., Kent, T. A., Lipscomb, J. D., Huynh, B. H., Le Gall, J., Xavier, A. V., and Münck, E., 1982, Conversion of [3Fe-4S] into [4Fe-4S] clusters: Mössbauer and EPR studies of *Desulfovibrio gigas* ferredoxin II, *J. Biol. Chem.* **257**:6259–6267.

Moura, J. J. G., Xavier, A. V., Hatchikian, E. C., and Le Gall, J., 1978, Structural control of the redox potentials and of the physiological activity by oligomerization of ferredoxin, *FEBS Lett.* **89**:177–179.

Mus-Veteau, I., Dolla, A., Guerlesquin, F., Payan, F., Czjzek, M., Haser, R., Bianco, P., Haladjian, J., Rapp-Giles, B. J., Wall, J. D., Voordouw, G., and Bruschi, M., 1992, Site-directed mutagenesis of tetraheme cytochrome c$_3$. *J. Biol. Chem.* **267**:16851–16858.

Natsoulis, G., Hilger, F., and Fink, G. R., 1986, The HTS1 gene encodes both the cytoplasmic and mitochondrial histidine tRNA synthetases of *S. crevisiae, Cell* **46**:235–243.

Newman, D., and Postgate, J. R., 1968, Rubredoxin from a nitrogen-fixing variety of *Desulfovibrio desulfuricans, Eur. J. Biochem.* **7**:45–50.

Nivière, V., Bernadac, A., Forget, N., Fernàndez, V. M., and Hatchikian, C. E., 1991, Localization of hydrogenase in *Desulfovibrio gigas* cells, *Arch. Microbiol.* **155**:579–586.

Odom, J. M., Bruschi, M., Peck, H. D., Jr., and Le Gall, J., 1976, Structure-function relationships among rubredoxins: comparative reactivities with NADH:rubredoxin oxidoreductase from *Desulfovibrio gigas, Proc. Fed. Am. Soc. Exp. Biol.* **35**:1360.

Odom, J. M., and Peck, H. D., Jr., 1981, Hydrogen cycling as a general mechanism for energy coupling in the sulfate reducing bacteria, *Desulfovibrio* sp., *FEMS Microbiol. Lett.* **12**:47–50.

Odom, J. M., and Peck, H. D., Jr., 1984, Hydrogenase, electron-transfer proteins, and energy coupling in the sulfate-reducing bacteria *Desulfovibrio*, *Ann. Rev. Microbiol.* **38:**551–592.

Odom, J. M., and Singleton, R., Jr., 1993, *The Sulfate-Reducing Bacteria: Contemporary Perspectives.* Springer-Verlag, New York.

Ogata, M., Kondo, S., Okawara, N., and Yagi, T., 1988, Purification and characterization of ferredoxin from *Desulfovibrio vulgaris* Miyazaki, *J. Biochem.* (Tokyo) **103:**121–125.

Okawara, N., Ogata, M., Yagi, T., Wakabayashi, S., and Matsubara, M., 1988a, Amino acid sequence of ferredoxin I from *Desulfovibrio vulgaris* Miyazaki, *J. Biochem.* (Tokyo) **104:**196–199.

Okawara, N., Ogata, M., Yagi, T., Wakabayashi, S., and Matsubara, M., 1988b, Characterization and complete amino acid sequence of ferredoxin II from *Desulfovibrio vulgaris* Miyazaki, *Biochimie* **70:**1815–1820.

Palma, P. N., Moura, I., Le Gall, J., Van Beeumen, J., Wampler, J. E., and Moura, J. J. G., 1994, Evidence for a ternary complex formed between flavodoxin and cytochrome $c_3$, *Biochemistry* **33:**6394–6407.

Papaefthymiou, V., Girerd, J.-J., Moura, I., Moura, J. J. G., and Münck, E., 1987, Mössbauer study of *D. gigas* ferredoxin II and spin-coupling model for the $Fe_3S_4$ cluster with valence, *J. Am. Chem. Soc.* **109:**4703–4710.

Park, J. S., Kano, K., Morimoto, Y., Higuchi, Y., Yasuoka, N., Ogata, M., Niki, K., and Akutsu, H., 1991, $^1H$ NMR studies on ferricytochrome $c_3$ from *Desulfovibrio vulgaris* Miyazaki F and its interaction with ferredoxin I. *J. Biomol. NMR* **1:**271–282.

Peck, H. D., Jr., 1966, Phosphorylation coupled with electron transfer in extracts of the sulfate reducing bacterium, *Desulfovibrio gigas, Biochem. Biophys. Res. Commun.* **22:**112–118.

Peck, H. D., Jr., Deacon, T. E., and Davidson, J. T., 1965, Studies on adenosine 5'-phosphosulfate reductase from *Desulfovibrio desulfuricans* and *Thiobacillus thioparus, Biochim. Biophys. Acta* **96:**429–446.

Peck, H. D., Jr., and Le Gall, J., 1982, Biochemistry of dissimilatory sulfate reduction. *Phil. Trans. R. Soc.* **B298:**433–466.

Petitdemange, H., Marczak, R., and Gay, R., 1979, NADH-flavodoxin oxidoreductase activity in *Clostridium tyrobutiricum, FEMS Lett.* **5:**291–294.

Pierik, A. J., Wolbert, R. B., Portier, G. L., Verhagen, M. F., Hagen, W. R., 1993, Nigerythrin and rubrerythrin from *Desulfovibrio vulgaris* each contain two mononuclear iron centers and two dinuclear iron clusters, *Eur. J. Biochem.* **212:**237–245.

Pierrot, M., Haser, R., Frey, M., Payan, F., and Astier, J. P., 1982, Crystal structure and electron transfer properties of cytochrome $c_3$, *J. Biol. Chem.* **257:**14341–14348.

Pollock, W. B. R., Loutfi, M., Bruschi, M., Rapp-Giles, B. J., Wall, J. D., and Voordouw, G., 1991, Cloning, sequencing, and expression of the gene encoding the high-molecular-weight cytochrome c from *Desulfovibrio vulgaris* Hildenborough, *J. Bacteriol.* **173:**220–228.

Postgate, J. R., 1954, Presence of cytochrome in an obligate anaerobe, *Biochem. J.* **56:**xi.

Postgate, J. R., 1984, *The Sulphate-reducing Bacteria*, 2nd edn., Cambridge University Press, Cambridge, England, pp. 10–82.

Prickril, B. C., Czechowski, M. H., Przybyla, A. E., Peck, H. D., Jr., and Le Gall, J., 1986, Putative signal peptide on the small subunit of the periplasmic hydrogenase from *Desulfovibrio vulgaris, J. Bacteriol.* **167:**722–725.

Prickril, B. C., He, S. H., Li, C., Menon, N., Choi, E.-S., Przybyla, A. E., DerVartanian, D. V., Peck, H. D., Jr., Fauque, G., Le Gall, J., Teixeira, M., Moura, I., Moura, J. J. G., Patil, D., and Huynh, B. H., 1987, Identification of three classes of hydrogenases in the genus *Desulfovibrio, Biochem. Biophys. Res. Commun.* **149:**369–377.

Prickril, B. C., Kurtz, D. M., Jr., Le Gall, J., and Voordouw, G., 1991, Cloning and sequencing of the gene for rubrerythrin from *Desulfovibrio vulgaris* (Hildenborough), *Biochemistry* **30:**11118–11123.

Ragsdale, S. W., Ljungdahl, L. G., and DerVartanian, D. V., 1983, Isolation of carbon monoxide dehydrogenase from *Acetobacterium woodii* and comparison of its properties with those of the *Clostridium thermoaceticum* enzyme. *J. Bacteriol.* **155:**1224–1237.

Robbins, A. H., and Stout, C. D., 1989, Structure of activated aconitase: formation of the [4Fe-4S] cluster in the crystal. *Proc. Natl. Acad. Sci. USA* **86:**3639–3643.

Rossi, M., Pollock, W. B. R., Reij, M. W., Keon, R. G., Fu, R., and Voordouw, G., 1993, The *hmc* operon of *Desulfovibrio vulgaris* subsp. *vulgaris* Hildenborough encodes a potential transmembrane redox protein complex, *J. Bacteriol.* **175:**4699–4711.

Saint-Martin, P., Lespinat, P. A., Fauque, G., Berlier, Y., Le Gall, J., Moura, I., Teixeira, M., Xavier, A. V., and Moura, J. J. G., 1988, Hydrogen production and deuterium-proton exchange reactions catalyzed by *Desulfovibrio* nickel (II)-substituted rubredoxins, *Proc. Natl. Acad. Sci. USA* **85:**9378–9380.

Samain, E., Albagnac, G., and Le Gall, J., 1986, Redox studies of the tetraheme cytochrome $c_3$ isolated from the propionate oxidizing, sulfate reducing bacterium, *Desulfobulbus elongatus*, *FEBS Lett.* **204:**247–250.

Samain, E., Patil, D. S., DerVartanian, D. V., Albagnac, G., and Le Gall, J., 1987, Isolation of succinate dehydrogenase from *Desulfobulbus elongatus*, a propionate oxidizing, sulfate-reducing bacterium, *FEBS Lett.* **216:**140–144.

Santos, H., Faraleira, P., Xavier, A. V., Chen, L., Liu, M.-Y., and Le Gall, J., 1993, Anaerobic metabolism of carbon reserves by the "obligate anaerobe" *Desulfovibrio gigas*, *Biochem. Biophys. Res. Commun.* **195:**551–557.

Seki, S., Ikeda, A., and Ishimoto, M., 1988, Rubredoxin as an intermediary electron carrier for nitrate reduction by NAD(P)H in *Clostridium perfringens*, *J. Biochem.* **103:**583–584.

Seki, Y., Seki, S., Satoh, M., Ikeda, A., and Ishimoto, M., 1989, Rubredoxin from *Clostridium perfringens:* complete amino acid sequence and participation in nitrate reduction, *J. Biochem.* **106:**336–341.

Shimizu, F., Ogata, M., Yagi, T., Wakabayashi, S., and Matsubara, H., 1989, Amino acid sequence and function of rubredoxin from *Desulfovibrio vulgaris* Miyazaki, *Biochimie* **71:**1171–1177.

Sieker, L. C., Adman, E. T., Jensen, L. H., and Le Gall, J., 1984, Crystallization and preliminary X-ray diffraction study of the 3-Fe ferredoxin II from the bacterium *Desulfovibrio gigas*, *J. Mol. Biol.* **179:**151–155.

Sieker, L. C., Jensen, L. H., and Le Gall, J., 1986a, Preliminary X-ray studies of the tetraheme cytochrome $c_3$ from *Desulfovibrio gigas*, *FEBS Lett.* **209:**261–264.

Sieker, L. C., Stenkamp, P. E., Jansen, L. H., Prickril, B., and Le Gall, J., 1986b, Structure of rubredoxin from the bacterium *Desulfovibrio desulfuricans*, *FEBS Lett.* **208:**73–76.

Sieker, L. C., Jensen, L. H., Prickril, B., and Le Gall, J., 1983, Crystallographic study of rubredoxin from *Desulfovibrio desulfuricans* strain 27774, *J. Mol. Biol.* **171:**101–103.

Stenkamp, R. E., Sieker, L. C., Jensen, L. H., 1990, The structure of rubredoxin from *Desulfovibrio desulfuricans* strain 27774 at 1.5 Å resolution, *Proteins* **8:**352–364.

Stewart, D. E., Le Gall, J., Moura, I., Moura, J. J. G., Peck, H. D., Jr., Xavier, A. V., Weiner, P. K., and Wampler, J. E., 1988, A hypothetical model of the flavodoxin-tetraheme cytochrome $c_3$ complex of sulfate-reducing bacteria, *Biochemistry* **27:**2444–2450.

Stewart, D. E., Le Gall, J., Moura, I., Moura, J. J., Peck, H. D., Jr., Xavier, A. V., Weiner, P. K., and Wampler, J. E., 1989, Electron transport in sulfate-reducing bacteria. Molecular modeling and NMR studies of the rubredoxin-tetraheme-cytochrome-$c_3$ complex, *Eur. J. Biochem.* **185:**695–700.

Stewart, D. E., and Wampler, J. E., 1991, Molecular dynamics simulations at the cytochrome $c_3$-rubredoxin complex from Desulfovibrio vulgaris, Proteins **11:**142–152.

Stout, C. D., 1988, 7-iron ferredoxin revisited, J. Biol. Chem. **263:**9256–9260.

Stout, C. D., 1989, Refinement of the 7 Fe ferredoxin from Azotobacter vinelandii at 1.9 Å resolution, J. Mol. Biol. **205:**545–555.

Stout, G. H., Turley, S., Sieker, L. C., and Jensen, L. H., 1988, Structure of ferredoxin I from Azotobacter vinelandii, Proc. Natl. Acad. Sci. USA **85:**1020–1027.

Suh, B., and Akagi, J. M., 1969, Formation of thiosulfate from sulfite by Desulfovibrio vulgaris. J. Bacteriol. **99:**210–215.

Surerus, K., Münck, E., Moura, I., Moura, J. J. G., and Le Gall, J., 1987, Evidence for the formation of a $ZnFe_3S_4$ cluster in Desulfovibrio gigas, J. Am. Chem. Soc. **109:**3805–3807.

Tan, J. A., and Cowan, A., 1990, Coordination and redox properties of a novel triheme cytochrome from Desulfovibrio vulgaris (Hildenborough), Biochemistry **29:**4886–4892.

Tavares, P., Ravi, N., Moura, J. J. G., Le Gall, J., Huang, Y.-H., Crouse, B. R., Johnson, M. K., Huynh, B. H., and Moura, I., 1994, Spectroscopic properties of desulfoferrodoxin from Desulfovibrio desulfuricans (ATCC 27774), J. Biol. Chem. **269:**10504–10510.

Thomson, A. J., Breton, J., Butt, J. N., Hatchikian, E. C., and Armstrong, F. A., 1992, Iron-sulphur clusters with labile metal ions, J. Inorg. Biochem. **47:**197–207.

Thomson, A. J., Robinson, A. E., Johnson, M. K., Moura, J. J. G., Moura, I., Xavier, A. V., and Le Gall, J., 1981, The three iron cluster in a ferredoxin from Desulfovibrio gigas, a low-temperature magnetic circular dichroism study, Biochim. Biophys. Acta **670:**93–100.

Van-Beeumen, J. J., Van-Driessche, G., Liu, M.-Y., and Le Gall, J., 1991, The primary structure of rubrerythrin, a protein with inorganic pyrophosphatase activity from Desulfovibrio vulgaris. Comparison with hemerythrin and rubredoxin, J. Biol. Chem. **266:**20645–20653.

van den Berg, W. A. M., Stokkermans, J. P. W. G., and van Dongen, M. A. M., 1993, The operon for the Fe-hydrogenase in Desulfovibrio vulgaris (Hildenborough): mapping of the transcript and regulation of expression, FEMS Microbiol. Lett. **110:**85–90.

Van Rooijen, G. J. H., Bruschi, M., and Voordouw, G., 1989, Cloning and sequencing of the gene encoding cytochrome $c_{553}$ from Desulfovibrio vulgaris Hildenborough, J. Bacteriol. **171:**3575–3578.

Vervoort, J., Muller, F., Le Gall, J., Bacher, A., and Sedlmaier, H., 1985, Carbon-13 and nitrogen-15 nuclear-magnetic-resonance investigation on Desulfovibrio vulgaris flavodoxin, Eur. J. Biochem. **151:**49–57.

Vervourt, J., van Berkel, W. J. H., Mayhew, S. G., Muller, F., Bacher, A., Nielsen, P., Le Gall, J., 1986, Properties of the complexes of riboflavin 3′,5′-biphosphate and the apoproteins form Megasphaera elsdenii and Desulfovibrio vulgaris, Eur. J. Biochem. **161:**749–756.

Visser, A. J. G. W., Carreira, L. A., Le Gall, J., and Lee, J., 1980, Coherent anti-stokes raman spectroscopy of flavodoxin in solution and crystal phases, J. Phys. Chem. **84:**3344–3346.

Vogel, H., Bruschi, M., and Le Gall, J., 1977, Phylogenetic studies of two rubredoxins from sulfate-reducing bacteria, J. Mol. Evol. **9:**111–119.

Voordouw, G., 1988, Cloning of genes encoding redox proteins of known amino acid sequence from a library of the Desulfovibrio vulgaris (Hildenborough) genome, Gene **69:**75–83.

Voordouw, G., and Brenner, S., 1986, Cloning and sequencing of the gene encoding cytochrome $c_3$ from Desulfovibrio vulgaris (Hildenborough), Eur. J. Biochem. **159:**347–351.

Voordouw, G., Pollock, W. B. R., Bruschi, M., Guerlesquin, F., Rapp-Giles, B. J., and Wall, J. D., 1990a, Functional expression of *Desulfovibrio vulgaris* Hildenborough after conjugational gene transfer from *Escherichia coli, J. Bacteriol.* **172:**6122–6126.

Voordouw, G., Niviere, V., Ferris, F. G., Fedorak, P. M., and Westlake, D. W. S., 1990b, Distribution of hydrogenase genes in *Desulfovibrio* spp. and their use in identification of species from the oil field environment, *Appl. Environ. Microbiol.* **56:**3748–3754.

Watenpaugh, K. D., Sieker, L. G., Jensen, L. H., Le Gall, J., and Dubourdieu, M., 1972, Structure of the oxidized form of a flavodoxin at 2.5 Å resolution. *Proc. Natl. Acad. Sci. USA* **69:**3185–3188.

Watt, W., Tulinsky, A., Swenson, R. P., and Watenpaugh, K. D., 1991, Comparison of the crystal structures of a flavodoxin in its three oxidation states at cryogenic temperatures, *J. Mol. Biol.* **218:**195–208.

Wu, M., and Tzagoloff, A., 1987, Mitochondrial and cytoplasmic fumarases in *Saccharomyces cerevisiae* are encoded by a single nuclear gene FUM 1, *J. Biol. Chem.* **262:**12275–12282.

Yagi, T., 1979, Purification and properties of cytochrome $c_{553}$, an electron acceptor for formate dehydrogenase of *Desulfovibrio vulgaris* Miyazaki, *Biochim. Biophys. Acta* **548:**96–105.

Zubieta, J. A., Mason, R., and Postgate, J. R., 1973, A four-iron ferredoxin from *Desulfovibrio desulfuricans, Biochem. J.* **133:**851-854.

# Solute Transport and Cell Energetics

# 6

## HERIBERT CYPIONKA

## 1. INTRODUCTION

The main energy problems of a sulfate-reducing bacterium were described by Wood (1978), Stouthamer (1988), and Thauer (1989). In this chapter I try to develop a simplified unifying scheme of sulfate reduction. After some general considerations on the bioenergetics of sulfate reduction, we will follow the fate of a sulfate molecule metabolized by a sulfate-reducing bacterium, from uptake until excretion as $H_2S$. Sulfate transport is discussed in detail, although no complete compilation of data is given. Each step of sulfate reduction will be discussed, mainly with hydrogen gas as a model electron donor. In the last part, we shall regard alternative processes related to energy conservation in sulfate-reducing bacteria (SRB), i.e., fermentation of organic compounds, disproportionation, and even oxidation of inorganic sulfur compounds, and use of alternate electron acceptors such as nitrate and molecular oxygen. Many of these processes have been detected recently and are unique for SRB. They demonstrate the metabolic flexibility of the energy metabolism of SRB, and will hopefully provide tools for the elucidation of the energy conservation during sulfate reduction.

### 1.1. Thermodynamics of Dissimilatory Sulfate Reduction

The dissimilatory SRB have little energy available to them. The upper limits of energy conservation from sulfate reduction are set by thermodynamics. If a very potent electron donor like $H_2$ ($E_0' = -420$

HERIBERT CYPIONKA • Carl von Ossietzky Universität Oldenburg, Institut für Chemie und Biologies des Meeres, D-26111 Oldenburg, Germany.

*Sulfate-Reducing Bacteria*, edited by Larry L. Barton. Plenum Press, New York, 1995.

mV) is oxidized, the free energy change of the overall reaction under standard conditions at neutral pH is $-155$ kJ/mole [Eq. (1)] and thus sixfold lower than with $O_2$ as electron acceptor [Eq. (2)].

$$4 \ H_2 + SO_4^{2-} + 1.5 \ H^+ \rightarrow 0.5 \ HS^- + 0.5 \ H_2S$$
$$+ \ 4 \ H_2O \ \Delta G^{o\prime} = -155 \ kJ/mole \tag{1}$$

$$4 \ H_2 + 2 \ O_2 \rightarrow 4 \ H_2O \ \Delta G^{o\prime} = -949 \ kJ/mole \tag{2}$$

About $-70$ kJ/mole are required for ATP synthesis in growing cells, (Thauer and Morris, 1984; Schink, 1992). Thus, not more than 2 ATP per sulfate reduced can be conserved. Even less can be expected, because thermodynamics does not regard transport processes and biochemical pathways. If those include critical steps, increasing portions of the free energy change might not be available for energy conservation. Indeed, several observations, such as the ATP requirement for sulfate activation (Peck, 1959; Ishimoto, 1959), energy requirement for assimilatory sulfate reduction, and (for a long time) a restricted spectrum of known substrates that were only incompletely oxidized to acetate, might have supported the impression that sulfate reduction is not coupled to energy conservation at all. Although oxidative phosphorylation has been demonstrated (Peck, 1960, 1966), the spore-forming sulfate reducers were regarded as fermenting bacteria that use sulfate as an electron "sink" for excess electrons produced during fermentation (Peck and LeGall, 1982).

However, the demonstration of lithotrophic growth with $H_2$ and sulfate (Badziong et al., 1978; Badziong and Thauer, 1978; Klemps et al., 1985) has proven that sulfate reduction is a true respiration process that can accomplish net energy conservation. $H_2$ is oxidized in a single step that does not allow substrate-level phosphorylation. Therefore, growth with $H_2$ as the only electron donor must be coupled to chemiosmotic energy conservation during reduction of the electron acceptor.

## 1.2. Some Relevant Properties of Sulfate, Sulfide, and Sulfite

Sulfate reduction entails several problems not encountered during aerobic respiration. Some of these can be easily deduced by comparing eqs. 1 and 2.

While in the second reaction only membrane-permeable compounds, gases, and water, are involved, sulfate reduction involves consumption and production of ions. Sulfate even carries two negative charges which are repelled by the membrane protential, i.e., an excess of negative charges at the inner side of the membrane. This demonstrates the problem of sulfate transport.

While the end product of aerobic respiration is water, sulfate reduction produces $H_2S$, which is not only malodorous, but toxic, even for SRB at concentrations above 5 mM (Klemps *et al.*, 1985).

Furthermore, sulfate reduction tends to change the pH, since protons are consumed by formation of $H_2S$ from the sulfuric acid. Hydrogen sulfide is a weak acid only partially dissociated at neutral pH. In this review, the term sulfide is used for the sum of hydrogen sulfide ($H_2S$) and bisulfide ($HS^-$). Because the first dissociation constant ($pk_1$) of $H_2S$ is about 7.0, $H_2S$ and $HS^-$ are both present within the pH range where SRB are active. The second dissociation constant ($pK_2$) of $H_2S$ is around 17 to 19 (Myers, 1986). This means that the sulfide ion ($S^{2-}$) is a stronger base than $OH^-$. Therefore, free $S^{2-}$ cannot exist in aqueous solutions, because an $S^{2-}$ ion, even in strongly alkaline solution, immediately deprotonizes water to form $OH^-$ and $HS^-$.

While sulfate and thiosulfate can be regarded as almost completely dissociated, the second pK of sulfurous acid ($H_2SO_3$) is 6.9. This means that sulfite is present as bisulfite ($HSO_3^-$) and sulfite ($SO_3^{2-}$) in approximately equal amounts at neutral pH.

## 1.3. Scalar and Vectorial Processes

The smallest unit of cell energetics is not ATP, but a single proton, any other ion, or even an electron that is transported across the cell membrane by various chemiosmotically relevant processes. A conversion factor of three protons translocated per ATP (Thauer and Morris, 1984) is assumed in this review, without having been tested in SRB. For other ions, such as $Na^+$, the conversion factor depends on the transport mechanisms involved. Furthermore, uptake of substrates and excretion of end products includes energetic aspects. Thus, transport processes and cell energetics are closely related. Generally, one has to distinguish between *scalar* and *vectorial* processes. Scalar changes are those in which the net amount of a compound is changed. For example, the alkalinization during sulfate reduction to sulfide [Eq. (1)] is a scalar process. The change is chemically analyzable in a system with no compartments (cells, vesicles etc.), but also in a bacterial culture, since the inner cell volume makes up a very small percentage [e.g., 1.4 μl per mg dry mass (Varma *et al.*, 1983a; Kreke and Cypionka, 1992a)] of the culture volume. For the quantification of scalar pH changes, one has to regard the medium pH and the dissociation constants of the compounds metabolized. It would be easier to read and formally correct to write as

$$4 H_2 + SO_4^{2-} \rightarrow S^{2-} + 4 H_2O \quad \Delta G^{\circ\prime} = -118 \text{ kJ/mole} \tag{3}$$

or

$$4 H_2 + SO_4^{2-} + H^+ \rightarrow HS^- + 4 H_2O \quad \Delta G^{\circ\prime} = -152 \text{ kJ/mole} \tag{4}$$

But this would not correctly describe the scalar proton disappearance in neutral medium. The mixed forms of sulfide (and sulfite) and the different numbers of protons also result in slight differences in the free energy changes (all calculated after Thauer et al., 1977).

Although often so stated in the literature it must be pointed out that scalar processes cannot be used for energy conservation. If acid is added to a cell suspension the resulting *scalar* acidification cannot be utilized for ATP conservation by the cells. Only very few protons will enter the cell, because the scalar change affects only one side of the membrane. The change in the transmembrane $\Delta pH$ will be balanced by a change in the electrical membrane potential. For the same reason, the net scalar alkalinization caused by sulfate reduction is not expected to diminish the energy yield, as long as the pH stress is not too hard. Therefore, terms such as "scalar proton transport" as found sometimes in the literature make no sense.

For chemiosmotic energy coupling *vectorial* processes are required, which are coupled to concentration changes on both sides of the cytoplasmic membrane. If a proton (or another ion) is transported from inside the cell across the cell membrane, the total amount of $H^+$ in the assay is not changed. However, only such a vectorial transport generates a force (composed of the transmembrane electrical and chemical concentration difference), pulling the proton back into the cell.

Movement of the proton may be coupled to a chemical reaction (e.g., reduction of an electron acceptor or ATP synthesis); this is then classified as a *primary transport system*. Alternatively, the proton may be transported in symport or antiport with another compound. In this case, called a *secondary transport system*, the driving force depends on the transmembrane concentration differences of both compounds. (The terms *active* and *passive transport* are less suitable for differentiating between primary and secondary transport systems.) If transport results in a net transfer of charges across the membrane, it is called *electrogenic*. In this case the electrical membrane potential has to be taken into account, which is not necessary for *electroneutral transport* systems.

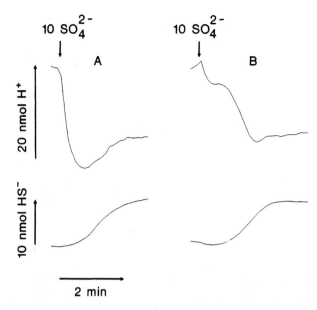

**Figure 1.** pH and sulfide traces recorded during sulfate reduction by *Desulfovibrio desulfuricans*. Washed cells were incubated in $H_2$-saturated 150 mM KCl. The concentrations were followed by means of electrodes. (A) Upon the addition of 10 nmoles of sulfate, protons disappear due to proton–sulfate symport, which precedes sulfide formation. Release of sulfide is then accompanied by a slight acidification due to partial dissociation of $H_2S$. The final result is a scalar alkalinization according to eq. (1). The shoulder in the second experiment (B) indicates vectorial proton translocation that does not change the scalar result.

## 1.4. A First Look at Proton Effects Coupled to Sulfate Reduction

Both scalar and vectorial proton movements during sulfate reduction according to eq. (1) can be observed with a pH electrode (Fig. 1A). Upon addition of sulfate (a small pulse of 2.5 μM) to cell suspensions incubated in nonbuffered KCl under $H_2$, the first event is an alkalinization. The rapid disappearance of protons preceeds the formation of sulfide and indicates vectorial proton uptake in symport with sulfate. Then, the transient alkalinization is compensated by an acidification phase in parallel with the release of $H_2S$, which outside the cell partially dissociates according to

$$H_2S \rightarrow 0.5\ H_2S + 0.5\ HS^- + 0.5\ H^+ \quad \Delta G^{\circ\prime} = 3\ kJ/mol \qquad (5)$$

The final result is a scalar disappearance of 1.5 $H^+$ per sulfate according to eq. (1). The purpose of sulfate reduction by the bacterium—

generation of a proton motive force across the cytoplasmic membrane—is not so easily seen in this experiment. Sometimes one can observe a shoulder that can be attributed to proton translocation [or $H^+$ release by a periplasmic hydrogenase (Fig. 1B)], but barely allows quantitative calculations. Translocated protons (or those set free by a periplasmic hydrogenase) are rapidly taken up for sulfate transport and for ATP synthesis. The latter might be especially important in SRB because ATP is consumed for the activation of sulfate. Another shortcome of pH measurement under these conditions is that only the electroneutral exodus of protons can be seen as long as the cells have an electrical membrane potential (West and Mitchell, 1974). Therefore, special experimental conditions must be set in order to study each of the different steps (see below).

## 2. SULFATE TRANSPORT

### 2.1. Different Feasible Mechanisms of Sulfate Transport

Because sulfate has to be activated by ATP, which is available within cytoplasm only, the precondition of assimilatory and dissimilatory reduction of sulfate is its uptake into the cell. The capacity of assimilatory sulfate reduction is common to many plants, fungi, and bacteria. Sulfate uptake for assimilatory purposes is often achieved by *primary transport systems*. In enterobacteria (Ames, 1988; Hryniewicz et al., 1990) and cyanobacteria (Jeanjean and Broda, 1978), periplasmic sulfate-binding proteins are involved in sulfate uptake, and the uptake is driven by the hydrolysis of ATP. Those systems are unidirectional, and thus prevent loss of intracellular sulfate.

In dissimilatory SRB, ATP hydrolyzation for sulfate uptake would consume about half of the free energy of sulfate reduction. Therefore, it is not surprising that more economic mechanisms for sulfate uptake have been developed.

A feasible transport mechanism would be group translocation comparable to the phosphotransferase systems for sugar uptake. The transported solute is chemically modified during uptake and transformed into a form that can be further metabolized, but does not permeate the cytoplasmic membrane. As with sugars, the sulfate molecule has to be activated by ATP (although bound to an adenyl instead of a phosphate group). However, adenylylation of sulfate during transport has not been found and appears improbable since the ATP sulfurylase reaction is, in spite of ATP consumption, strongly endergonic (see below) and anything but suited to drive sulfate accumulation.

Instead, secondary sulfate transport systems have been found in SRB. Accumulation of sulfate is driven by preexisting gradients of protons or sodium ions. Sulfate reducers have a very high affinity to sulfate. The $K_m$ or $K_s$ values of growing cells (Nethe-Jaenchen and Thauer, 1984) or sulfate-reducing cell suspensions (Ingvorsen *et al.*, 1984; Ingvorsen and Jørgensen, 1984) are in the range of 5 to 200 µM.

## 2.2. Sulfate Accumulation by Symport with Protons

In order to study sulfate transport in detail, it is necessary to prevent sulfate reduction and immediate excretion of $H_2S$. For this purpose, the cell suspension is either chilled to 0 °C or simply exposed to air (Warthmann and Cypionka, 1990; Cypionka, 1994). Under these conditions, the alkalinization can be separated from sulfide formation, as shown in Fig. 2A. The amount of two protons that disappear per sulfate added can be calculated by comparison with equimolar HCl calibration pulses.

By use of radiolabeled sulfate it could be demonstrated that the

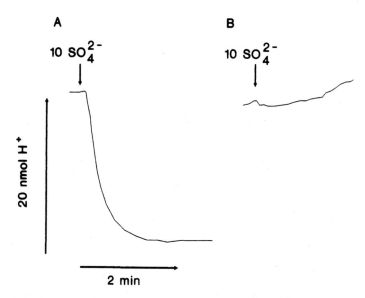

**Figure 2.** Proton traces during sulfate uptake by a freshwater species (*Desulfobulbus propionicus*) and a marine sulfate reducer (*Desulfococcus multivorans*). In this experiment sulfate reduction was prevented by oxygen. Two protons per sulfate disappear with the freshwater strain (A), while the marine strain causes little alkalinization (B).

proton uptake is accompanied by sulfate accumulation. More than 90% of the sulfate added (1.25 $\mu$M) was taken up by the cells. The residual extracellular concentration was about 0.1 $\mu$M, while the intracellular concentration amounted to about 0.5 mM (Cypionka, 1987; Kreke and Cypionka, 1992). Thus, the cells are able to accumulate sulfate more than a thousandfold.

## 2.3. Sodium-Dependent Sulfate Accumulation by Marine Sulfate Reducers

Freshwater usually contains low sulfate concentrations, and the sediments, the main habitat of SRB, are typically sulfate-depleted some millimeters below the surface. Therefore, an adaptation to low sulfate concentrations and the capacity to accumulate sulfate were to be expected for freshwater species. However, most marine species that stem from environments with high sulfate concentrations are also able to express high-accumulating sulfate transport systems. There is, however, a significant difference in the transport mechanism. Very little pH effect occurs during sulfate uptake (Fig. 2B). Instead, sulfate accumulation depends on the presence of sodium ions, which can partially be replaced by lithium ions (Warthmann and Cypionka, 1990; Stahlmann et al., 1991).

## 2.4. Calculation of Steady State Sulfate Accumulation

Accumulated (labeled) sulfate can be washed out of the cells by an excess of (unlabeled) sulfate very rapidly. This indicates that there is no irreversible step in sulfate uptake. Sulfate is not immediately withdrawn from equilibrium by further chemical reactions, and transport is reversible. Reversibility is the precondition for the application of chemiosmotic equations that describe steady states of gradients across the cytoplasmic membrane. Generally, the steady state accumulation of an anion by symport with a cation can be described as

$$\log (c_i/c_o) = -(m + n) \, \Delta\psi/z + n \log(x_o/x_i) \qquad (6)$$

where $c_i$, $c_o$ = concentrations of the anion inside and outside the cell, m = charge of the anion, n = charge(s) of the symported cation(s), $\Delta\psi$ = membrane potential (mV), $Z = 2.3 \, RT/F \approx 60$ mV at 30 °C, and $x_i$, $x_o$ = concentrations of the symported cation inside and outside the cell.

In sulfate symport with two protons (as shown in Fig. 2A) accumulation would be independent of the membrane potential. This would be electroneutral symport (no net charges are transferred, m + n = 0). The only driving force would be $\Delta pH$. In order to calculate the proton gradient required for a thousandfold sulfate accumulation, eq. (6) can be simplified to

$$\log (c_i/c_o) = 2 * \Delta pH. \qquad (7)$$

The pH gradient required for a thousandfold steady state accumulation would be 1.5 (log 1000 = 3), higher than known from any neutrophilic bacterium.

Further studies have revealed that high sulfate accumulation is not independent of the membrane potential (Cypionka, 1989; Warthmann and Cypionka, 1990). It has been shown that sulfate accumulation is sensitive to inhibitors known to affect the membrane potential (valinomycin + $K^+$, TCS) and the pH gradient (monensin + $Na^+$, nigericin + $K^+$). Obviously, high sulfate accumulation depends on both, $\Delta pH$ and $\Delta \psi$, and is electrogenic, while the pH electrode detects only electroneutral proton movements.

## 2.5. Correlation of Proton Potential and Sulfate Accumulation

After determination of $\Delta pH$ and $\Delta \psi$, quantitative calculation of the number of protons symported was possible.

The transmembrane pH gradient of SRB has been studied by two different methods, $^{13}P$-NMR (Kroder et al., 1991) and distribution of membrane permeant weak acids (Kreke and Cypionka, 1992). Both methods gave similar results: At neutral external pH the cells keep a $\Delta pH$ of about 0.5 units (inside alkaline). $\Delta pH$ increased to 1.2 pH units at pH 5.9 in the external medium. At alkaline pH (above pH 7.7), there was no pH gradient in Desulfovibrio desulfuricans (Kroder et al., 1991; Kreke and Cypionka, 1992). The use of membrane-permeable radiolabeled probes ($^{14}C$-benzoate) gave $\Delta pH$ values between 0.25 and 0.8 in ten freshwater strains tested (Kreke and Cypionka, 1992).

The determination of the electrical membrane potential was more difficult due to unspecific binding of the probe (Thauer, 1989). Only with four of ten freshwater strains studied were reliable values in the range of −80 to −140 mV obtained. The proton potential in these strains was in the range between −110 and −155 mV (Kreke and Cypionka, 1992).

Assuming that sulfate accumulation was in equilibrium with the proton-motive force, even the highest accumulation factors observed could be explained by a stoichiometry of three protons per sulfate.

## 2.6. Stoichiometry of Sulfate Symport with Sodium Ions

In the marine strains studied in detail (*Desulfococcus multivorans* and *Desulfovibrio salexigens*), sulfate accumulation was sensitive to inhibitors that affect the transmembrane sodium gradient (monensin, amiloride, ETH157). As in freshwater strains a stoichiometry of 3 cations symported with sulfate was found for the conditions of maximum sulfate accumulation (Kreke and Cypionka, 1994).

Of course, marine sulfate reducers are adapted to the high sodium concentrations (about 450 mM $Na^+$) of seawater, and most of them require sodium ions for growth. However, with the exception of a sodium/proton antiporter in a freshwater species (Varma *et al.*, 1983a), specific functions of sodium ions for SRB had not been described before. The energy metabolism of *Desulfovibrio salexigens*, the only marine sulfate reducer investigated in this respect, is based on primary proton pumps and not on sodium ions (Kreke and Cypionka, 1994). A sodium ion-driven sulfate transport seems to be advantageous for the marine bacteria, since an $Na^+$ gradient is sustained and represents a form of energy. However, such an accumulation mechanism requires at the same time a perfectly balanced regulation, otherwise the cells would be flooded and deenergized by the two ions, which occur in high concentrations in their environment.

## 2.7. Regulation of Sulfate Transport

The features discussed so far refer to experiments with bacteria grown under sulfate limitation to which only micromolar concentrations of sulfate were added. The high accumulation under these conditions is, however, not always obtained. Instead at least two different sulfate transport mechanisms are found, and transport is regulated at the genetic and the activity level. Littlewood and Postgate (1957) had already observed that in nonmetabolizing cell suspensions of *Desulfovibrio* high sulfate concentrations added were not diluted by intracellular water. The cells appeared impermeable for sulfate. Similar results were obtained by Varma *et al.* (1983a) in plasmolysis and reswelling experiments. By contrast, Furusaka (1961) found that *Desulfovibrio* cells accumulated radiolabeled sulfate during its reduction. Obviously, sulfate transport is strictly regulated, which is not surprising for a molecule of such central importance.

Highest sulfate accumulation is observed in cells grown in chemostats under sulfate-limiting conditions for long periods. With cells grown under excess sulfate, accumulation scarcely exceeds one hundredfold.

Obviously, the cells possess more than one sulfate transport system. The high-accumulating transport system is expressed under sulfate limitation, only, while a low-accumulating system is constitutively present.

Besides regulation on the genetic level (derepression of the high-accumulating transport system by sulfate limitation) there also exists a fast regulatory mechanism on the activity level. If cells with the expressed high-accumulating system are exposed to increasing sulfate concentrations, accumulation decreases correspondingly. The cells are not deenergized by sulfate uptake, as indicated by constant ATP levels and proton potentials over a wide range of sulfate concentrations (Stahlmann et al., 1991; Kreke and Cypionka, 1992).

The correlation of the steady state sulfate accumulation with the protonmotive force gives decreasing stoichiometries. At external sulfate concentrations above 100 μM the calculated stoichiometry even decreased to values below 2. Since this would mean electrogenic transport with net uptake of a negative charge against the membrane potential, it must be concluded that in those cases a regulatory mechanism stops sulfate transport, before an equilibrium with the proton potential is reached. These observations are in agreement with those of the apparent impermeability for sulfate discussed above.

## 2.8. Transport of Thiosulfate and Other Sulfate Analogs

SRB are able to reduce various sulfur compounds other than sulfate. The most important of those are thiosulfate and sulfite. So far, only thiosulfate has been used for detailed transport studies. Thiosulfate is a structural analogue of sulfate, since it differs from sulfate only by an additional S atom that replaces an O atom. The characteristics of sulfate and thiosulfate transport were quite similar in nineteen freshwater and marine strains (Stahlmann et al., 1991). The capacity of high thiosulfate accumulation was induced by sulfate limitation. The washout of (labeled) accumulated sulfate by an excess of (unlabeled) sulfate could also be obtained with thiosulfate and vice versa. The cells appeared unable to distinguish between sulfate and thiosulfate. Obviously, thiosulfate is accumulated by the high-accumulating sulfate transport system in most of the strains tested. So far, *Desulfovibrio desulfuricans* strain Essex appears to be the only exception found in this respect (Cypionka, 1987). With this strain the proton uptake kinetics differed during uptake of sulfate and thiosulfate depending on the electron acceptor supplied during growth.

From several other structural sulfate analogues tested at 25-fold excess, molybdate (the classical inhibitor of sulfate reduction) and tungstate had little influence on sulfate accumulation (Cypionka, 1989; Warthmann and Cypionka, 1990). Chromate caused strong inhibition, while selenate, which can be reduced by some sulfate reducers (Zehr and Oremland, 1987; Cypionka, 1989), inhibited sulfate accumulation approximately by half.

### 2.9. Assessment of the Energy Requirement for Sulfate Transport

Generally, sulfate uptake in dissimilatory SRB occurs via secondary transport systems in symport with cations (Fig. 3). The transport systems are reversible (all transmembrane arrows in Fig. 3 must be regarded as two-directional). In freshwater species protons are symported, while marine species use sodium ions. High accumulation is achieved by electrogenic mechanisms (symport of three cations), which are found only after

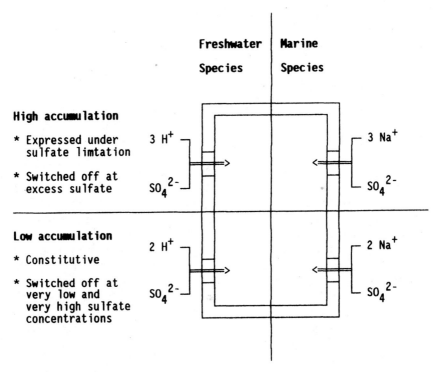

**Figure 3.** Mechanisms and regulation of sulfate transport in freshwater and marine sulfate-reducing bacteria.

growth under sulfate limitation, and are immediately switched off at increased sulfate concentration. In cells grown under excess sulfate, no capacity of high accumulation is found. Sulfate appears to be taken up by electroneutral systems, which are hardly characterized so far. The low-accumulating systems must be shut down at very low (to allow high accumulation by the other system) and at high sulfate concentrations (where accumulation stops before equilibrium with the driving forces is reached). Otherwise the cells would be deenergized by flooding with sulfate.

The energy requirement for sulfate transport can easily be estimated for symport with protons. If three protons are consumed per ATP regenerated by the membrane-bound ATPase, electrogenic symport with three protons would correspond to a consumption of 1 ATP per sulfate taken into the cell. Electroneutral symport with two protons would mean consumption of two-thirds ATP. Thus, at first glance, the energy investment for secondary sulfate transport seems to be as high as that of assimilatory sulfate reducers that hydrolyze ATP for sulfate uptake.

However, the energy for sulfate uptake is partially compensated by $H_2S$ excretion. In contrast to assimilatory sulfate reduction, $H_2S$, the end product of dissimilatory sulfate reduction, is immediately set free. Therefore, an estimation of the energy balance of sulfate transport must include the energetic aspects of sulfide release. $H_2S$ can diffuse as a gas through the cell membrane. Nothing is known about transport of the hydrogen sulfide ion ($HS^-$), which at an intracellular pH of about 7.5 has a higher concentration than $H_2S$. If one assumes that the anion $HS^-$ cannot permeate the cell membrane (or is transported by electroneutral symport with $H^+$) then sulfide release is energetically equivalent to export of two protons. Under these conditions electroneutral uptake of sulfate plus release of $H_2S$ would not consume energy. Electrogenic uptake with three protons would finally result in consumption of one-third ATP. Corresponding differences of the growth yields under conditions of sulfate excess and limitation have been found in the chemostat with *Desulfotomaculum orientis* (Cypionka, 1986; Cypionka and Pfennig, 1986). This strain revealed the capacity of high sulfate accumulation only after growth under sulfate limitation, when the growth yield was 20 percent lower than under sulfate excess.

The energy requirement for $Na^+$-dependent sulfate accumulation is more difficult to assess, since the mechanisms of the generation of the required sodium gradient must be known. In freshwater (Varma *et al.*, 1983a) and marine (Kreke and Cypionka, 1994) species, electrogenic sodium/proton antiport systems have been found that use more than one $H^+$ per sodium ion. If such mechanisms are involved, sulfate transport would consume more energy than proton-coupled systems.

## 3. ENERGETICS OF SULFATE ACTIVATION

In spite of their name, SRB are unable to reduce sulfate. Instead, sulfate must be activated first under consumption of ATP. The ATP sulfurylase (or adenylyl sulfatase) forms APS and pyrophosphate from ATP and sulfate

$$SO_4^{2-} + ATP + 2\ H^+ \rightarrow APS + PP_i \quad \Delta G^{o\prime} = +46\ kJ/mol \quad (8)$$

Although ATP is hydrolyzed the reaction is strongly endergonic, and must be pulled to completion by removal of the end products. Pyrophosphate can be hydrolyzed by pyrophosphatase

$$PP_i + H_2O \rightarrow 2\ P_i \quad \Delta G^{o\prime} = -22\ kJ/mol \quad (9)$$

Thus, a second phosphate ester is cleaved, and the energy effort of sulfate activation equals the hydrolyzation of two ATP molecules to ADP and phosphate. For example, catalyzed by adenylate kinase, the AMP formed can react with a second ATP to give two ADP without free energy change

$$AMP + ATP \rightarrow 2\ ADP \quad \Delta G^{o\prime} = 0\ kJ/mol \quad (10)$$

Removal of the second product of ATP sulfurylase, APS, is the first redox reaction

$$APS + H_2 \rightarrow HSO_3^- + AMP + H^+ \quad \Delta G^{o\prime} = -69\ kJ/mol \quad (11)$$

The reduction of APS to sulfite and AMP by APS reductase is more exergonic than pyrophosphate cleavage. With $H_2$ as electron donor the free energy change is about three times higher.

Equations 8 to 11 can be summarized as

$$SO_4^{2-} + H_2 + H^+ + 2\ ATP + H_2O \rightarrow HSO_3^- + 2\ ADP$$
$$+ 2\ P_i \quad \Delta G^{o\prime} = -45\ kJ/mol \quad (12)$$

The free energy change of this sequence is strongly negative, even at lower $H_2$ concentrations. It is questionable whether the dissimilatory sulfate reducers can afford such a luxurious energy investment or if either pyrophosphate hydrolyzation or APS reduction are coupled to energy-conserving mechanisms.

### 3.1.  Energy Coupling of Pyrophosphatase or APS Reductase

The hydrolyzation of pyrophosphate appears thermodynamically unnecessary since the APS reductase reaction should be sufficiently exergonic to pull sulfate activation. Several possibilities have been discussed for how pyrophosphatase could save energy. Pyrophosphatase activity in *Desulfovibrio* species and *Desulfotomaculum orientis* is stimulated by reductants such as dithionite (Ware and Postgate, 1971; Thebrath *et al.*, 1989). This was interpreted as a mechanism that prevents waste of energy by sulfate activation if no reducing substrates are available.

In several *Desulfotomaculum* strains no stimulation of pyrophosphatase activity by reductants was observed. Instead, there were some hints of a membrane-bound localization of the enzyme (Thebrath *et al.*, 1989). Thus, it cannot be excluded that pyrophosphatase is involved in energy-conservation by proton translocation as reported from anoxygenic phototrophic bacteria (Nore *et al.*, 1986; Nyrén and Strid, 1991; Romero *et al.*, 1991), plants (Martinoia, 1992) or yeast (Pereiradasilva *et al.*, 1993).

Furthermore, it was proposed that pyrophosphate could be used to form acetyl phosphate in *Desulfotomaculum* species (Liu and Peck, 1981; Peck and LeGall, 1982). However, a pyrophosphate-dependent acetate kinase, as described in the anaerobic protozoon *Entamoeba histolytica* (Reeves and Guthrie, 1975), was not demonstrated in *Desulfotomaculum*. Instead, part of the observations giving hints on the pyrophosphate-dependent acetyl phosphate formation could be explained as artifacts (Thebrath *et al.*, 1989).

Reports on growth with pyrophosphate as an external energy source (Liu *et al.*, 1982; Varma and Peck, 1983; Varma *et al.*, 1983b; Cruden *et al.*, 1983; Jones and Simon, 1984) also turned out to be based on an artifact. Pyrophosphate added to the growth medium causes slow precipitation of salts, but no protein increase (Klemps *et al.*, 1985; Thebrath *et al.*, 1989).

Since the reduction of APS is strongly exergonic [eq. (11)] it would be an advantage if APS reductase was coupled to energy conservation. However, in most of the strains studied so far, the enzyme appears to be localized within the cytoplasm (Kremer *et al.*, 1988). In spite of that, APS reduction would generate a protonmotive force if the electrons are delivered via a periplasmic hydrogenase (see below). Furthermore, an indication for energy coupling of APS reductase can be derived from studies on the mechanism of sulfite disproportionation (see below). Sulfite oxidation to APS in the presence of AMP is most probably driven by reverse electron transport and is thus energy-coupled.

At present there is no proof that the activation of sulfate consumes less than two ATP. This value was also derived from growth yield studies in the chemostat. Sulfite and thiosulfate, which require no activation, allowed a threefold higher maximum growth yield of *Desulfovibrio vulgaris* than sulfate (Nethe-Jaenchen and Thauer, 1984; Thauer, 1989). Smaller growth yield differences were obtained with *Desulfotomaculum orientis* in similar experiments (Cypionka, 1986; Cypionka and Pfennig, 1986). However, the data obtained could not be extrapolated to maximum growth yields, which restricts the possibilities of interpretation.

## 4. ENERGETICS OF SULFITE REDUCTION

The reduction of sulfite to sulfide has to compensate the energy investment of sulfate activation and yield additional ATP for growth. The standard free energy change of sulfite reduction to sulfide with $H_2$ as electron donor is $-174$ kJ/mol.

$$0.5 \ HSO_3^- + 0.5 \ SO_3{}^{2-} + 3 \ H_2 + H^+ \rightarrow 0.5 \ HS^- + 0.5 \ H_2S$$
$$+ 3 \ H_2O \quad \Delta G^{\circ\prime} = -174 \ kJ/mol \tag{13}$$

Thermodynamically, this could allow regeneration of at least 2 ATP. Although details of the biochemistry of sulfite reduction and electron transport are not to be discussed here, we have to compare the energetical consequences of a one-step reduction of sulfite to sulfide and the disputed trithionate pathway, which would form trithionate and thiosulfate as intermediates of sulfite reduction (Kobayashi *et al.*, 1969; Akagi, 1981; Kim and Akagi, 1985). The latter mechanism involves three enzymes, a sulfite reductase forming trithionate from three sulfite molecules [eq. (14)], a trithionate reductase forming thiosulfate and sulfite [eq. (15)], and a thiosulfate reductase forming sulfide and sulfite [eq. (16)].

$$1.5 \ SO_3^{2-} + 1.5 \ HSO_3^- + H_2 + 2.5 \ H^+ \rightarrow S_3O_6^{2-}$$
$$+ 3 \ H_2O \quad \Delta G^{\circ\prime} = -48 \ kJ/mol \tag{14}$$

$$S_3O_6^{2-} + H_2 \rightarrow S_2O_3^{2-} + 0.5 \ SO_3^{2-} + 0.5 \ HSO_3^-$$
$$+ 1.5 \ H^+ \quad \Delta G^{\circ\prime} = -122 \ kJ/mol \tag{15}$$

$$S_2O_3^{2-} + H_2 \rightarrow 0.5 \; HS^- + 0.5 \; H_2S + 0.5 \; HSO_3^-$$
$$+ \; 0.5 \; SO_3^{2-} \quad \Delta G^{\circ\prime} = -4 \; kJ/mol \qquad (16)$$

The redox potentials and the corresponding free energy changes of these steps are quite different. While the one-step reduction has a standard midpoint potential of $-116$ mV, the redox couples of the trithionate pathway have midpoint potentials of $-173$ mV, $+225$ mV, and $-402$ mV. The highest free energy change occurs during trithionate reduction, while the last and sulfide-forming reaction is hardly exergonic even at $H_2$ saturation.

The trithionate pathway would require electron transport to three different electron acceptors that are sequentially formed. Each of these electron acceptors would consume electrons as a terminal acceptor (in contrast to electron carriers that are reversibly reduced and reoxidized). It is unknown whether the different enzymes would be directly coupled to energy conservation (function as proton pumps) or whether chemiosmotic gradients are generated by the electron transport chain(s), only. So far, sulfite reductases have been described to be localized in the cytoplasm, although a membrane association is discussed (Badziong and Thauer, 1980; Kremer *et al.*, 1988; Thauer, 1989), and a membrane-associated sulfite reductase has now been purified from *Desulfovibrio desulfuricans* (Steuber *et al.*, 1995).

Without considering enzyme studies that of course are essential in order to clarify the mechanism of sulfite reduction, several observations with whole cells are in favor of a pathway that includes intermediates. Formation of thiosulfate and trithionate has been observed in growing cultures and with washed cells (Vainshtein *et al.*, 1980; Fitz and Cypionka, 1990; Sass *et al.*, 1992). As expected from thermodynamics, limitation of electron donor supply was a prerequisite. By contrast, Chambers and Trudinger (1975), who did not find indications for formation of intermediates in a study with labeled sulfate and thiosulfate, always had used excess concentrations of the electron donor.

Two chemiosmotic observations are also in favor of the trithionate pathway. First, sulfite reduction to $H_2S$ is inhibited by uncouplers and depends on an intact cell structure. This indicates an energy-dependent step (as the requirement of reverse electron transport in order to reduce thiosulfate). In case of a one-step reduction, one would expect a stimulation of sulfite reduction by uncouplers (as observed with $O_2$ reduction, see below). Second, the highest $H^+/e^-$ ratios upon micromolar $H_2$ additions were found when sulfite reduction remained incomplete and was not coupled to sulfide formation (Fitz and Cypionka, 1989, 1991). As expected, thiosulfate reduction resulted in little proton translocation in those experiments. Thus, it appears that the cells can use the electrons for the favorable reaction under electron donor limitation. Sulfite reduction stops at intermediates whose reduction to sulfide would consume electrons without energy coupling. The question arises as to why SRB do not always stop sulfite reduction at the level of thiosulfate, if sulfide formation consumes electrons without allowing energy conservation. Probably sulfide formation is completed, since, concomitantly, sulfite as electron acceptor (that allows further energy conservation) is regenerated. Otherwise, the next sulfate molecule would have to be activated under ATP consumption.

## 5. GENERATION OF A PROTONMOTIVE FORCE

Many SRB use molecular hydrogen as electron donor. In several species more than one hydrogenase has been found (Odom and Peck, 1984). Hydrogenases may be located in the intracellular as well as in the periplasmic space. If a periplasmic hydrogenase passes the electrons to a cytochrome and electron consumption occurs in the cytoplasm, this process generates inevitably a membrane potential and a transmembrane pH gradient without pumping protons across the cell membrane. This simple but elegant mechanism, called vectorial electron transport (Badziong and Thauer, 1980), is also found in other bacteria and with other substrates, e.g., formate (Hooper and DiSpirito, 1985; LeGall and Fauque, 1988). The protons produced do not remain in the periplasm as a scalar product of $H_2$ oxidation. Instead, they are finally consumed for sulfate reduction inside the cell [eq. (1)].

Alternatively or additionally, SRB may pump protons across the membrane, performing the classical type of vectorial proton translocation. Discrimination between the two mechanisms is difficult, since both result in an reversible acidification of the outer bulk phase. The distinction is possible, however, when the enzyme localization is known, and when the $H^+/H_2$ ratios exceed two, which is only possible by classical proton translocation.

### 5.1. Hydrogen Cycling

Since the generation of a proton motive force by vectorial electron transport appears so easy, it is tempting to speculate that such a mechanism could also be involved in energy conservation from substrates that are oxidized inside the cell. Due to the hypothesis of hydrogen cycling (Odom and Peck, 1981; Peck and Lissolo, 1988), $H_2$ is formed inside the cell by a cytoplasmic hydrogenase. Then, it diffuses into the periplasmic space where it is oxidized by the periplasmic hydrogenase. However, $H_2$ formation inside the cell is thermodynamically less favorable than outside due to the transmembrane pH gradient and the fact that at alkaline pH the redox potential of $H_2$ becomes more negative (for example, $H_2$ uptake with methyl viologen, $E_0' = -440$ mV, is usually tested at pH 9). Since it must be assumed that the two hydrogenases are "electrically connected" via reversible electron carriers, the hypothesis appears daring. Of course, SRB are able to produce $H_2$ from organic substrates (Bryant *et al.*, 1977; Pankhania *et al.*, 1988; Seitz *et al.*, 1990), and cells oxidizing CO (Lupton *et al.*, 1984) or pyruvate (Peck *et al.*, 1987) with sulfate have been shown to produce and consume $H_2$. However, the midpoint redox potentials of CO and pyruvate are more negative than that of $H_2$. With lactate, which has a midpoint potential of $-190$ mV, no $H_2$ cycling was demonstrated. Anyhow, hydrogen cycling cannot be obligate. Some sulfate reducers do not have hydrogenases (e.g. *Desulfovibrio sapovorans*). The spore-forming *Desulfotomaculum* species have a gram-positive cell wall type and no periplasmic hydrogenase (Cypionka and Dilling, 1986). In *Desulfovibrio vulgaris* the hydrogenase was inhibited by $CuCl_2$ or by $O_2$ concentrations greater than 2%, while the capacity of lactate oxidation remained unaffected (Fitz and Cypionka, 1991). Finally, a mutant of *D. desulfuricans*, unable to reduce sulfate with $H_2$, could grow normally with lactate (Odom and Wall, 1987).

## 5.2. Proton Release by Periplasmic Hydrogenase and by Translocation

The classical demonstration of vectorial proton translocation (Scholes and Mitchell, 1970) has been done with the oxidant pulse method. Small $O_2$ pulses are added to anoxically incubated cells. The short period of electron transport causes a reversible acidification of the outer bulk medium. Then, the translocated protons are taken up again and drive ATP regeneration. This experiment was successful with aerobic respiration since $O_2$ must not be transported, but can permeate the cell membrane. The same type of experiment with sulfate reducers and sulfate instead of $O_2$ gives a different result, as already described (Fig. 1). Alkalinization instead of acidification is observed, since sulfate must be transported into the cell. (The first study on proton translocation by Kobayashi *et al.*, 1982, who used the oxidation pulse method, certainly must be regarded with caution, since disappearance of protons due to transport and the scalar proton binding by $H_2S$ formation from sulfate were not observed.)

To study generation of the proton motive force by sulfate reduction this lab uses the reductant pulse method. Cells containing no or little endogenous substrates are preincubated with excess sulfate. Thus, transport of sulfate (or other electron acceptors to be tested) is unnecessary when electron transport is started by the addition of a small $H_2$ pulse. The disadvantage of this method is that nonreductive metabolism of the electron acceptors (like sulfate activation, disproportionation of sulfite or thiosulfate) prior to $H_2$ addition is not definitely excluded. Furthermore, since the assay is saturated with $N_2$ and no electron donors are present, the redox potential is more positive (which is an advantage in some cases) and the rates of electron transport are lower than after oxidant pulses.

In our experiments, pH traces typical for respiration-driven proton translocation (or $H^+$ formation by a periplasmic hydrogenase) have been obtained (Fitz and Cypionka, 1989, 1991). The detectable $H^+$ release was significantly increased by reagents that dissipate the membrane potential (thiocyanate, valinomycin + EDTA, methyl triphenylphosphonium ion) and was completely inhibited by uncouplers.

With sulfate as electron acceptor and *D. desulfuricans* strain Essex, appearance of 1.8 $H^+$ per $H_2$ was observed (Fig. 4A). With sulfite the

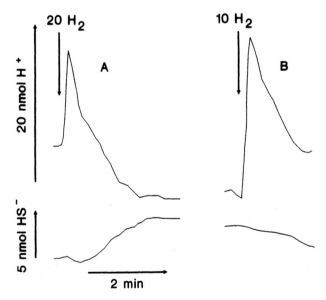

**Figure 4.** Proton translocation after $H_2$ pulses (reductant pulse method). Washed cells of *Desulfovibrio desulfuricans* were incubated in $N_2$-saturated 150 mM KCl with 150 mM KSCN and 10 mM sulfate. The first experiment (A), in which sulfide formation and scalar alkalinization occurred, revealed less proton translocation than the second (B), where sulfate reduction was incomplete and no sulfide was formed.

$H^+/H_2$ ratios were slightly higher, while with thiosulfate only 0.5 $H^+/H_2$ was observed. As already mentioned, considerably higher values (up to 3.1 with sulfate and 3.4 with sulfite) were obtained in experiments where no $H_2S$ formation occurred (Fig. 4B). Obviously, sulfate or sulfite could be incompletely reduced, a result not to be expected for an obligate one-step reaction. With thiosulfate, the ninefold increase of proton translocation in experiments without sulfide formation is far from clear. Since thiosulfate addition caused even disappearance of sulfide in several experiments, other transformations that have not been analyzed so far, may be involved (Fitz and Cypionka, 1989).

However, $H^+/H_2$ ratios of up to 4.4 allowed the unequivocal conclusion that *D. desulfuricans* generates a protonmotive force by classical proton translocation. This was confirmed by the fact that no indication for a periplasmic localization of the hydrogenase were obtained (Fitz and Cypionka, 1989).

*Desulfovibrio vulgaris* strain Marburg has a very active periplasmic hydrogenase (Badziong and Thauer, 1980), which was found to release less than 2 $H^+/H_2$ in experiments with sulfate, sulfite or thiosulfate as electron acceptors (Fitz and Cypionka, 1991). Correspondingly, proton translocation was sensitive to $CuCl_2$, which specifically inhibits the periplasmic hydrogenase, since it does not enter the cells (Cypionka and Dilling, 1986). Proton translocation with lactate and pyruvate, however, was not sensitive to $CuCl_2$ and thus obviously caused by classical proton translocation. $H^+/H_2$ ratios of up to 4 with $O_2$ as electron acceptor (see below) leave no doubt that the strain can make use of both vectorial electron transport and transmembrane proton translocation.

## 6. COMPREHENSIVE ASSESSMENT OF THE ENERGETICS OF SULFATE REDUCTION

Fifteen years after the assessment of Wood (1978) considerable progress has been made in our understanding of sulfate respiration. Sulfate transport, sulfate activation, vectorial electron transport, proton translocation, and the resulting transmembrane gradient have been studied in detail. However, central questions, such as energy coupling of pyrophosphatase and APS reductase, sites of proton translocation and the pathway of sulfite reduction still remain to be answered (Fig. 5).

With $H_2$ as electron donor, net energy conservation of about 1 ATP per sulfate reduced to sulfide is to be expected from energetical considerations (Thauer, 1989) and from growth yields found in chemostat experiments. Translated into the chemiosmotic currency, this equals three protons to be translocated across the cell membrane or released in

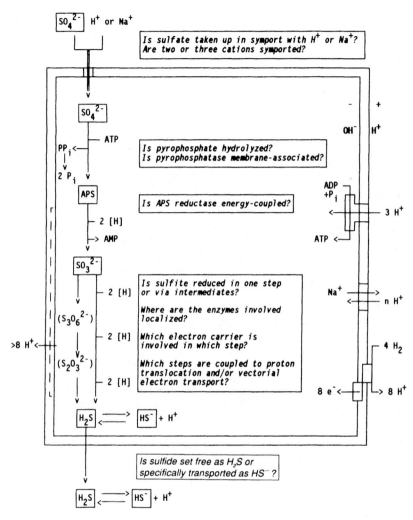

**Figure 5.** Generalized scheme of dissimilatory sulfate reduction with several questions that have or might have different answers in different sulfate-reducing bacteria.

the periplasm with concomitant electron transport into the cytoplasm. The scalar binding of protons due to $H_2S$ formation is not directly relevant, but has influence on the export of protons taken up together with sulfate.

At sufficient sulfate supply, sulfate uptake requires symport of two protons, equivalent to two-thirds ATP. Under sulfate limitation, the high-accumulating transport mechanism will be expressed, which consumes three protons or 1 ATP per sulfate.

As long as energy coupling of pyrophosphatase and/or APS reductase is not unequivocally demonstrated, we must assume that the activation of sulfate again requires hydrolyzation of two phosphate ester bounds.

The energy for sulfate activation is compensated during sulfite reduction. Although some hints of a stepwise sulfite reduction have been obtained, the obligate involvement of the trithionate pathway is still not clarified.

Two mechanisms of generation of the protonmotive force have been found: release of protons by periplasmic enzymes with vectorial electron transport into the cell, and classical proton translocation. A periplasmic hydrogenase alone could release 2 $H^+/H_2$ or 8 $H^+$ per sulfate, and compensate 2 ATP (corresponding to 6 $H^+$) consumed during sulfate activation and allow growth. However, periplasmic substrate oxidation is not found in all species, while the capacity of proton translocation appears to be always present (indicated by $H^+/H_2$ values above 2). The occurrence of hydrogen cycling (intracellular formation of $H_2$ which is oxidized by the periplasmic hydrogenase) has only been demonstrated for reactions with very negative redox potentials such as CO or pyruvate. The energy investment for sulfate uptake will be regained when sulfide finally leaves the cell as $H_2S$.

The energetics of marine sulfate reducers have not yet been studied intensively. The sodium requirement for sulfate transport increases the costs for sulfate transport if the sodium gradient is generated by electrogenic $Na^+/H^+$ antiport ($> 1 H^+/Na^+$). The energy metabolism of the only marine strain studied in this respect is based primarily on the protonmotive force, and not on primary sodium ion translocation (Kreke and Cypionka, 1994).

## 7. ENERGY CONSERVATION BY PROCESSES OTHER THAN SULFATE REDUCTION

SRB are not restricted to energy-conservation by sulfate reduction. They are also capable of fermentative growth or utilization of other electron acceptors. Reduction of sulfite and thiosulfate has been discussed already. Additionally, elemental sulfur may be reduced by several SRB (Biebl and Pfennig, 1977; Fauque et al., 1979; Cypionka, 1989). Several of the processes now discussed have been discovered only recently.

## 7.1. Fermentation of Organic Substrates

Many SRB are able to ferment organic substrates in the absence of sulfate (for review see Widdel, 1988, and Hansen, 1993). In the simplest case, pyruvate is fermented to acetate, carbon dioxide, and $H_2$. Energy conservation during this fermentation is achieved via acetate kinase by substrate level phosphorylation only. Similarly, lactate may be fermented to $H_2$, acetate, and $CO_2$ (Pankhania et al., 1988). This fermentation does, however, normally not allow growth, since lactate oxidation to pyruvate ($E_0' = -190$ mV) requires energy-dependent reverse electron transport. Accordingly, the reaction is catalyzed by a membrane-bound enzyme. It might be important to note that ATP formed by substrate-level phosphorylation can be used to energize reverse electron transport. In syntrophic cocultures, where the hydrogen partial pressure is kept sufficiently low by a $H_2$-consuming partner, growth by lactate fermentation is possible (Bryant et al., 1977).

Other sulfate reducers are able to grow by propionate fermentation. In this case, a chemiosmotic step is involved. Propionate formation follows the pathway known from propionibacteria and thus includes fumarate reductase, which can be energy-coupled (Barton et al., 1970; Stams et al., 1984). The same reaction is involved in the fermentation of fumarate and malate (Miller and Wakerly, 1966).

Another fermentation that includes chemiosmotic aspects is homoacetate fermentation, which is carried out by *Desulfotomaculum orientis* (Klemps et al., 1985; Cypionka and Pfennig, 1986). Homoacetate fermentation can also be regarded as carbonate respiration, and must be coupled to chemiosmotic energy conservation if electron donors that do not allow substrate level phosphorylation are utilized. *D. orientis* formed acetate from $H_2 + CO_2$, formate, ethanol or lactate.

## 7.2. Disproportionation of Sulfur Compounds

Many sulfate reducers can perform a unique fermentation of inorganic sulfur compounds (Bak and Cypionka, 1987; Bak and Pfennig, 1987), which are disproportionated to sulfate and sulfide [Eqs. (17 and 18)]. For example, thiosulfate is transformed to equal amounts of sulfate and sulfide

$$S_2O_3^{2-} + H_2O \rightarrow SO_4^{2-} + 0.5\ H_2S + 0.5\ HS^- + 0.5\ H^+\ \Delta G^{o\prime}$$
$$= -25\ kJ/mole. \tag{17}$$

Sulfite is disproportionated to three-fourths sulfate and one-fourth sulfide

$$2 \; HSO_3^- + 2 \; SO_3^{2-} \rightarrow 3 \; SO_4^{2-} + 0.5 \; H_2S + 0.5 \; HS^-$$
$$+ \; 0.5 \; H^+ \; \Delta G^{o\prime} = -235 \; kJ/mol. \quad\quad (18)$$

About half of 19 SRB tested were capable of at least one of these transformations, while this capacity so far has not been found in colorless or phototrophic sulfur bacteria (Krämer and Cypionka, 1989). The free energy change of thiosulfate disproportionation ($-25$ kJ/mole) is low and cannot always be utilized for growth. Sulfite disproportionation was not so widespread, but could be utilized for growth (Krämer and Cypionka, 1989). Even the disproportionation of elemental sulfur is possible, if the sulfide formed is removed by chemical reaction with iron or manganese (Thamdrup et al., 1993).

The capacity of disproportionation is constitutively expressed. The enzymes required for the disproportionation appear to be the same as for sulfate reduction. Experiments with cell extracts and inhibitors have shown that sulfate formation is obtained by a reversal of sulfate activation. ATP regeneration includes ATP sulfurylase, a mechanism analogous to substrate level phosphorylation as typical for fermentations (Fig. 6).

However, two endergonic steps have to be overcome. First, APS reductase releases electrons with a standard redox potential of $-60$ mV. The reduction of thiosulfate to sulfide and sulfite, however, requires a more negative redox potential ($E_0' = -402$ mV). Therefore, the electrons from APS reduction must be lowered in their redox potential by reverse electron transport. This means that APS reduction could be coupled to chemiosmiotic processes and thus perhaps conserve energy during sulfate reduction. Secondly, the supply of pyrophosphate for the ATP sulfurylase is problematic. The sulfate reducers studied did not show ADP sulfurylase activity, which would have allowed formation of ADP from phosphate and AMP. As discussed above, energy coupling of pyrophatase could allow its reversal.

In any case, most of the ATP regenerated by ATP sulfurylase must be consumed for the endergonic steps. Thermodynamics would not allow a yield of 1 ATP per thiosulfate disproportionated.

## 7.3. Energy Conservation by Reduction of Nitrate

Some *Desulfovibrio* species and *Desulfobulbus propionicus* are able to utilize nitrate as electron acceptor (Widdel and Pfennig, 1982; Keith and Herbert, 1983; McCready et al., 1983; Mitchell et al., 1986; Seitz and Cypionka, 1986). Ammonia, but not $N_2$ is formed as end product

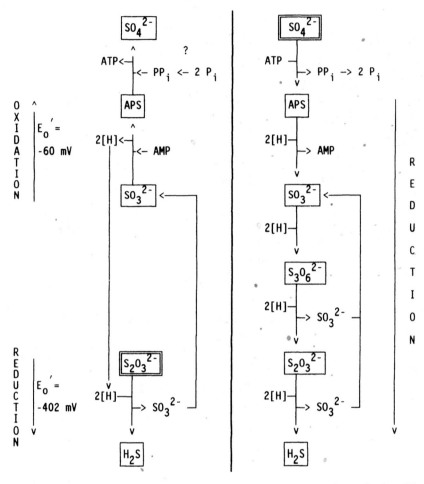

**Figure 6.** Pathway of thiosulfate disproportionation compared to sulfate reduction. The same enzymes can catalyze both processes. However, thiosulfate disproportionation requires the reversal of sulfate activation. Critical steps are the formation of pyrophosphate and reverse electron transport required for the reduction of thiosulfate.

$$4 \, H_2 + NO_3^- + 2 \, H^+ \rightarrow NH_4^+ + 3 \, H_2O \; \Delta G^{\circ\prime}$$
$$= -600 \; kJ/mole. \tag{19}$$

Nitrite is formed as an intermediate of nitrate reduction and can be reduced by many sulfate reducers unable to reduce nitrate.

$$3 \, H_2 + NO_2^- + 2 \, H^+ \rightarrow NH_4^+ + 2 \, H_2O \; \Delta G^{\circ\prime}$$
$$= -437 \; kJ/mole. \tag{20}$$

While the capacity of nitrite reduction appears to be present constitutively, nitrate reductase is expressed only if nitrate is offered as an electron acceptor. Depending on the species, in the presence of both electron acceptors either sulfate (Widdel and Pfennig, 1982) or nitrate (Seitz and Cypionka, 1986) may be the preferred electron acceptor, or both electron acceptors can be reduced concomitantly (McCready et al., 1986).

Vectorial proton translocation during nitrite reduction has been demonstrated with whole cells (Steenkamp and Peck, 1981; Fitz and Cypionka, 1989). ATP production coupled to nitrite reduction was also obtained in membrane preparations (Barton et al., 1983). Since hydrogenase and nitrite reductase were found to be located in the periplasmic space, and $H^+/H_2$ ratios above two were obtained, proton translocation must be responsible for energy coupling of nitrite respiration.

Proton translocation during reduction of nitrate has not been observed. However, from the growth yields in chemostat experiments (Seitz and Cypionka, 1986) it could be concluded that both nitrate reduction to nitrite and nitrite reduction to ammonia are coupled to energy conservation. The thermodynamic efficiency of nitrate and nitrite ammonification is far lower than that of sulfate reduction. The growth yields in the chemostat were only slightly higher than with sulfate, although the free energy change per $H_2$ oxidized is about four times higher. Obviously, the energy conservation of SRB is optimized to reduce sulfate rather than nitrate.

## 7.4. Energy Conservation by Aerobic Respiration

Recently it has been discovered that molecular oxygen can be reduced by most SRB (Postgate, 1984; Abdollahi and Wimpenny, 1990; Dilling and Cypionka, 1990; Dannenberg et al., 1992; Marschall et al., 1993). The stoichiometries (for example 2 $H_2$ consumed per $O_2$ reduced [eq. 2]) indicate that $O_2$ can be completely reduced to water. Respiration rates comparable to those of aerobic organisms were observed with Desulfovibrio species. Respiration is coupled to proton translocation with higher $H^+/2e^-$ ratios than found in sulfur compounds or nitrite. $O_2$ respiration results in ATP formation. It is stimulated by uncouplers that prevent ATP formation at the same time.

All these criteria demonstrate that sulfate reducers can carry out true aerobic respiration. However, usually the cells do not grow with $O_2$ as electron acceptor. Not more than one doubling has been observed in homogeneously aerated systems (Marschall et al., 1993). Motility and viability of the cells decrease upon prolonged incubation. The cells are microaerophilic and reveal a very high affinity to $O_2$. Some strains show

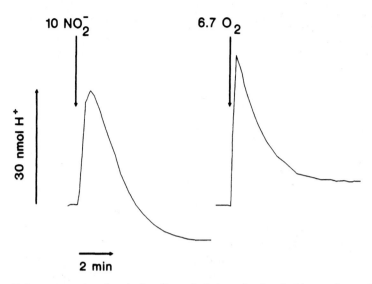

**Figure 7.** Proton translocation during $O_2$ and nitrite reduction (oxidant pulse method). Washed cells of *Desulfovibrio desulfuricans* were incubated in $H_2$-saturated KCl with 150 mM KSCN. Nitrite reduction to ammonia results in scalar alkalinization. Oxygen reduction results in fast proton translocation and high $H^+/2\ e^-$ ratios.

increasing respiration rates with decreasing $O_2$ concentrations. However, a lower sensitivity towards $O_2$ is observed when the cells are not incubated in homogeneously aerated suspension but growing in stabilized oxygen-sulfide gradients (Cypionka *et al.*, 1985; Gottschal and Szewzyk, 1985; Marschall *et al.*, 1993).

Sulfate reducers obviously possess terminal oxidases different from aerobic organisms. Respiration is not sensitive to classical inhibitors such as cyanide or azide. Besides cytochrome $c_3$ (Postgate, 1984), another oxygen-reducing enzyme has recently been described (Chen *et al.*, 1993). An FAD-containing hemoprotein was purified that catalyzed the oxidation of rubredoxin with $O_2$ in *Desulfovibrio gigas*. However, since the turnover number in activity tests was only $2\ s^{-1}$ it must be questioned whether this protein can be the only terminal oxidase.

Generally, the same substrates may be used for aerobic respiration or sulfate reduction (Dannenberg *et al.*, 1992). However, oxidation of some substrates is sensitive to $O_2$ (e.g., hydrogenase of *Desulfovibrio vulgaris*, Fitz and Cypionka, 1991). Furthermore, with oxygen (or nitrate) many SRB can oxidize inorganic sulfur compounds (Dannenberg *et al.*, 1992). Sulfite, thiosulfate, polysulfide, and even sulfide can serve as electron donors for aerobic respiration. Depending on the strain, these compounds are incompletely or completely oxidized to sulfate. The oxidation of sulfur compounds with $O_2$ (or even nitrate and nitrite) appears to be a logical consequence of the disproportionation of sulfur compounds. It proves that sulfate reduction and electron transport in SRB is not a one-way street.

## 8. CONCLUSION

After a period of exciting discoveries concerning the versatility of carbon metabolism in SRB (Widdel and Pfennig, 1977; Widdel, 1988), sulfur and energy metabolism have also turned out to be copious. However, central bioenergy questions are still unresolved. Although the diversity of SRB is becoming more and more evident, most studies use *Desulfovibrio* species and are restricted to freshwater strains. It gets increasingly clear that even small chances offered by thermodynamics are utilized. Growth coupled to disproportionation of inorganic sulfur compounds demonstrates an excellent energy efficiency. On the other hand, the dissimilatory sulfate reducers remain experts in sulfur transformation. Apparently, more favorable possibilities, such as the use of nitrate or molecular oxygen, are possible but less exploited. The complete reversal of sulfate reduction demonstrates the complexity of sulfur metabolism, the limited state of our knowledge, and that we need to get the biochemistry and bioenergetics of sulfate reduction clarified.

ACKNOWLEDGMENTS. I thank Prof. Dr. Ulrich Fischer, Bernd Kreke, and Rolf Warthmann for critically reading the manuscript. Our studies on sulfate-reducing bacteria have been supported by the Deutsche Forschungsgemeinschaft.

## REFERENCES

Abdollahi, H., and Wimpenny, J. W. T., 1990, Effects of oxygen on the growth of *Desulfovibrio desulfuricans*, *J. Gen. Microbiol.* **136**:1025–1030.
Akagi, J. M., 1981, Dissimilatory sulfate reduction, mechanistic aspects, in: *Biology of Inorganic Nitrogen and Sulfur* (H. Bothe and A. Trebst, eds.) Springer, Heidelberg, pp. 169–177.
Ames, G. F.-L., 1988, Structure and mechanism of bacterial periplasmic transport systems, *J. Bioenerget. Biomembr.* **20**:1–18.
Badziong, W., and Thauer, R. K., 1978, Growth yields and growth rates of *Desulfovibrio vulgaris* (Marburg) growing on hydrogen plus sulfate and hydrogen plus thiosulfate as the sole energy source, *Arch. Microbiol.* **117**:209–214.
Badziong, W., Thauer, R. K., and Zeikus, J. G., 1978, Isolation and characterization of

*Desulfovibrio* growing on hydrogen plus sulfate as the sole energy source, *Arch. Microbiol.* **117**:209–214.

Badziong, W., and Thauer, R. K., 1980, Vectorial electron transport in *Desulfovibrio vulgaris* (Marburg), growing on hydrogen plus sulfate as sole energy source, *Arch. Microbiol.* **125**:167–174.

Bak, F., and Cypionka, H., 1987, A novel type of energy metabolism involving fermentation of inorganic sulphur compounds, *Nature* **326**:891–892.

Bak, F., and Pfennig, N., 1987, Chemolithotrophic growth of *Desulfovibrio sulfodismutans* sp. nov. by disproportionation of inorganic sulfur compounds, *Arch. Microbiol.* **147**:184–189.

Barton, L. L., Le Gall, J., and Peck, H. D., Jr., 1970, Phosphorylation coupled to oxidation of hydrogen with fumarate in extracts of the sulfate reducing bacterium, *Desulfovibrio gigas, Biochem. Biophys. Res. Commun.* **41**:1036–1042.

Barton, L. L., Le Gall, J., Odom, J. M., and Peck, H. D., 1983, Energy coupling to nitrite respiration in the sulfate-reducing bacterium *Desulfovibrio gigas, J. Bacteriol.* **153**:867–871.

Biebl, H., and Pfennig, N., 1977, Growth of sulfate-reducing bacteria with sulfur as electron acceptor, *Arch. Microbiol.* **112**:115–117.

Bryant, M. P., Campbell, L. L., Reddy, C. A., and Crabill, M. R., 1977, Growth of *Desulfovibrio* in lactate or ethanol media low in sulfate in association with $H_2$-utilizing methanogenic bacteria, *Appl. Environ. Microbiol.* **33**:1162–1169.

Chambers, L. A., and Trudinger, P. A., 1975, Are thiosulfate and trithionate intermediates in dissimilatory sulfate reduction? *J. Bacteriol.* **123**:36–40.

Chen, L., Liu, M. Y., Le Gall, J., Fareleira, P., Santos, H., and Xavier, A. V., 1993, Rubredoxin oxidase, a new flavo-hemo-protein, is the site of oxygen reduction to water by the strict anaerobe *Desulfovibrio gigas, Biochem. Biophys. Res. Commun.* **193**:100–105.

Cruden, D. L., Durbin, W. E., and Markovetz, A. J., 1983, Utilization of $PP_i$ as an energy source by *Clostridium* sp., *Appl. Environ. Microbiol.* **46**:1403–1408.

Cypionka, H., 1986, Sulfide-controlled continuous culture of sulfate-reducing bacteria, *J. Microbiol. Meth.* **5**:1–9.

Cypionka, H., 1987, Uptake of sulfate, sulfite and thiosulfate by proton-anion symport in *Desulfovibrio desulfuricans, Arch. Microbiol.* **148**:144–149.

Cypionka, H., 1989, Characterization of sulfate transport in *Desulfovibrio desulfuricans, Arch. Microbiol.* **152**:237–243.

Cypionka, H., 1994, Sulfate transport, in: *Methods in Enzymology* **243**:3–14.

Cypionka, H., and Dilling, W., 1986, Intracellular localization of the hydrogenase in *Desulfotomaculum orientis, FEMS Microbiol. Lett.* **36**:257–260.

Cypionka, H., and Pfennig N., 1986, Growth yields of *Desulfotomaculum orientis* with hydrogen in chemostat culture, *Arch. Microbiol.* **143**:366–369.

Cypionka, H., Widdel, F., and Pfennig, N., 1985, Survival of sulfate-reducing bacteria after oxygen stress, and growth in sulfate-free oxygen-sulfide gradients, *FEMS Microbiol. Ecol.* **31**:39–45.

Dannenberg, S., Kroder, M., Dilling, W., and Cypionka, H., 1992, Oxidation of $H_2$, organic compounds and inorganic sulfur compounds coupled to reduction of $O_2$ or nitrate by sulfate-reducing bacteria, *Arch. Microbiol.* **158**:93–99.

Dilling, W., and Cypionka, H., 1990, Aerobic respiration in sulfate-reducing bacteria, *FEMS Microbiol. Lett.* **71**:123–128.

Fauque, G., Herve, D., and LeGall, J., 1979, Structure-function relationship in hemoproteins: The role of cytochrome $c_3$ in the reduction of colloidal sulfur by sulfate-reducing bacteria, *Arch. Microbiol.* **121**:261–264.

Fitz, R. M., and Cypionka, H., 1989, A study on electron transport-driven proton translocation in *Desulfovibrio desulfuricans*, *Arch. Microbiol.* **152:**369–376.

Fitz, R. M., and Cypionka, H., 1990, Formation of thiosulfate and trithionate during sulfite reduction by washed cells of *Desulfovibrio desulfuricans*, *Arch. Microbiol.* **154:**400–406.

Fitz, R. M., and Cypionka, H., 1991, Generation of a proton gradient in *Desulfovibrio vulgaris*, *Arch. Microbiol.* **155:**444–448.

Furusaka, C., 1961, Sulphate transport and metabolism by *Desulphovibrio desulphuricans*, *Nature* **192:**427–429.

Gottschal, J. C., and Szewzyk, R., 1985, Growth of a facultative anaerobe under oxygen-limiting conditions in pure culture and in co-culture with a sulfate-reducing bacterium. *FEMS Microbiol. Ecol.* **31:**159–170.

Hansen, T. A., 1993, Carbon metabolism of sulfate-reducing bacteria, in: *The sulfate-reducing bacteria: Contemporary perspectives,* (J. M. Odom and R. Singleton, Jr., eds.), Springer, New York, pp. 21–40.

Hooper, A. B., and DiSpirito, A. A., 1985, In bacteria which grow on simple reductants, generation of a proton gradient involves extracytoplasmic oxidation of substrate, *Microbiol. Rev.* **49:**140–157.

Hryniewicz, M., Sirko, A., Palucha, A., Böck, A., and Huilanicka, D., 1990, Sulfate and thiosulfate transport in *Escherichia coli* K12—Identification of a gene encoding a novel protein involved in thiosulfate binding, *J. Bacteriol.* **172:**3358–3366.

Ingvorsen, K., and Jørgensen, B. B., 1984, Kinetics of sulfate uptake by freshwater and marine species of *Desulfovibrio*, *Arch. Microbiol.* **139:**61–66.

Ingvorsen, K., Zehnder, A. J. B., and Jørgensen, B. B., 1984, Kinetics of sulfate and acetate uptake by *Desulfobacter postgatei*, *Appl. Environ. Microbiol.* **47:**403–408.

Ishimoto, M., 1959, Sulfate reduction in cell-free extracts of *Desulfovibrio*, *J. Biochem.* **46:**105–106.

Jeanjean, R., and Broda, E., 1977, Dependence of sulphate uptake by *Anacystis nidulans* on energy, on osmotic shock and on sulphate starvation, *Arch. Microbiol.* **114:**19–23.

Jones, J. G., and Simon, B. M., 1984, The presence and activity of *Desulfotomaculum* spp. in sulphate-limited freshwater sediments, *FEMS Microbiol. Lett.* **21:**47–50.

Keith, S. M., and Herbert, R. A., 1983, Dissimilatory nitrate reduction by a strain of *Desulfovibrio desulfuricans*, *FEMS Microbiol. Lett.* **18:**55–59.

Kim, J. H., and Akagi, J. M., 1985, Characterization of trithionate reductase system from *Desulfovibrio vulgaris*, *J. Bacteriol.* **163:**472–475.

Klemps, R., Cypionka, H., Widdel, F., and Pfennig, N., 1985, Growth with hydrogen, and further physiological characteristics of *Desulfotomaculum* species, *Arch. Microbiol.* **143:**203–208.

Kobayashi, K., Tachibana, S., and Ishimoto, M., 1969, Intermediary formation of trithionate in sulfite reduction by sulfate-reducing bacteria, *J. Biochem.* **65:**155–157.

Kobayashi, K., Hasegawa, H., Takagi, M., and Ishimoto, M., 1982, Proton translocation associated with sulfite reduction in a sulfate-reducing bacterium *Desulfovibrio vulgaris,* *FEBS Lett.* **142:**235–237.

Krämer, M., and Cypionka, H., 1989, Sulfate formation via ATP sulfurylase in thiosulfate- and sulfite-disproportionating bacteria, *Arch. Microbiol.* **151:**232–237.

Kreke, B., and Cypionka, H., 1992, Protonmotive force in freshwater sulfate-reducing bacteria, and its role in sulfate accumulation in *Desulfobulbus propionicus*, *Arch. Microbiol.* **158:**183–187.

Kreke, B., and Cypionka, H., 1994, Role of sodium ions for sulfate transport and energy metabolism in *Desulfovibrio salexigens*, *Arch. Microbiol.* **161:**55–61.

Kremer, D. R., Veenhuis, M., Fauque, G., Peck, H. D., LeGall, J., Lampreia, J., Moura,

J. J. G., and Hansen, T. A., 1988, Immunocytochemical localization of APS reductase and bisulfite reductase in three *Desulfovibrio* species, *Arch. Microbiol.* **150:**296–301.

Kroder, M., Kroneck, P. M. H., and Cypionka, H., 1991, Determination of the transmembrane proton gradient in the anaerobic bacterium *Desulfovibrio desulfuricans* by $^{31}P$ nuclear magnetic resonance, *Arch. Microbiol.* **156:**145–147.

LeGall, J., and Fauque, G., 1988, Dissimilatory reduction of sulfur compounds, in: *Biology of anaerobic microorganisms,* (A. J. B. Zehnder, ed.), Wiley & Sons, New York, pp. 587–639.

Littlewood, D., and Postgate, J. R., 1957, On the osmotic behaviour of *Desulphovibrio desulphuricans, J. Gen. Microbiol.* **16:**596–603.

Liu, C.-L., and Peck, H. D., 1981, Comparative bioenergetics of sulfate reduction in *Desulfovibrio* and *Desulfotomaculum* spp., *J. Bacteriol.* **145:**966–973.

Liu, C.-L., Hart, N., and Peck, H. D., 1982, Inorganic pyrophosphate: energy source for sulfate-reducing bacteria of the genus *Desulfotomaculum, Science* **217:**363–364.

Lupton, F. S., Conrad, R., and Zeikus, J. G., 1984, Physiological function of hydrogen metabolism during growth of sulfidogenic bacteria on organic substrates, *J. Bacteriol.* **159:**843–849.

Marschall, C., Frenzel, P., and Cypionka, H., 1993, Influence of oxygen on sulfate reduction and growth of sulfate-reducing bacteria, *Arch. Microbiol.* **159:**168–173.

Martinoia, E., 1992, Transport processes in vacuoles of higher plants, *Bot. Acta* **105:**232–245.

McCready, R. G. L., Gould, W. D., and Barendregt, R. W., 1983, Nitrogen isotope fractionation during the reduction of $NO_3^-$ to $NH_4^+$ by *Desulfovibrio* sp., *Can. J. Microbiol.* **29:**231–234.

Miller, J. D. A., and Wakerly, D. S., 1966, Growth of sulphate-reducing bacteria by fumarate dismutation, *J. Gen. Microbiol.* **43:**101–107.

Mitchell, G. J., Jones, J. G., and Cole, J. A., 1986, Distribution and regulation of nitrate and nitrite reduction by *Desulfovibrio* and *Desulfotomaculum* species, *Arch. Microbiol.* **144:**35–40.

Myers, R. J., 1986, The new low value for the second dissociation constant for $H_2S$. Its history, its best value, and its impact on the teaching of sulfide equilibria, *J. Chem. Education* **63:**687–690.

Nethe-Jaenchen, R., and Thauer, R. K., 1984, Growth yields and saturation constant of *Desulfovibrio vulgaris* in chemostat culture, *Arch. Microbiol.* **137:**236–240.

Nore, B. F., Husain, I., Nyrén, P., and Baltscheffsky, M., 1986, Synthesis of pyrophosphate coupled to the reverse energy-linked transhydrogenase reaction in *Rhodospirillum rubrum* chromatophores, *FEBS Lett.* **200:**133–138.

Nyrén, P., and Strid, A., 1991, Hypothesis—The physiological role of the membrane-bound proton-translocating pyrophosphatase in some phototrophic bacteria, *FEMS Microbiol. Lett.* **77:**265–269.

Odom, J. M., and Peck, H. D., 1981, Hydrogen cycling as a general mechanism for energy coupling in the sulfate-reducing bacteria, *Desulfovibrio* sp., *FEMS Microbiol. Lett.* **12:**47–50.

Odom, J. M., and Peck, H. D., 1984, Hydrogenases, electron-transfer proteins, and energy coupling in the sulfate-reducing bacteria *Desulfovibrio, Ann. Rev. Microbiol.* **38:**551–592.

Odom, J. M., and Wall, J. D., 1987, Properties of a hydrogen-inhibited mutant of *Desulfovibrio desulfuricans* ATCC 27774, *J. Bacteriol.* **169:**1335–1337.

Pankhania, I. P., Spormann, A. M., Hamilton, W. A., and Thauer, R. K., 1988, Lactate conversion to acetate, $CO_2$ and $H_2$ in cell suspensions of *Desulfovibrio vulgaris* (Marburg): Indications for the involvement of an energy driven step, *Arch. Microbiol.* **150**:26–31.

Peck, H. D., 1959, The ATP-dependent reduction of sulfate with hydrogen in extracts of *Desulfovibrio desulfuricans, Proc. Natl. Acad. Sci.* USA **45**:701–708.

Peck, H. D., 1960, Evidence for oxidative phosphorylation during the reduction of sulfate with hydrogen by *Desulfovibrio desulfuricans, J. Biol. Chem.* **235**:2734–2738.

Peck, H. D., 1966, Phosphorylation coupled with electron transfer in extracts of the sulfate-reducing bacterium *Desulfovibrio gigas, Biochem. Biophys. Res. Commun.* **22**:112–118.

Peck, H. D., and LeGall, J., 1982, Biochemistry of dissimilatory sulphate reduction, *Phil. Trans. Roy. Soc. London* B **298**:443–466.

Peck, H. D., LeGall, J., Lespinat, P. A., Berlier, Y., and Fauque, G., 1987, A direct demonstration of hydrogen cycling by *Desulfovibrio vulgaris* employing membrane-inlet mass spectrometry, *FEMS Microbiol. Lett.* **40**:295–299.

Peck, H. D., and Lissolo, T., 1988, Assimilatory and dissimilatory sulphate reduction: enzymology and bioenergetics, in: *The nitrogen and sulphur cycles*, (J. A. Cole and S. J. Ferguson, eds.), Cambridge University Press, pp. 99–132.

Pereiradasilva, L., Sherman, M., Lundin, M., and Baltscheffsky, H., 1993, Inorganic pyrophosphate gives a membrane potential in yeast mitochondria, as measured with the permeant cation tetraphenylphosphonium, *Arch. Biochem. Biophys.* **304**:310–313.

Postgate, J. R., 1984, The sulphate-reducing bacteria. Cambridge University Press, 2nd edition.

Reeves, R. E., and Guthrie, J. D., 1975, Acetate kinase (pyrophosphate). A fourth pyrophosphate-dependent kinase from *Entamoeba histolytica, Biochem. Biophys. Res. Commun.* **66**:1389–1395.

Romero, I., Gomezpriego, A., and Celis, H., 1991, A membrane-bound pyrophosphatase from respiratory membranes of *Rhodospirillum rubrum, J. Gen. Microbiol.* **137**:2611–2616.

Sass, H., Steuber, J., Kroder, M., Kroneck, P. M. H., and Cypionka, H., 1992, Formation of thionates by freshwater and marine strains of sulfate-reducing bacteria, *Arch. Microbiol.* **158**:418–421.

Schink, B., 1992, Syntrophism among prokaryotes, in: *The prokaryotes. A handbook on the biology of bacteria: Ecophysiology, isolation, identification, applications*, (A. Balows, H. G. Trüper, M. Dworkin, W. Harder, K. H. Schleifer, eds.), Springer, New York, pp. 276–299.

Scholes, P., and Mitchell, P., 1970, Respiration-driven proton translocation in *Micrococcus denitrificans, Bioenergetics* **1**:309–323.

Seitz, H.-J., and Cypionka, H., 1986, Chemolithotrophic growth of *Desulfovibrio desulfuricans* with hydrogen coupled to ammonification of nitrate or nitrite, *Arch. Microbiol.* **146**:63–67.

Seitz, H.-J., Schink, B., Pfennig, N., and Conrad, R., 1990, Energetics of syntrophic ethanol oxidation in defined chemostat cocultures. 1. Energy requirement of $H_2$ production and $H_2$ oxidation, *Arch. Microbiol.* **155**:89–93.

Stahlmann, J., Warthmann, R., and Cypionka, H., 1991, $Na^+$-dependent accumulation of sulfate and thiosulfate in marine sulfate-reducing bacteria, *Arch. Microbiol.* **155**:554–558.

Stams, A. J. M., Kremer, D. R., Nicolay, K., Weenk, G. H., and Hansen, T. A., 1984, Pathway of propionate formation in *Desulfobulbus propionicus, Arch. Microbiol.* **139**:167–173.

Steenkamp, D. J., and Peck, H. D., 1981, Proton translocation associated with nitrite respiration in *Desulfovibrio desulfuricans, J. Biol. Chem.* **256**:5450–5458.

Steuber, J., Cypionka, H., Kroneck, P. M. H., 1994, Mechanism of dissimilatory sulfite reduction by *Desulfovibrio desulfuricans:* purification of a membrane-bound sulfite reductase and coupling with cytochrome $c_3$ and hydrogenase, *Arch. Microbiol.* **162:**255–260.

Stouthamer, A. H., 1988, Bioenergetics and yields with electron acceptors other than oxygen, in: *Handbook on anaerobic fermentations,* (L. E. Erickson and D. Y.-C. Fung, eds.), Marcel Decker, New York, Basel, pp. 345–437.

Thamdrup, B., Finster, K., Hansen, J. W., and Bak, F., 1993, Bacterial disproportionation of elemental sulfur coupled to chemical reduction of iron or manganese, *Appl. Environ. Microbiol.* **59:**101–108.

Thauer, R. K., 1989, Energy metabolism of sulfate-reducing bacteria, in: *Autotrophic bacteria,* (H. G. Schlegel and B. Bowien, eds.), Science Tech Publishers, Madison, pp. 397–413.

Thauer, R. K., Jungermann, K., and Decker, K., 1977, Energy conservation in chemotrophic anaerobic bacteria, *Bacteriol. Rev.* **41:**100–180.

Thauer, R. K., and Morris, J. G., 1984, Metabolism of chemotrophic anaerobes: Old views and new aspects, in: *The Microbe 1984: Part II, Prokaryotes and eukaryotes,* (D. P. Kelly and N. G. Carr, eds.), Cambridge University Press, pp. 123–168.

Thebrath, B., Dilling, W., and Cypionka, H., 1989, Sulfate activation in *Desulfotomaculum, Arch. Microbiol.* **152:**296–301.

Vainshtein, M. B., Matrosov, A. G., Baskunov, V. P., Zyakun, A. M., and Ivanov, M. V., 1980, Thiosulfate as an intermediate product of bacterial sulfate reduction, *Microbiology* (engl. Transl.) **49:**672–675.

Varma, A. K., and Peck, H. D., 1983, Utilization of short and long-chain polyphosphates as energy sources for the anaerobic growth of bacteria, *FEMS Microbiol. Lett.* **16:**281–285.

Varma, A., Schönheit, P., and Thauer, R. K., 1983a, Electrogenic sodium ion/proton antiport in *Desulfovibrio vulgaris, Arch. Microbiol.* **136:**69–73.

Varma, A. K., Rigsby, W., and Jordan, D. C., 1983b, A new inorganic pyrophosphate utilizing bacterium from a stagnant lake, *Can. J. Microbiol.* **29:**1470–1474.

Ware, D. A., and Postgate, J. R., 1971, Physiological and chemical properties of reductant-activated inorganic pyrophosphatase from *Desulfovibrio desulfuricans, J. Gen. Microbiol.* **67:**145–160.

Warthmann, R., and Cypionka, H., 1990, Sulfate transport in *Desulfobulbus propionicus* and *Desulfococcus multivorans, Arch. Microbiol.* **154:**144–149.

West, I. C., and Mitchell, P., 1974, Proton/sodium ion antiport in *Escherichia coli, Biochem. J.* **144:**87–90.

Widdel, F., 1988, Microbiology and ecology of sulfate- and sulfur-reducing bacteria, in: *Biology of anaerobic microorganisms,* (A. J. B. Zehnder, ed.) Wiley & Sons, New York, pp. 469–585.

Widdel, F., and Pfennig, N., 1977, A new anaerobic, sporing, acetate-oxidizing, sulfate-reducing bacterium, *Desulfotomaculum* (emend.) *acetoxidans, Arch. Microbiol.* **112:**119–122.

Widdel, F., and Pfennig, N., 1982, Studies on dissimilatory sulfate-reducing bacteria that decompose fatty acids. II. Incomplete oxidation of propionate by *Desulfobulbus propionicus* gen. nov., sp. nov., *Arch. Microbiol.* **131:**360–365.

Wood, P. M., 1978, A chemiosmotic model for sulphate respiration, *FEBS Lett.* **95:**12–18.

Zehr, J. P., and Oremland, R. S., 1987, Reduction of selenate to selenide by sulfate-respiring bacteria: Experiments with cell suspensions and estuarine sediments, *Appl. Environ. Microbiol.* **53:**1365–1369.

# Molecular Biology of Redox-Active Metal Proteins from *Desulfovibrio*

7

WALTER M.A.M. VAN DONGEN

## 1. INTRODUCTION

Anaerobe sulfate reducers are a rich source of proteins with redox-active metal groups. In the past two decades, a variety of such proteins, especially from *Desulfovibrio* species, has been analyzed, and many novel metal centers have been described, some functioning in electron transport, others as the sites for catalytic activity. Sulfate reducers attracted the attention of molecular biologists almost 10 years ago, when the potential relevance of some of these proteins for technological application was realized (e.g., hydrogenases for production of $H_2$ as an energy source). Therefore, it is not surprising that the first gene, cloned from a sulfate reducer in 1985, encoded a hydrogenase from *Desulfovibrio vulgaris* Hildenborough with a very high specific activity (Voordouw *et al.*, 1985), soon followed by that for cytochrome $c_3$, the putative electron donor (Voordouw and Brenner, 1986). Subsequent research has shown the complexity of hydrogenases and has slightly tempered the high expectations of that time about short-term technological application. Nevertheless, many other genes have been isolated from *Desulfovibrios* since then, most of them encoding metal proteins (hydrogenases, sulfite reductase, cytochromes, rubredoxin, rubrerythrin). Comparative analysis of the aminoacid sequences, derived from the nucleotide sequence of the genes, has proven to be an invaluable tool for identification of metal

WALTER M.A.M. VAN DONGEN • Department of Biochemistry, Agricultural University, Dreyenlaan 3, 6703 HA Wageningen, The Netherlands.

*Sulfate-Reducing Bacteria*, edited by Larry L. Barton. Plenum Press, New York, 1995.

binding sites in these proteins (for example, for analysis of nickel-containing hydrogenases, Przybyla *et al.*, 1992).

However, attempts to investigate how the structure of these proteins was related to their properties (coordination of the metal, redox behavior) by introducing deliberate changes in the structure by site-directed mutagenesis of the cloned genes, initially failed. It soon became evident that expression of many of these genes (for example, those encoding hydrogenases) in the *Escherichia coli* hosts usually applied for this purpose, resulted only in the production of apoproteins, without the metal clusters. Probably, *E. coli* lacks appropriate factors required for biosynthesis and processing of these proteins (Voordouw *et al.*, 1987; van Dongen *et al.*, 1988). This prompted the search for genetic methods aimed at the use of *Desulfovibrios* as hosts for expression of (mutated) genes. In 1989, the group of Dr. Mergeay in Belgium (working on the resistance against heavy metals) and my group reported the first methods for plasmid transfer to *Desulfovibrio* (Powell *et al.*, 1989; van den Berg *et al.*, 1989). Since then, expression systems have been developed, allowing overproduction of (mutated) proteins in *Desulfovibrio* (Mus-Veteau *et al.*, 1992, Stokkermans *et al.*, 1992b).

Many techniques belonging to the standard toolbox of *E. coli* geneticists are not yet available on a regular scale for *Desulfovibrio* (for example, transposon mutagenesis and other techniques to disrupt genes). Introduction of these techniques is expected to be a matter of time.

In this chapter, genetics and molecular biology of sulfate reducers will be restricted to *Desulfovibrio* spp. *Desulfovibrios* are till now the only sulfate reducers from which genes have been cloned and sequenced (apart from sequences of 16S ribosomal RNA from various sulfate reducers for determination of phylogenetic relationships, Devereux *et al.*, 1989). They are also the only sulfate reducers for which an, although primitive, system for genetic manipulation has been developed. This chapter opens with a very brief summary of the little that is known about *Desulfovibrio* genetics, to indicate constituents of the *Desulfovibrio* genome that may be applied for future development of the system for genetic manipulation. Next, gene transfer to *Desulfovibrio* with broad-host-range vectors is described. The final section gives examples on the use of gene cloning and sequencing as one of the tools for the analysis of metal proteins from *Desulfovibrio* (hydrogenases, cytochromes), and on the application of gene cloning in *Desulfovibrio* for production of mutated proteins (to investigate how structural changes affect properties) or to introduce changes in the *Desulfovibrio* genome.

## 2. *DESULFOVIBRIO* GENETICS

### 2.1. The Genome

In the classification system of Woese (1987), based on analysis of 16S rRNA sequences, most nonspore-forming sulfate reducers are placed in the delta subdivision of the purple bacteria. Within the delta subdivision, *Desulfovibrios* are classified as a single, heterogeneous group (Devereux *et al.*, 1989). The G + C content of the DNA varies between 49 and 65% for different species (Skyring and Jones, 1972). One of the manifestations of this heterogeneity may be in the different capabilities for plasmid uptake and replication of various strains (section 3).

Genomes of *Desulfovibrio* spp. are small: approximately 1630 kb for *D. gigas* and 1720 kb for *D. vulgaris*, against 3950 kb for *E. coli* (Postgate *et al.*, 1984). These small sizes may reflect the limited metabolic capabilities of *Desulfovibrios*. *Desulfovibrio* cells may contain several copies of the genome: 9–17 copies per cell have been found in *D. gigas* and four copies per cell in *D. vulgaris* (Postgate *et al.*, 1984).

Several libraries of random *D. vulgaris* DNA fragments cloned in plasmid or lambda vectors have been constructed (Voordouw, 1988; Stokkermans, unpublished). Construction of a genomic map of *D. vulgaris* by chromosome walking is in progress (Deckers and Voordouw, 1994).

### 2.2. Plasmids

Megaplasmids, between 60 and 195 kb in size, have been found in six of 16 investigated *Desulfovibrio* strains (Postgate *et al.*, 1984, 1988; Powell, 1989). Copy numbers of the plasmids have not been determined, but are probably low. Plasmids account for approximately 10% of the total DNA in *D. gigas* and *D. vulgaris* Hildenborough, if present at only one copy per chromosome (Postgate *et al.*, 1984).

Except for the gene encoding nitrogenase component 2 in *D. vulgaris* strains, no other proteins are known to be encoded by plasmid-located genes (Postgate *et al.*, 1988).

Efforts to isolate these megaplasmids have been only successful with *D. gigas* (Postgate *et al.*, 1988). Further investigation of properties of these plasmids for exploitation in the construction of cloning vectors for *Desulfovibrio* (determination of the host-range, mapping of the basic replicon and of regions required for maintenance and, possibly, transfer) has not been accomplished.

Only recently a small plasmid has been found in a *Desulfovibrio* strain (Wall *et al.*, 1993). This plasmid, pBG1, is only 2.3 kb in size and occurs in approximately 20 copies in *D. desulfuricans* G100A. No functions have yet been attributed to pBG1. The plasmid apparently does not replicate in *E. coli*, but a conjugable shuttle vector has been constructed by fusion of pBG1 with an *E. coli* vector (see also section 3.1).

### 2.3. Phages

Induction of phagelike particles from *D. vulgaris* with mitomycin C or UV light was first demonstrated in 1973 (Handley *et al.*, 1973), but the first phage was isolated only in 1987 by Rapp and Wall from culture supernatant of *D. desulfuricans* ATCC 27774. This nonplaque-forming phage contained double-stranded DNA of a uniform size (13.6 kb), but analysis of restriction digests of isolated phage DNA on agarose gels showed smears of DNA rather than discrete fragments, presumably by random packaging of bacterial sequences into the phage heads.

Contrary to the ATCC 2774 phages, phages isolated from *D. salexigens*, *D. vulgaris* Hildenborough and *D. desulfuricans* ATCC 13541 contained DNA that showed discrete restriction fragments upon digestion (Kamimura and Araki, 1989; Seyedirashti *et al.*, 1991). The Hildenborough and ATCC 13541 phages both contained 40–45 kb of DNA; lack of cross hybridization of their DNAs suggests that they are not closely related. The Hildenborough phage did not replicate in the *D. desulfuricans* strains ATCC 13541 and Norway 4 (Seyedirashti *et al.*, 1991). The *D. salexigens* phage is the only one for which plaque formation has been demonstrated (Kamimura and Araki, 1989).

Application of *Desulfovibrio* phages for construction of vectors for gene cloning and genetic transfer has not yet been reported.

## 3. TRANSFER OF BROAD-HOST-RANGE PLASMIDS TO *DESULFOVIBRIO:* A BASIC STEP FOR GENETIC MANIPULATION

### 3.1. IncQ Plasmids for Gene Cloning in *Desulfovibrio*

The delta subdivision of purple bacteria, to which *Desulfovibrio* belongs, has long been the only one of the four subdivisions for which no plasmid transfer systems were described. Only in 1989, two groups reported, for the first time, transfer of broad-host-range vectors to *D. vulgaris* Hildenborough, *Desulfovibrio* sp. Holland SH-1 and *D. desulfuri-*

*cans* Norway 4 (Powell *et al.*, 1989; van den Berg *et al.*, 1989). Subsequently, plasmid transfer to *D. desulfuricans* strain G200 and *D. fructosovorans* has been described (Voordouw *et al.*, 1990b; Rousset *et al.*, 1991).

Several procedures have been described for transformation of *Desulfovibrio*, either by conjugal transfer from an *E. coli* host (Powell *et al.*, 1989; van den Berg *et al.*, 1989; Argyle *et al.* 1992; van Dongen *et al.*, 1994) or by electroporation (Rousset *et al.*, 1991). Conjugation with *E. coli* is still the favorite technique. In principle, these procedures are not very different from those described for transformation of other gram-negatives, but the anaerobic nature and the high level of resistance of *Desulfovibrio* strains against antibiotics commonly used as selection makers, cause extra complications. Genetic experiments require that bacteria can be grown on agar plates as colonies derived from single cells. Therefore, highly efficient and reliable plating procedures have been developed (van den Berg *et al.*, 1989; Argyle *et al.*, 1992).

Broad-host-range cloning vectors have been developed from plasmids belonging to incompatibility groups P, Q, and W. Such vectors contain at least an origin of replication, genes involved in plasmid maintenance and conjugal transfer, one or more antibiotic resistance genes and, ideally, also a multiple cloning site with recognition sequences for a number of restriction enzymes. Only for IncQ vectors conjugal transfer to various *Desulfovibrio* species (*D. vulgaris* Hildenborough, *Desulfovibrio* sp. Holland SH-1 and *D. desulfuricans* strains G200 and Norway 4) has been firmly established, at transfer frequencies ranging from $10^{-4}$ to 1 (van den Berg *et al.*, 1989; Powell *et al.*, 1989; Voordouw *et al.*, 1990b; Argyle *et al.*, 1992). IncQ plasmids are stably maintained in these species. Van den Berg and colleagues (1989) cultivated *D. vulgaris* Hildenborough harboring IncQ vectors pSUP104 or pSUP104Ap during 32 generations in the absence of selective antibiotics; accumulation of cells that had lost the plasmid encoded selective markers did not occur in this period. Copy numbers of these plasmids were approximately 12 per cell, with a slight tendency to decrease when they contained large inserts. IncQ vectors pSUP104 (Priefer *et al.*, 1984) and pJRD215 (Davison *et al.*, 1987) have been applied as expression vectors in *D. vulgaris* Hildenborough and *D. desulfuricans* G200, respectively, allowing overproduction of proteins from cloned genes (van den Berg *et al.*, 1989; Voordouw *et al.*, 1990; Stokkermans *et al.*, 1992b). The promoter of the tetracycline resistance gene in pSUP104 has been shown to produce high levels of transcription of cloned genes in *D. vulgaris* (van den Berg *et al.*, 1991; Stokkermans *et al.*, 1992b).

A serious problem for transformation of *Desulfovibrio* is the limited availability of suitable markers (*e.g.* genes conferring resistance against antibiotics) for selection of cells that have incorporated plasmids (Powell, 1989; Argyle *et al.* 1992). Applications which require expression of a single selection marker (transformation of cells with stably maintained, high-copy number plasmids) are now possible for the strains mentioned, but applications in which expression of more than one marker is desired, are still difficult (for example marker exchange, in which genes in the genome are replaced by plasmid encoded copies that are inactivated by insertion of a gene encoding a selective marker, or transposon mutagenesis). A systematic screening for different markers (less common antibiotic resistance genes, genes that complement auxotrophic mutations) is clearly required to extend the toolbox for *Desulfovibrio* genetics. The only reported case of marker exchange till now is for *D. fructosovorans;* in this strain, an IncQ vector with genes for the NiFe-hydrogenase interrupted with a gene for kanamycin resistance, has been used for inactivation of the genomic copies of these genes (Rousset *et al.*, 1991). Apparently, the IncQ vector did not replicate in *D. fructosovorans.*

D. desulfuricans ATCC 27774 (Argyle *et al.*, 1992), *D. gigas* and *D. vulgaris* oxamicus ssp. Monticello (van den Berg, unpublished observations) have resisted uptake of IncQ plasmids by conjugation. It is not clear, whether these strains are deficient in conjugal uptake or in replication of IncQ plasmids.

Besides IncQ plasmids, a shuttle vector constructed by fusion of the *Desulfovibrio* plasmid pBG1 (section 2.2) with an *E. coli* vector could also be transferred from *E. coli* to *D. desulfuricans* G100A and *D. fructosovorans* by conjugation (Wall *et al.*, 1993).

### 3.2. Applications of Gene Cloning in *Desulfovibrio*

Gene cloning in *Desulfovibrio* has mainly been used for overproduction of proteins and enzymes with redox-active heme groups or iron-sulfur clusters. Expression of genes for several *Desulfovibrio* heme- and iron-sulfur proteins (cytochromes, hydrogenase, prismane protein) in *E. coli* has been shown to result only in the accumulation of apoproteins (Voordouw *et al.*, 1987; Pollock *et al.*, 1989; Stokkermans *et al.*, 1992b), but many of these proteins are overproduced in their native states, with the metal centers present, when a *Desulfovibrio* strain is used as a host (Voordouw *et al.*, 1990b; Bruschi *et al.*, 1992; Stokkermans *et al.*, 1992b). Inserting genes at the proper sites in cloning vectors pJRD215 or pSUP104 allows 5 to 30-fold overproduction of the proteins (van den Berg *et al.*, 1989; Voordouw *et al.*, 1990b; Stokkermans *et al.*, 1992b). Recently *Desulfovibrio* has been used as a host for expression of genes for metal proteins with deliberate changes in the primary structure, made by site-directed mutagenesis, to modify the redox behavior (Mus-Veteau *et al.*, 1992).

Disruption of the genes for the NiFe-hydrogenase in the genome of *D. fructosovorans* is still the only example of marker exchange in *Desulfovibrio* (Rousset *et al.*, 1991). Marker exchange has been proven very useful in a range of bacteria, for elucidation of the role of specific proteins or enzymes in a metabolic pathway by gene disruption or for introduction of new properties in a bacterium by stable incorporation of genes from a different species. Lack of suitable selective markers has prevented wider application of this powerful technique in *Desulfovibrio*. Instead of gene disruption by marker exchange, van den Berg *et al.* (1991) used plasmids from which hydrogenase antisense RNA was transcribed to reduce the amount of Fe-hydrogenase in *D. vulgaris* in order to analyze the metabolic role of this enzyme. Although the antisense RNA technique may give less unequivocal results than gene disruption (production of a protein is reduced but not completely abolished), it might be a proper choice for analyzing the function of genes whose disruption is lethal for the cell.

## 4. MOLECULAR BIOLOGY AS A TOOL FOR ANALYSIS OF REDOX-ACTIVE METAL PROTEINS

Besides heme-containing proteins: several *c*-type cytochromes, and sulfite reductase (Liu *et al.*, 1988; Peck and Le Gall, 1982), *Desulfovibrios* have proteins with nonsulfur di-iron sites: rubrerythrin, and nigerythrin (Le Gall *et al.*, 1988; Pierik *et al.*, 1993), with mononuclear $FeS_4$ centers: rubredoxin, rubrerythrin, and nigerythrin (Pierrot *et al.*, 1976; Adman *et al.*, 1977; LeGall *et al.*, 1988; Pierik *et al.*, 1993), [3Fe-4S] clusters: ferredoxin II from *D. gigas* (Huynh *et al.*, 1980) and with [4Fe-4S] clusters: ferredoxins, and hydrogenase (Bruschi and Guerlesquin, 1988; Voordouw and Brenner, 1985). All these metal centers are usually involved in single-electron transfer and cycle between two redox states. Very recently, a new species was added to this spectrum by the isolation of a protein from *D. vulgaris* with a novel type of iron-sulfur center, probably [6Fe-6S] with four possible redox states (Hagen *et al.*, 1989; Pierik *et al.*, 1992a,b).

For some of these proteins, mainly small electron carriers that are stable under aerobic conditions (*e.g.* ferredoxins, cytochrome $c_3$), determination of the three-dimensional protein structure with X-ray diffraction has yielded information about the structure and ligation of the metal centers. For a variety of reasons (oxygen sensitivity, difficulty to isolate the proteins with fully intact metal clusters), this has often not yet been possible for enzymes with catalytic metal centers. Isolation and sequencing of genes encoding such enzymes for determination of the primary structure and comparing these primary structures with those of homologous proteins or proteins with similar types of centers has, together with analysis of data obtained by spectroscopic techniques (EPR, EXAFS, Mössbauer), contributed to our understanding of how these metals might be ligated in more complex enzymes.

## 4.1. Hydrogenases

Hydrogenases are abundant in *Desulfovibrios*. Three classes are distinguished by the composition of the metal centers: hydrogenases that contain only Fe-S clusters and no other metals ('Fe-only' or Fe-hydrogenases); hydrogenases that have next to Fe-S clusters also nickel (NiFe-hydrogenases); and, hydrogenases with Fe-S clusters, nickel, and selenium (NiFeSe-hydrogenases). All hydrogenases catalyze *in vitro* the reversible production of $H_2$ with low-potential artificial electron donors (*e.g.* viologen dyes), but considerable differences exist in the specific activities in $H_2$ evolution and $H_2$ uptake, the ratio between these activities, and the sensitivity towards inhibitors of metal enzymes like CO, NO, and acetylene (Lissolo *et al.*, 1986; Prickril *et al.*, 1987; He *et al.*, 1989b; Pierik *et al.*, 1992c).

### 4.1.1. The Fe-Hydrogenases

The periplasmic Fe-hydrogenases from *D. vulgaris* Hildenborough (van der Westen *et al.*, 1978) and *D. desulfuricans* ATCC 7757 (Hatchikian *et al.*, 1992) have the highest specific activity of all hydrogenases known so far (approximately 9000 U/mg and 55,000 U/mg in the $H_2$-production and -consumption assays, respectively, with methyl viologen as electron donor/acceptor). Contrary to other Fe-hydrogenases, the *Desulfovibrio* enzymes can be isolated in a "resting" form, which is insensitive to $O_2$-inactivation. Transformation to the active form occur under reducing conditions.

Both Fe-hydrogenases are αβ heterodimers with subunits of 42 (46) and 10 (11) kDa (Voordouw *et al.*, 1985; Voordouw and Brenner, 1985; Hatchikian *et al.*, 1992). Extensive analysis of the *D. vulgaris* enzyme demonstrated that it contains no other metals than iron (Huynh *et al.*, 1984, Pierik *et al.*, 1992c). The iron and sulfide content were determined at 13–15 and 12–14 moles/mole of protein, respectively (Hagen *et al.*,

1986b). EPR spectra of reduced enzyme show a complex signal analogous to that found in ferredoxins with two weakly interacting [4Fe-4S] clusters (Grande *et al.*, 1983). Indeed, the amino acid sequence of the large α subunit has in the N-terminal region twice the motif (Cys-(Xaa)$_2$-Cys-(Xaa)$_2$-Cys-(Xaa)$_3$-Cys) which is conserved in ferredoxins; the cysteines are known to coordinate the irons of the two cubanes (Fig 1, the F-region). The [4Fe-4S] clusters are involved in electron transport to the active site of the enzyme. A third cluster with a midpoint potential for reduction of −300 mV at neutral pH, probably the site of H$_2$ activation, showed spectral (EPR and Mössbauer) properties that were different from those of [2Fe-2S] or [4Fe-4S] clusters (Hagen *et al.*, 1986a, Pierik *et al.*, 1992c). Based on the total iron content of the hydrogenase (13–15

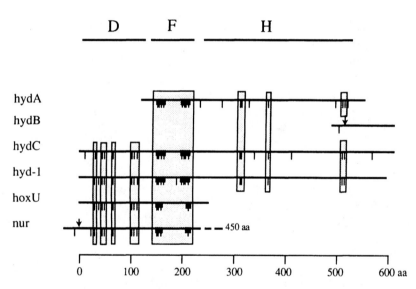

**Figure 1.** Schematic representation of the homology between the α and β subunits of the Fe-hydrogenase (hydA and hydB) and hydC from *D. vulgaris,* hydrogenase-I from *C. pasteurianum* (hyd-1), the δ subunit of NAD-reducing hydrogenase of *A. eutrophus* H16 (hoxU), and the N-terminal sequence of the 75-kDa subunit of beef heart mitochondrial NADH-dehydrogenase (nur). Small gaps required for optimal alignment of the sequences are not shown. Downward pointing arrows indicate splicing sites for leader peptidases in hydB and nur. Downward ticks represent positions of cysteines; conserved cysteines are in shaded boxes. Black boxes indicate cysteine motifs for coordination of [4Fe-4S] ferredoxin-like cubanes (note the presence of two such motifs in hydA, hydC, and hyd-1; only one in nur and hoxU). D, F, and H represent the regions predicted to be involved in coordination of [2Fe-2S] clusters, ferredoxin-like [4Fe-4S] cubanes, and the putative [6Fe-6S] active-site cluster of hydrogenase, respectively.

moles/mole protein) and the amount supposed to be present in the two ferredoxin-like cubanes (8 Fe), Hagen *et al.* (1986b) proposed a—then revolutionary—[6Fe-6S] cluster for the site of $H_2$ activation, a concept that has been used since then to explain some of the spectral data on clostridial Fe-hydrogenases as well (Adams, 1990). Although the spectroscopy appears as yet too complicated for a definite proof of a [6Fe-6S] active-site cluster in hydrogenase, evidence for such Fe-S superclusters in proteins has been obtained subsequently in the 'prismane protein', also isolated from *D. vulgaris* (section 4.2). How this complex cluster is ligated into hydrogenase is unknown; comparison of the sequence of the *D. vulgaris* Hildenborough hydrogenase with those of other Fe-hydrogenases (from *D. vulgaris* subsp. *oxamicus* Monticello, Voordouw *et al.*, 1989b, and from *Clostridium pasteurianum*, Meyer and Gagnon, 1991) indicated relatively great similarity. Besides the eight cysteines in the ferredoxin-like cysteine motifs, nine of the 10 remaining Cys-residues in the Hildenborough large subunit are conserved in the Monticello enzyme and five in the clostridial hydrogenase I. However, ligation of the active-site Fe-S cluster to other residues other than cysteine cannot be excluded.

Downstream from the *hydAB* genes encoding the dimeric Fe-hydrogenase in *D. vulgaris* Hildenborough, a gene has been found that might encode another Fe-hydrogenase (Stokkermans *et al.*, 1989). The primary structure of the 63 kDa polypeptide encoded by this *hydC* gene has homology with the *D. vulgaris* Fe-hydrogenase and especially with the clostridial hydrogenase I (Meyer and Gagnon, 1991). Figure 1 shows the alignment of the sequences and the position of the conserved cysteines. Interestingly, the central part and the C-terminal parts of the HydC sequence are homologous to the α- and β-subunit sequences of the dimeric Fe-hydrogenase, respectively (exclusive the leader peptide of the β-subunit which is removed upon export across the cytoplasmic membrane). Probably, the current *hydC* and *hydAB* genes derive from a common ancestral gene, which was duplicated, followed by splicing of one of the copies and attachment of a leader sequence to the small subunit (see section 4.1.3 for the role of the small subunit in export of the enzyme). Splicing apparently did not occur in the clostridial hydrogenase.

Both HydC and the clostridial hydrogenase have a cysteine-rich N-terminal extension that is not found in the *D. vulgaris* Fe-hydrogenase (region D in Fig. 1). Region D also has a high degree of sequence conservation with a subunit of a completely different hydrogenase, the γ-subunit of the diaphorase protein of the NAD-reducing NiFe-hydrogenase from *Alcaligenes eutrophus* (Tran-Betcke *et al.*, 1990) and with part of a 75-kDa subunit of bovine NADH-ubiquinone oxidoreductase, the first enzyme in the mitochondrial respiratory chain (Runswick *et al.*, 1989). Cysteines in region D may coordinate a [2Fe-2S] cluster for electron transport (Stokkermans, 1993): spectroscopic evidence for [2Fe-2S] clusters (obtained with EPR and EXAFS) exists for NADH-dehydrogenase (Bakker and Albracht, 1986, Krishnamoorthy and Hinkle, 1988), the diaphorase subunit of NAD-dependent hydrogenase (Schneider *et al.*, 1984) and for the clostridial hydrogenase I (Adams, 1990). [The NAD-reducing hydrogenase from *Alcaligenes* and bovine NADH-ubiquinone oxidoreductase share another region of homology, which is not found, however, in HydC or clostridial hydrogenase: The α-subunit of the NAD-reducing hydrogenase, which is the FMN-containing subunit where NAD is reduced, has high similarity to the 24-kDa subunit of the FMN-containing, NADH oxidizing subfraction of NADH dehydrogenase (Pilkington and Walker, 1989). Surprisingly, these three enzymes with very different functions (Fe-hydrogenase: synthesis of $H_2$ in strictly anaerobic bacteria; NAD-dependent NiFe hydrogenase: reduction of NAD by $H_2$ in an aerobe; NADH-dehydrogenase: oxidation of NADH coupled to ATP synthesis in eukaryotes) contain highly similar elements, of which one (region D) has not yet been found in any other enzyme. Probably, part of the electron transport route within these proteins is highly conserved.

Unfortunately, HydC from *D. vulgaris* has not yet been purified, and the function of this protein and the conditions which induce its synthesis are not known.

### 4.1.2. The NiFe- and NiFeSe-Hydrogenases

NiFe- and NiFeSe-hydrogenases contain nickel and iron-sulfur clusters as redox-active metals. NiFe-hydrogenases have been isolated from a variety of microorganisms, including strict aerobes; NiFeSe-hydrogenases have been found only in a few *Desulfovibrio* strains: *D. baculatus*, *D. salexigens* and *D. vulgaris* Hildenborough (Rieder *et al.*, 1984, Teixeira *et al.*, 1986, Lissolo *et al.*, 1986). Most eubacterial NiFe- and NiFeSe-hydrogenases consist of two subunits, a large subunit of 50–65 kDa and a small subunit of 25–40 kDa; some contain additional subunits, *e.g.* the NAD- or F420-reducing hydrogenases. The active site is contained in the large subunit: "monomeric hydrogenases," consisting of only a large subunit with activity in the *in vitro* assay with viologen dyes, have been extracted (as indicated in the review by Przybyla *et al.* (1992), which clearly demonstrates the usefulness of a careful compilation of data available on primary structure, enzymology, and spectroscopy).

The selenium in the NiFeSe-hydrogenases is part of a recently discovered new amino acid, selenocysteine. The codon for selenocysteine is

TGA, normally a stop codon. Whether TGA is read as stop codon or codon for selenocysteine is determined by the structure of the environing DNA. In *E. coli*, the *sel* operon encodes proteins involved in selenosysteine biosynthesis (Sawers *et al.*, 1991). Selenocysteine replaces in these hydrogenases a cysteine residue which is conserved in the NiFe-hydrogenases without selenium.

*In vitro* activities of NiFe- and NiFeSe-hydrogenases in hydrogen uptake or production with artificial electron donors/acceptors are low compared to those of the Fe-hydrogenases: 440 and 470 U/mg in the $H_2$-production assay and 1500 and 120 U/mg in the $H_2$-consumption reaction for the *D. gigas* NiFe-hydrogenase and the *D. baculatus* NiFeSe-hydrogenase, respectively (Prickril *et al.*, 1987).

In all nickel containing hydrogenases, nickel in the large subunit is the site for $H_2$ activation. NiFe-hydrogenases occur in one of three states with distinct spectroscopy of the nickel: resting or "unready," "ready," and "active." Determination of the formal valence of nickel in the different states of the enzyme proved extremely difficult, and considerable uncertainty still exists (van der Zwaan *et al.*, 1985). Aerobic isolation yields inactive or unready enzyme with nickel as Ni(III). Ni(III) in unready enzyme can be titrated to Ni(II), with midpoint potentials $-200-$ $-50$ mV at pH 7 for most NiFe-hydrogenases. Unready hydrogenases are converted to the active state by incubation under $H_2$. This reductive activation is slow and strongly dependent on pH and temperature. Evidence has been presented for Ni(II)/Ni(I) and Ni(I)/'Ni(0)' redox transitions in active *Methanobacterium* enzyme with midpoint potentials of $-235$ and $-270$ mV, respectively (Coremans, 1991). $H_2$-production in the active site may proceed by reduction of a proton to a hydride at the nickel, followed by extraction of a proton. Anaerobic oxidation of active enzyme at very high redox potentials yields the ready state. In the ready state, nickel is Ni(III) like in the unready enzyme, but with a different EPR spectrum; activation of ready hydrogenase with $H_2$ is fast (Fernandez *et al.*, 1986). Ready hydrogenase can convert to the unready state, spontaneously or after exposure to oxygen. This conversion apparently does not involve a redox transition, but may be caused by a change in the protein conformation and/or nickel coordination (Coremans *et al.*, 1989).

EXAFS studies indicated, that nickel is penta- or hexacoordinated and has a mixture of sulfur and nitrogen and/or oxygen ligands. In NiFeSe-hydrogenases selenium is a ligand for nickel (Eidsness *et al.*, 1989, He *et al.*, 1989a). Comparison of aminoacid sequences of 17 large subunits, from *Desulfovibrios* and other bacteria, showed that, although the homology between subunits from different species may be as low as

30%, they all have two strictly conserved regions, one N-terminal (Arg-Xaa-Cys-(Xaa)$_2$-Cys), the other C-terminal (Asp-Pro-Cys-(Xaa)$_2$-Cys), the only variation being replacement of the first Cys in the C-terminal conserved region by selenocysteine in NiFeSe-hydrogenases (Przybyla *et al.*, 1992). These authors replaced six of these seven conserved residues in *E. coli* hydrogenase. Each mutation resulted in complete loss of hydrogenase activity. They suggested the four cysteines, arginine and aspartic acid as ligands for the nickel. However, ligation of metals to arginine is questionable and without precedent.

Some NiFe-hydrogenases have a [3Fe-4S] cluster, probably in the large subunit, such as the hydrogenase of *D. gigas*. This cluster is not required for hydrogenase activity (not even with cytochrome $c_3$ as electron donor) and does not interact with the nickel (Teixeira *et al.*, 1989).

Finally, NiFe-hydrogenases have two interacting [4Fe-4S] clusters; Mössbauer experiments suggest the irons to be coordinated by cysteines. Hydrogenases that can be isolated in a monomeric form do not have these clusters, suggesting that they are localized in the small subunit. All small subunit sequences have 10 conserved cysteines, although not in the typical motifs known from ferredoxins with two [4Fe-4S] clusters (Tran-Betcke *et al.*, 1990). Probably, these clusters are involved in electron transport between the active site nickel and the physiological electron carriers, but are not strictly required with artificial acceptors/donors.

### 4.1.3. Biosynthesis of Hydrogenases

The first indications that biosynthesis of active hydrogenase requires specific factors that are not present in every bacterial host came from the work of Voordouw *et al.* (1987), who used *E. coli* as a host for production of *D. vulgaris* Fe-hydrogenase from the cloned genes. The hydrogenase αβ dimer produced in *E. coli* was purified and was shown to contain only the two ferredoxin-like [4Fe-4S] clusters, but not the active-site cluster. Moreover, only a minor part of the subunits produced in *E. coli* was exported to the periplasm and assembled into the dimer. The majority of the small subunits accumulated as precursor molecules associated with the membrane, while most of the large subunits remained in the cytoplasm. If one of the subunits was not present in *E. coli*, translocation of the other subunit was entirely blocked, indicating that translocation might be coupled to assembly of the subunits. Translocation of the subunits across the membrane probably requires a specific system that is also involved in assembly of the enzyme and that is not or in insufficient amounts provided for by the *E. coli* host (van Dongen *et al.*, 1988).

Proteins to be secreted to the periplasmic space are generally made as precursors with a presequence of approximately 25 amino acids. These leader sequences have N-terminally a few positively charged residues, followed by a stretch of hydrophobic residues that form a transmembrane helix during translocation; after translocation leaders are removed with a specific signal peptidase (von Heijne, 1985). Comparison of the N-terminal aminoacid sequences of the hydrogenase subunits with those derived from the nucleotide sequences of the genes indicated that only the small (β), but not the large (α) subunit is synthesized with a leader sequence (Prickril *et al.*, 1986; van Dongen *et al.*, 1988). Moreover, the β-subunit leader has characteristics that differ from those of common bacterial leaders: it is much longer, has an unusual short hydrophobic core, and has positively charged residues throughout large part of the sequence (Fig. 2). Subsequently, it was found that also NiFe-hydrogenases have leader sequences only at the small subunits that are similar to that of the Fe-hydrogenase small subunit. These leaders all contain the consensus sequence Arg-Arg-Xaa-Phe-Xaa-Lys (Voordouw *et al.*, 1989b) that may serve as signal sequence for recognition by a specific translocation factor.

It was concluded that for expression of active hydrogenase, besides the subunits, specific factors for export across the cytoplasmic membrane, assembly, and insertion of the active-site metal cluster are necessary (Voordouw *et al.*, 1987; van Dongen *et al.*, 1988). Apparently, these factors were not or in insufficient amounts available in the *E. coli* host and it was suggested that biosynthesis of active hydrogenase requires a set of helper proteins.

Fe-hydrogenases have only been found in bacteria which are not yet very accessible for the genetic techniques required to identify genes for such helper proteins (*Desulfovibrio*, *Clostridium* and *Megasphaera elsdenii*). However, such genes have subsequently been found for NiFe-hydrogenases in various bacteria, although not yet in *Desulfovibrios* (Przybyla *et al.*, 1992). In most species, but apparently not in *Desulfovibrios*, genes for the structural subunits and helper genes are clustered in the genome. Analysis of the genes in these clusters and the function they have in biosynthesis, proceeds most rapidly in *E. coli* and *Alcaligenes eutrophus*. These species have three and two different NiFe-hydrogenases, respectively. In both organisms, genes for the structural subunits of each of the hydrogenases are clustered with a set of 4–7 helper genes (only two helper genes have been found for the soluble hydrogenase of *Alcaligenes*), which are required for biosynthesis of the

| Hydrogenase | Type | Localization | Leader β-subunit |
|---|---|---|---|
| | | | -50        -40 |
| A. D.baculatus a) | NiFeSe | Periplasm | MSLS |
| B. D.gigas b) | NiFe | Periplasm | MKCIYGRGKDQVEERLERR-GVS |
| C. D.vulgaris (My) c) | NiFe | Inner membrane | MKISIGLGKEGVEERLAER-GVS |
| D. E.coli d) | NiFe | Inner membrane | MNNEETFYQAMRRQGVT |
| E. Br.japonicum e) | NiFe | Inner membrane | MGAATETFYSVIRRQGIT |
| F. Rh.capsulatus f) | NiFe | Inner membrane | MQLSDIETFYDVMRRQGIT |
| G. A.chroococcum g) | NiFe | Inner membrane | MRRQGIT |
| H. D.vulgaris (M) h) | Fe | Periplasm ?? | MQIVNLT |
| I. D.vulgaris (H) i) | Fe | Periplasm | MQIASIT |
| **Consensus** | | | **-----------mrr-g-t** |
| | | | |
| E. coli ß-lactamase j) | | Periplasm | MSIQHFRVALIPFFAAFCLPVFA |

```
                    -30          -20          -10          -1  S +1     +10
D. baculatus     RREFVKLSC-AGVAGLGISQIYH-PGIVHA  /  MTEGAKKAPVIWVQ..
D. gigas         RRDFMK-FCTAVAVAMGMGPAFA-PKVAEA  /  LT--AKKRPSVVYL..
D. vulgaris (My) RRDFLK-FCTAIAVTMGMGPAFA-PEVARA  /  LMG--PRRPSVVYL..
E. coli          RRSFLK-YCSLAATSLGLGAGMA-PKIAWA  /  LEN--KPRIPVVWI..
Br. japonicum    RRSFHK-FCSLTATSLGLGPSLAASRIANA  /  LET--KPRVPVIWM..
Rh. capsulatus   RRSFMKSVRSPQHV-LGLGPSFV-PKIGEA  /  MET--KPRTPVVWV..
A. chroococcum   RRSFLK-YCSLTGRP-GLGPTFA-PQIAHA  /  MET--RPPTPVVWL..
D. vulgaris (M)  RRGFLKAACVVTAAALISIRMT--GKAVA   /  AAKQLKDYMDRIN..
D. vulgaris (H)  RRGFLKVACVTTGAALIGIRMT---GKAVA  /  AVKQIKDYMLDRIN..
Consensus        RR-Flk--c------lg-gp--a-p-ia-A  /

E. coli ß-lactamase      MSIQHFRVALIPFFAAFCLPVFA  /  HPE..
```

**Figure 2.** Leader sequences of the small subunits of Fe- and NiFe-hydrogenases. Residues conserved in at least 50% of the sequences are shown bold and given in the consensus sequence. Invariant residues are shown in capitals in the consensus. 'S' is the splice site. a): Menon et al., 1987; Voordouw et al., 1989a; b): Li et al., 1987; c): My = Myazaki; Deckers et al., 1990; d): Menon et al., 1990; e): Sayavedra-Soto et al., 1988; f): Leclerc et al., 1988; g): Ford et al., 1990; h): M = Monticello; Voordouw et al., 1989b; i): H = Hildenborough; Voordouw and Brenner, 1985; Prickril et al., 1986; j): E. coli β-lactamase leader peptide as a typical bacterial leader, Sjöström et al., 1987.

specific hydrogenase encoded in the same cluster (Friedrich, 1990; Eberz and Friedrich, 1991; Menon *et al.*, 1991; Sauter *et al.*, 1992; Przybyla *et al.*, 1992). In *E. coli*, each of the three sets of helper genes is remarkably dissimilar to the other two sets. On the other hand, the set of helper genes in the *E. coli* hydrogenase-1 gene cluster is similar to those found in hydrogenase gene clusters in several other bacteria (for example, encoding the membrane-bound *Alcaligenes* hydrogenase). By analysis of mutants in which specific helper genes have been deleted, the function of each of these genes is being elucidated now. Some genes encode proteins involved in regulation of expression, other genes encode electron carrier proteins, that are probably *in vivo* involved in electron transport to the enzymes. In *Wolinella succinogenes,* one of these genes was shown to encode a *b*-type cytochrome which couples $H_2$ oxidation to reduction of quinone in the respiratory chain. After careful extraction of hydrogenase from the membranes with Triton, this cytochrome was present as a third subunit of the enzyme (Dross *et al.*, 1992). This gene had been found before in hydrogenase gene clusters from other species; it had been indicated as encoding the membrane anchor. One of the proteins encoded by the *E. coli* hydrogenase-1 gene cluster (hyaF) enhances nickel incorporation into the enzyme; deletion of another gene (*hyaD*) resulted in the accumulation of unprocessed subunits in the membrane.

Evidently, only the large subunit of *E. coli* hydrogenase-1: as monomeric enzyme, containing only the large subunit, can be extracted and is still active with artificial electron donors (*in vivo,* the situation may be different). However, during synthesis of hydrogenase-1, the small subunit has to be present for localizing the large subunit at the membrane, and processing and activating it; only after biosynthesis, the subunits might be separated to give active monomeric enzyme. On the other hand, similar to the Fe-hydrogenase, processing of the small subunit (removal of the leader sequence) occurs only when the large subunit is present. In the absence of nickel, or after deletion of the presumed nickel binding site, the large subunit did not localize at the membrane, while the small subunit accumulated at the membrane as a precursor (Menon *et al.*, 1991). Apparently, nickel and the small subunit are required for localization of the large subunit at the membrane, and the small subunit, *hyaD,* and probably, *hyaE* for processing and activation at the membrane.

Besides the helper genes in the different hydrogenase gene clusters that are only involved in biosynthesis of the specific hydrogenase encoded by the respective cluster, both *E. coli* and *Alcaligenes* have a homologous set of genes which have a pleiotropic effect on the biosynthesis of all hydrogenases in the respective species (Lutz *et al.*, 1991; Friedrich, 1991). This cluster contains, besides a regulatory gene, a gene required for nickel sequestration and transport (*hypB, hoxN*, Eberz *et al.*, 1989) and two genes that affect the size of the large subunits (Lutz *et al.*, 1991). Mutation of the latter two genes results in a higher apparent molecular mass of the large subunits on SDS-gels. The large subunits have no N-terminal leader sequence which can be removed, but for *A. vinelandii* NiFe-hydrogenase evidence has been presented by mass spectral analysis for removal of 14 C-terminal residues during biosynthesis (Gollin *et al.*, 1992). It has been proposed that this C-terminal fragment is required to keep the enzyme in a conformation that allows nickel incorporation, and that its removal allows assembly of the Ni-site.

Although similar sets of helper genes for hydrogenase biosynthesis have not yet been isolated from *Desulfovibrio* spp., there is no reason to believe that processes involved in localization and metal incorporation are very different from those in *E. coli* or *Alcaligenes*.

### 4.1.4. Function of Hydrogenases

Most *Desulfovibrios* have more than one hydrogenase, but the distribution and localization of the various types differs from one species to another: *D. vulgaris* Hildenborough has a periplasmic Fe-hydrogenase, membrane localized NiFe− and NiFeSe-hydrogenases; *D. baculatus* has, besides a membrane localized NiFe-hydrogenase, NiFeSe-hydrogenases in all three cell compartments; and *D. gigas* has a periplasmic and a cytoplasmic NiFe−, but no membrane localized hydrogenase (Lissolo *et al.*, 1986; Fauque *et al.*, 1988; Rohde *et al.*, 1990). The functions of each of the hydrogenases in the different species are by no means clear and are subject for continuing research. An uptake hydrogenase is required during growth of *Desulfovibrios* on $H_2$ as an energy source. Furthermore, production of $H_2$ occurs during growth on lactate or pyruvate as substrates; it has been suggested, that $H_2$ production might either be involved in removal of excess, reducing equivalents from the cell to control the redox state of internal electron carriers (Lupton *et al.*, 1984), or in a cycling mechanism for generation of proton-motive force and ATP production (Odom and Peck, 1981, 1984). In this latter model, a hydrogenase with the active site in the cytoplasm couples $H_2$ production to lactate or pyruvate oxidation; $H_2$ diffuses to the periplasm, where it is oxidized by a periplasmic hydrogenase, generating a proton gradient.

Van den Berg *et al.* (1991) and Rousset *et al.* (1991) attempted a genetic approach to obtain insight in the function of the periplasmic Fe-hydrogenase of *D. vulgaris* Hildenborough and the periplasmic NiFe-hydrogenase of *D. fructosovorans,* respectively. Van den Berg and coworkers were able to specifically suppress synthesis of the Fe-hydrogenase in *D. vulgaris* with a plasmid from which hydrogenase antisense RNA was produced, resulting in a 2–3 fold decrease in the cellular hydrogenase activity. This rather slight decrease in activity resulted in 2–3 fold lower growth rates of the mutant strains in lactate medium; furthermore, the mutants accumulated much less $H_2$ in the culture headspace than did the wild-type strain. Although these experiments indicated an important role for the Fe-hydrogenase for growth of *D. vulgaris* during the metabolization of lactate, and suggested a role in $H_2$ production, they gave no direct clue for the process in which the enzyme is involved. Recent experiments indicate that the amount of Fe-hydrogenase in *D. vulgaris* cells is variable and increases when the amount of soluble iron in a culture becomes low (Bryant *et al.,* 1993). This suggests a role in iron metabolism, probably acquisition.

Although a highly active periplasmic Fe-hydrogenase is very important for *D. vulgaris* Hildenborough (and accumulates in high amounts in this strain, to approximately 1% of the soluble protein), this is apparently not the case for every *Desulfovibrio* strain. Out of 25 *Desulfovibrio* strains tested, 13 do not have genes for a Fe-hydrogenase (Voordouw *et al.,* 1990a), and one of the strains that has the genes (*D. vulgaris* subspecies *oxamicus* Monticello) does not produce detectable amounts of the enzyme (Stokkermans, unpublished results).

Rousset and colleagues deleted the genes for the periplasmic NiFe-hydrogenase of *D. fructosovorans* by marker exchange. The deletion resulted in a 90% decreased hydrogenase activity (the remaining 10% accounted for the activity of a Fe-hydrogenase). Surprisingly, the mutant strain had normal growth rates on several organic compounds or $H_2$ as substrates, and in co-culture with a $H_2$-producing bacterium, and produced sufficient $H_2$ from lactate to support normal growth of a methanogen. The only physiological effect was an extended lag period of the mutant during growth with $H_2$ (Rousset *et al,*. 1991; Rousset, 1991).

Apparently, these two periplasmic hydrogenases have a different physiological function in *D. vulgaris* and *D. fructosovorans.* Elimination of all hydrogenase genes in a particular strain, alone and in combination, may be attempted to determine the role of each of the enzymes, but may not be successful if a vital function is affected.

### 4.2. Proteins with Novel Fe-S Clusters: Prismane Protein and Dissimilatory Sulfite Reductase

Although direct evidence for a catalytic [6Fe-6S] cluster in the Fe-hydrogenase of *D. vulgaris* is not yet available, proof for this cluster-type has been obtained in the 'prismane protein' from the same species, first described by Hagen *et al.* (1989). This protein, for which no function has been determined yet, contains only 6–7 irons and acid-labile sulfurs (Pierik *et al.*, 1992a; Stokkermans, 1992b). In the "as-isolated" state, the protein has signals in the EPR spectrum at an unusually low magnetic field strength. Applying principles of quantum mechanics, these resonances were ascribed to irons in $S = 9/2$ spin–system (Hagen *et al.*, 1989; Pierik *et al.*, 1992b). Such "superspins" have never been found in proteins with [2Fe-2S] or [4Fe-4S] clusters. Moreover the EPR-spectrum of the dithionite-reduced protein was similar to those of anorganic model compounds with a [6Fe-6S] prismane core (Pierik *et al.*, 1992b). Based on these observations, a novel type of Fe-S cluster, [6Fe-6S], was proposed for this protein. This Fe-S cluster could be titrated into four different redox states, with putative formal charges of 6+ (*i.e.* all irons oxidized), 5+ (the 'as isolated' state), 4+ and 3+ and midpoint potentials of the transitions of +285 mV (6+ → 5+), −5 mV (5+ → 4+) and −165 mV (4+ → 3+). These observations suggested that, contrary to [4Fe-4S] or [2Fe-2S] clusters, which normally serve as one-electron donors/acceptors in electron transport, this novel cluster can accept/donate multiple electrons at different potentials. It might be a prototype of a active-site Fe-cluster in enzymes catalyzing redox reactions involving multiple-electron transfer.

Indeed, subsequent to the discovery and characterization of the prismane protein, other proteins with established enzymatic activity in redox reactions were also shown to exhibit EPR spectra consistent with iron in a $S = 9/2$ spin system: carbon monoxide dehydrogenase from the methanogen *Methanotrix soehngenii* and dissimilatory sulfite reductase (desulfoviridin), also from *D. vulgaris* (Jetten *et al.*, 1991; Pierik and Hagen, 1991). Although analysis of the clusters in these enzymes is more complicated than in the prismane protein (they contain 20–30 Fe-atoms, arranged in multiple, different Fe-S clusters and heme-iron), the "superspins" in these proteins were also believed to derive from Fe-S clusters that have a more complex composition than the [4Fe-4S] cubane. For the Fe-hydrogenase of *D. vulgaris,* which was assumed to contain a [6Fe-6S] active-site cluster from the chemically determined amount of iron and sulfur, spectroscopic data consistent with $S = 9/2$ system are not yet available.

Mössbauer spectroscopy of the prismane protein in the different redox states suggested, that not all irons in the cluster are equivalent: upon reduction from the 5+ to the 3+-state, the extra electrons appeared to be localized primarily at two of the six irons, supposedly at opposite ends of the cluster (Pierik *et al.*, 1992b). It has been proposed that four 'core' irons are coordinated by cysteines and the two irons at the opposite ends by a more polar group, for example histidine (Pierik *et al.*, 1992b).

There is evidence that prismane protein occurs in several sulfate reducers (Stokkermans *et al.*, 1992a) and has been purified from *D. desulfuricans* (Moura *et al.*, 1992). Genes for the *D. vulgaris* and *D. desulfuricans* prismane proteins have been cloned and sequenced. Directly at the N-terminus they have a cysteine-rich region (Cys-(Xaa)$_2$-Cys-(Xaa)$_8$-Cys-(Xaa)$_3$-Gly-Xaa-Cys-Gly) that is partially conserved in the *M. soehngenii* carbon monoxide dehydrogenase (Cys-(Xaa)$_2$-Cys-(Xaa)$_4$-Cys-(Xaa)$_7$-Gly-Xaa-Cys-Gly) (Stokkermans *et al.*, 1992a, c), but is not found in the Fe-hydrogenase. The 4 cysteines in this cluster might ligate the four core-irons of this novel cluster. Experiments to assign with site-directed mutagenesis the residues involved in coordination of the Fe-S cluster in prismane protein, and eventually to modify the redox properties, are in progress. As this unusual Fe-S cluster is not made in *E. coli*, a highly efficient expression system for this protein in *Desulfovibrio* has been developed (Stokkermans *et al.*, 1992b; van den Berg *et al.*, 1994).

### 4.3. Cytochromes

*Desulfovibrio* species contain several types of *c*-type cytochromes that function as electron carriers. Four have been isolated from *D. vulgaris:* the monoheme cytochrome $c_{553}$ (9 kDa), tetraheme cytochrome $c_3$ (12 kDa), octaheme cytochrome $cc_3$ (26 kDa) and a high-molecular-weight cytochrome *c* (HMC) with 16 hemes (75 kDa) (Pollock *et al.*, 1991). All four cytochromes are found in the periplasm. There is no direct evidence for the metabolic routes they are involved in, but cytochrome $c_3$ is an efficient electron donor/acceptor in *in vitro* experiments with several enzymes, among these the Fe-hydrogenase. Kinetic experiments have shown that *in vitro* cytochrome $c_{553}$ can efficiently replace cytochrome $c_3$ as electron donor for the Fe-hydrogenase (Verhagen *et al.*, 1994). The gene for HMC is found in an operon containing several genes encoding (trans)membrane proteins with considerable similarity to subunits of NADH dehydrogenase or cytochrome *c* reductase; some of these have binding motifs for ferredoxin-like [4Fe-4S] clusters. It has been proposed that these proteins form a transmembrane complex for electron transfer between the periplasmic hydrogenase and the cytoplasmic enzymes involved in sulfate reduction (Rossi *et al.*, 1993).

The primary structures of the small cytochromes $c_{553}$ and $c_3$ from several *Desulfovibrio* species have been determined by peptide sequencing and nucleotide sequencing of the genes; the 3D-structures of cytochromes $c_3$ from *D. baculatus, D. vulgaris* and *D. gigas* have been established by x-ray-diffraction (Pierrot *et al.*, 1982; Higuchi *et al.*, 1984; Matias *et al.*, 1993). Overall structure and the arrangement of the hemes in the *D. vulgaris* and *D. baculatus* proteins is conserved, although they have less than 25% conserved aminoacids (Coutinho *et al.*, 1992; Turner *et al.*, 1992). Also the gene for *D. vulgaris* HMC has been isolated and sequenced (Pollock *et al.*, 1991). These authors convincingly demonstrated, by comparing the derived aminoacid sequence with the partial sequence of cytochrome $cc_3$, determined by peptide sequencing, that HMC and cytochrome $cc_3$ derive from a single gene. The octaheme cytochrome probably arises *in vivo* by limited proteolysis of the hexadeka heme HMC.

The hemes in the *c*-type cytochromes are all covalently linked to the protein by two cysteinyl residues which occur in the primary structure of the protein in the typical motifs (Cys-(Xaa)$_2$-Cys-His) and (Cys-(Xaa)$_4$-Cys-His). Both motifs occur twice in tetraheme cytochrome $c_3$ (Voordouw and Brenner, 1986). The histidine is the fifth ligand for the heme iron; the sixth ligand is either a methionine (in cytochrome $c_{553}$) or another histidine (all four hemes in cytochrome $c_3$, Pierrot *et al.*, 1982; Higuchi *et al.*, 1984). HMC has 16 (Cys-(Xaa)$_2$-Cys-His) heme binding motifs. Pollock *et al.* (1991) distinguished four regions in the primary structure of HMC, which may form separate domains in the 3D structure (Fig. 3). Three of the four regions each contain four heme-binding motifs (regions B, C, and D). Each of these three regions contains four additional histidines at similar positions and spacing as the histidines that form the sixth ligands of iron in cytochrome $c_3$. The fourth region, A, contains only three heme-binding motifs and three additional histi-

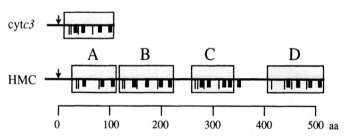

**Figure 3.** Schematic representation of the arrangement of histidines and Cys-Xaa$_{2/4}$-Cys-His heme-binding motifs in the primary structure of *D. vulgaris* tetraheme cytochrome $c_3$ (cyt$c_3$) and high-molecular-weight hexadeka heme cytochrome (HMC) (Pollock *et al.*, 1991). Heme binding motifs are represented by black boxes, the remaining histidines by downward ticks. The three complete (B, C, D) and one incomplete (A) cytochrome $c_3$-like regions in HMC are shaded. Arrows indicate leader peptidase splicing sites.

dines. The sixteenth motif is located outside one of these four regions. As HMC contains only 31 histidines, the iron in this heme probably does not have a histidine as the sixth ligand. EPR spectra indicate that 14–15 of the heme-irons are low-spin (as found in other bis(histidinyl) coordinated hemes), and 1–2 are high-spin (Bruschi et al., 1992). These authors suggested that the sixth Fe-ligand position of the sixteenth heme may be either a methionine or vacant.

The axial coordination of heme-iron has a distinct effect on the midpoint potential of the redox transition. The midpoint potential of the histidinyl-methionyl coordinated iron in the heme of cytochrome $c_{553}$ is −50 mV. Midpoint potentials of (bis)histidinyl coordinated heme irons are lower. The four hemes in cytochrome $c_3$ have different potentials, ranging between −200 and −350 mV for the D. gigas protein. Reduction of the hemes in cytochrome $c_3$ is cooperative, the midpoint potential of each of the hemes being dependent on the oxidation state of each of the others (Santos et al., 1984). Furthermore, midpoint potentials are pH-dependent, suggesting proton-linked conformational changes in the protein (Catarino et al., 1991). A thermodynamic model, accounting for both the heme-heme interactions and a pH-dependent equilibrium between two conformations, has been developed to describe the complex cooperative redox behavior (Coletta et al., 1991). 2D NMR is currently applied to study at the level of the protein structure the interactions between the closely spaced hemes (minimal distances less than 4Å).

The influence of the axial coordination of the four heme irons in D. vulgaris cytochrome $c_3$ on the midpoint potential of reduction was recently investigated with site-directed mutagenesis (Mus-Veteau et al., 1992; Dolla et al. 1994). Four single mutant proteins were constructed, each containing a methionine instead of a histidine at the sixth axial ligand position of one of the four heme-irons. [As E. coli does not incorporate hemes in cytochrome $c_3$ and accumulates the apoprotein, mainly as precursor (Pollock et al., 1989), the mutant proteins were produced in D. desulfuricans G200, resulting in a 3–12-fold overproduction of the cytochrome with the full complement of hemes (Voordouw et al., 1990; Bruschi et al., 1992)]. Replacement of the histidines ligated to the iron of hemes 3 and 4 resulted in an increase of the midpoint potential of the respective heme-iron to −90 and −65 mV, respectively, i.e., comparable to the midpoint potential of the histidinyl–methionyl coordinated iron in cytochrome $c_{553}$. Surprisingly, no such increase in midpoint potential was found when the histidine ligands of hemes 1 and 2 were mutated into methionines; furthermore, the replacing methionines in these two mutated proteins were oxidized to methionine sulfone (Dolla et al., 1994).

In HMC, midpoint potentials of 0, $-100$ and $-250$ mV have been found with cyclic voltammetry (Bruschi *et al.*, 1992). Although values obtained with cyclic voltammetry have to be considered with caution and no independent determination by EPR redox-titration was performed, the value of $-250$ mV is in the range found for the bis(histidinyl) coordinated hemes in cytochromes $c_3$.

HMC studies demonstrate how comparative analysis of the primary structure of this complex cytochrome can be used to predict some of the properties. Nevertheless, to understand the redox behavior, it is too simple to consider HMC as merely composed of four cytochrome $c_3$ domains (three complete and one incomplete). Differences in the folding of HMC and cytochrome $c_3$ were suggested by differences in the Soret signals in circular dichroism spectra of both proteins (Bruschi *et al.*, 1992). The redox behavior of the four hemes in cytochrome $c_3$ demonstrates that redox properties are strongly influenced by small changes in the surrounding protein structure.

## 5. PERSPECTIVES

Above examples show how cloning of genes for metal proteins and comparative analysis of the aminoacid sequences has contributed, together with the enzymology and spectroscopy, to our understanding of this important class of proteins and the metals they contain. Using *Desulfovibrio* as a host for cloning allows production of these proteins after modification by site-directed mutagenesis. This may further increase our understanding about the influence by protein structure on redox behavior of the metals. The abundance of different proteins with a variety of metal centers, together with the now available genetic techniques, will make *Desulfovibrio* one of the organisms of choice for this type of research.

A more technologically related offset of *Desulfovibrio* genetics is the application in the Reversed Sample Genome Probing technique for the identification of *Desulfovibrios* in environmental samples (for example: from an oil field, where *Desulfovibrios* are a main source of corrosion of *e.g.* pipelines) or in microbial consortia (for example: anaerobic sewage treatment) by DNA hybridization techniques (Voordouw *et al,*. 1991).

Marker exchange or other techniques that allow the construction of designed modifications in the genome are not yet applied on a regular base. Development of such techniques requires the elimination of genes in the genome, which may contribute to the unraveling of metabolic routes (for example, it is not well known, in which routes the numerous electron carrier proteins are involved). Such techniques will also make stable integration of genes from different species in the genome possible, creating new properties in the bacterium. One of the properties of *Desulfovibrios* is their ability to precipitate heavy metals as metal sulfides. Unfortunately, *Desulfovibrios* are not particularly resistant against these metals. Incorporation of metal resistance genes in the genome might make them less sensitive and improve their applicability for this purpose.

## REFERENCES

Adams, M. W. W., 1990, The structure and mechanism of iron-hydrogenases, *Biochim. Biophys. Acta* **1020:**115–145.

Adman, E. T., Sieker, L. C., Jensen, L. H., Bruschi, M., and LeGall, J., 1977, A structural model of rubredoxin from *Desulfovibrio vulgaris* at 2 Å resolution, *J. Mol. Biol.* **112:**113–120.

Argyle, J. L., Rapp-Giles, B. J., and Wall, J. D., 1992, Plasmid transfer by conjugation in *Desulfovibrio desulfuricans, FEMS Microbiol. Lett* **94:**255–262.

Bakker, P. T. A., and Albracht, S. P. J., 1986, Evidence for two independent pathways of electron transport in mitochondrial NADH:Q oxidoreductase, *Biochim. Biophys. Acta* **850:**413–422.

Bruschi, M., and Guerlesquin, F., 1988, Structure, function and evolution of bacterial ferredoxins, *FEMS Microbiol. Rev.* **54:**155–176.

Bruschi, M., Bertrand, P., More, C., Leroy, G., Bonicel, J., Haladjian, J., Chottard, G., Pollock, W. B. R., and Voordouw, G., 1992, Biochemical and spectroscopic characterization of the high molecular weight cytochrome *c* from *Desulfovibrio vulgaris* Hildenborough expressed in *Desulfovibrio desulfuricans* G200, *Biochemistry* **31:**3281–3288.

Bryant, R. D., van Ommen Kloeke, F., and Laishley, E. J., 1993, Regulation of the periplasmic [Fe] hydrogenase by ferrous iron in *Desulfovibrio vulgaris* (Hildenborough), *Appl. Environ. Microbiol.* **59:**491–495.

Catarino, T., Coletta, M., LeGall, J., and Xavier, A. V., 1991, Kinetic study of the reduction mechanism of *Desulfovibrio gigas* cytochrome $c_3$, *Eur. J. Biochem.* **202:**1107–1113.

Coletta, M., Catarino, T., LeGall, J., and Xavier, A. V., 1991, A thermodynamic model for the cooperative functional properties of the tetraheme cytochrome $c_3$ from *Desulfovibrio gigas, Eur. J. Biochem.* **202:**1101–1106.

Coremans, J. M. C. C., 1991, *Redox properties of hydrogenase,*Academic Thesis, University of Amsterdam, The Netherlands.

Coremans, J. M. C. C., van der Zwaan, J. W., and Albracht, S. P. J., 1989, Redox behavior of nickel in hydrogenase from *Methanobacterium thermoautotrophicum* (strain Marburg), *Biochim. Biophys. Acta* **997:**256–267.

Coutinho, I. B., Turner, D. L., LeGall, J., and Xavier, A. V., 1992, Revision of the heme-core architecture in the tetraheme cytochrome $c_3$ from *Desulfovibrio baculatus* by two-dimensional $^1$H NMR, *Eur. J. Biochem.* **209:**329–333.

Davison, J., Heuterspreute, M., Chevalier, N., Vinh, H. T., and Brunel, F., 1987, Vectors with restriction site banks. V. pJRD215, a wide-host range cosmid vector with multiple cloning sites, *Gene* **51:**275–280.

Deckers, H. M., and Voordouw, G., 1994, Identification of a large family of genes for putative chemoreceptor proteins in an ordered library of the *Desulfovibrio vulgaris* Hildenborough genome, *J. Bacteriol.* **176:**351–358.

Deckers, H. M., Wilson, F. R., and Voordouw, G., 1990, Cloning and sequencing of a [NiFe] hydrogenase operon from *Desulfovibrio vulgaris* Miyazaki F, *J. Gen. Microbiol.* **136:**2021–2028.

Devereux, R., Delaney, M., Widdel, F., and Stahl, D. A., 1989, Natural relationships among sulfate-reducing eubacteria, *J. Bacteriol.* **171:**6689–6695.

Dolla, A., Florens, L., Bianco, P., Haladjian, J., Voordouw, G., Forest, E., Wall, J., Guerlesquin, F., and Bruschi, M., 1994, Characterization and oxidoreduction properties of cytochrome $c_3$ after heme axial ligand replacements, *J. Biol. Chem.* **269:**6340–6346.

Dross, F., Geisler, V., Lenger, R., Theis, F., Krafft, T., Fahrenholz, F., Duchene, A., Tripier, D., Juvenal, K., and Kröger, A., 1992, The quinone-reactive Ni/Fe-hydrogenase of *Wolinella succinogenes*, *Eur. J. Biochem.* **106:**93–102.

Eberz, G., and Friedrich, B., 1991, Three *trans*-acting regulatory functions control hydrogenase synthesis in *Alcaligenes eutrophus*, *J. Bacteriol.* **173:**1845–1854.

Eberz, G., Eitinger, T., and Friedrich, B., 1989, Genetic determinants of a nickel-specific transport system are part of the plasmid-encoded hydrogenase gene cluster in *Alcaligenes eutrophus*, *J. Bacteriol.* **171:**1340–1345.

Eidsness, M. K., Scott, R. A., Prickril, B., DerVartanian, D. V., LeGall, J., Moura, I., Moura, J. J. G., and Peck Jr., H. D., 1989, Evidence for selenocysteine coordination to the active site nickel in the (NiFeSe) hydrogenase from *Desulfovibrio baculatus*, *Proc. Natl. Acad. Sci. USA* **86:**147–151.

Fauque, G., Peck Jr., H. D., Moura, J. J. G., Huynh, B.-H., Berlier, Y., DerVartanian, D. V., Teixeira, M., Przybyla, A. E., Lespinat, P. A., Moura, I., and LeGall, J., 1988, The three classes of hydrogenases from sulfate-reducing bacteria of the genus *Desulfovibrio*, *FEMS Microbiol. Rev.* **54:**299–344.

Fernandez, V. M., Hatchikian, E. C., Patil, D. S., and Cammack, R., 1986, ESR detectable nickel and iron-sulfur centers in relation to the reversible activation of *Desulfovibrio gigas* hydrogenase, *Biochim. Biophys. Acta* **883:**145–154.

Ford, C. M., Garg, N., Garg, R. P., Tibelius, K. H., Yates, M. G., Arp, D. J., and Seefeldt, L. C., 1990, The identification, characterization, sequencing, and mutagenesis of the genes (*hupSL*) encoding the small and large subunits of the $H_2$-uptake hydrogenase of *Azotobacter chroococcum*, *Mol. Microbiol.* **4:**999–1008.

Friedrich, B., 1990, The plasmid-encoded hydrogenase gene cluster in *Alcaligenes eutrophus*, *FEMS Microbiol. Rev.* **87:**425–430.

Friedrich, B., 1991, Genes involved in the formation of catalytically active Ni-containing hydrogenase of *Alcaligenes eutrophus*, In: *Abstracts on the 3rd International Conference on the Molecular Biology of Hydrogenases*, Troia, Portugal, pp. 113–116.

Gollin, D. J., Mortenson, L. E., and Robson, R. L., 1992, Carboxyl-terminal processing may be essential for production of active NiFe hydrogenase in *Azotobacter vinelandii*, *FEBS Lett.* **309:**371–375.

Grande, H. J., Dunham, W. R., Averill, B., van Dijk, C., and Sands, R. H., 1983, Electron paramagnetic resonance and other properties of hydrogenases isolated from *Desulfovibrio vulgaris* (strain Hildenborough) and *Megasphaera elsdenii*, *Eur. J. Biochem.* **136:**201–207.

Hagen, W. R., van Berkel-Arts, A., Krüse-Wolters, K. M., Dunham, W. R., and Veeger, C., 1986a, EPR of a novel high-spin component in activated hydrogenase from *Desulfovibrio vulgaris* (Hildenborough), *FEBS Lett.* **201:**158–162.

Hagen, W. R., van Berkel-Arts, A., Krüse-Wolters, K. M., Voordouw, G., and Veeger, C., 1986b, The iron-sulfur composition of the active site of hydrogenase from *Desulfovibrio vulgaris* (Hildenborough) deduced from its subunit structure and total iron-sulfur content, *FEBS Lett.* **203:**59–62.

Hagen, W. R., Pierik, A. J., and Veeger, C., 1989, Novel electron paramagnetic resonance signals from an Fe/S protein containing six iron atoms, *J. Chem. Soc. Faraday Trans.* I **85**:4083–4090.

Handley, J., Adams, V., and Akagi, J. M., 1973, Morphology of bacteriophage-like particles from *Desulfovibrio vulgaris, J. Bacteriol.* **115**:1205–1207.

Hatchikian, E. C., Forget, N., Fernandez, V. M., Williams, R., and Cammack, R., 1992, Further characterization of the [Fe]-hydrogenase from *Desulfovibrio desulfuricans* ATCC 7757, *Eur. J. Biochem.* **209**:357–365.

He, S.-H., Teixeira, M., LeGall, J., Patil, D. S., DerVartanian, D. V., Huynh, B. H., and Peck Jr., H. D., 1989a, EPR studies with $^{77}Se$ enriched (NiFeSe) hydrogenase of *Desulfovibrio baculatus.* Evidence for a selenium ligand to the active-site nickel, *J. Biol. Chem.* **264**:2678–2682.

He, S.-H., Woo, S. B., DerVartanian, D. V., LeGall, J., and Peck Jr., H. D., 1989b, Effects of acetylene on hydrogenases from the sulfate reducing and methanogenic bacteria, *Biochem. Biophys. Res. Comm.* **161**:127–133.

Higuchi, Y., Kusunoki, M., Matsuura, Y., Yasuoka, W., and Kakudo, M., 1984, Refined structure of cytochrome $c_3$ at 1.8 Å resolution, *J. Mol. Biol.* **172**:109–139.

Huynh, B.-H., Moura, J. J. G., Moura, I., Kent, T. A., LeGall, J., Xavier, A. V., and Munck, E., 1980, Evidence for a three-iron center in a ferredoxin from *Desulfovibrio gigas.* Mössbauer and EPR studies, *J. Biol. Chem.* **255**:3242–3244.

Huynh, B.-H., Czechowski, M. H., Krüger, H.-J., DerVartanian, D. V., Peck Jr., H. D., and LeGall, J., 1984, *Desulfovibrio vulgaris* hydrogenase: a non-heme iron enzyme lacking nickel that exhibits anomalous EPR and Mössbauer spectra, *Proc. Natl. Acad. Sci. USA* **81**:3728–3732.

Jetten, M. S. M., Pierik, A. J., and Hagen, W. R., 1991, EPR characterization of a high-spin system in carbon monoxide dehydrogenase from *Methanothrix soehngenii, Eur. J. Biochem.* **202**:1291–1297.

Kamimura, K., and Araki, M., 1989, Isolation and characterization of a bacteriophage lytic for *Desulfovibrio salexigens*, a salt-requiring, sulfate-reducing bacterium, *Appl. Environ. Microbiol.* **55**:645–648.

Krishnamoorthy, G., and Hinkle, P. C., 1988, Studies on the electron transfer pathway, topography of iron-sulfur centers, and site of coupling in NADH-Q oxidoreductase, *J. Biol. Chem.* **263**:17566–17575.

Leclerc, N., Colbeau, A., Cauvin, B., and Vignais, P. M., 1988, Cloning and sequencing of the genes encoding the large and small subunits of the $H_2$ uptake hydrogenase (*hup*) of *Rhodobacter capsulatus, Mol. Gen. Genet.* **214**:97–107.

Le Gall, J., Prickril, B. C., Moura, I., Xavier, A. V., Moura, J. J. G., and Huynh, B.-H., 1988, Isolation and characterization of rubrerythrin, a non-heme iron protein from *Desulfovibrio vulgaris* that contains rubredoxin centers and a hemerythrin-like binuclear iron cluster, *Biochemistry* **27**:1636–1642.

Li, C., Peck Jr., H. D., LeGall, J., and Przybyla, A. E., 1987, Cloning, characterization, and sequencing of the genes encoding the large and small subunits of the periplasmic [NiFe] hydrogenase of *Desulfovibrio gigas, DNA* **6**:539–551.

Lissolo, T., Choi, E.-S., LeGall, J., and Peck Jr., H. D., 1986, The presence of multiple intrinsic membrane nickel-containing hydrogenases in *Desulfovibrio vulgaris* (Hildenborough), *Biochem. Biophys. Res. Comm.* **139**:701–708.

Liu, M.-C., Costa, C., Coutinho, I. B., Moura, J. J. G., Moura, I., Xavier, A. V., and LeGall, J., 1988, Cytochrome components of nitrate- and sulfate-respiring *Desulfovibrio desulfuricans* ATCC 27774, *J. Bacteriol.* **170**:5545–5551.

Lupton, F. S., Conrad, R., and Zeikus, J. G., 1984, Physiological function of hydrogen metabolism during growth of sulfidogenic bacteria on organic substrates, *J. Bacteriol.* **159:**843–849.

Lutz, S., Jacobi, A., Schlensog, V., Böhm, R., Sawers, G., and Böck, A., 1991, Molecular characterization of an operon (*hyp*) necessary for the activity of the three hydrogenase isoenzymes in *Escherichia coli, Mol. Microbiol.* **5:**123–135.

Matias, P. M., Frazão, C., Morais, J., Coll, M., and Carrondo, M. A., 1993, Structure analysis of cytochrome $c_3$ from *Desulfovibrio vulgaris* Hildenborough at 1.9 Å resolution, *J. Mol. Biol.* **234:**680–699.

Menon, N. K., Peck Jr., H. D., LeGall, J., and Przybyla, A. E., 1987, Cloning and sequencing of the genes encoding the large and small subunits of the periplasmic [NiFeSe] hydrogenase of *Desulfovibrio baculatus, J. Bacteriol.* **169:**5401–5407 (Erratum, 1988, *J. Bacteriol.* **170:**4429).

Menon, N. K., Robbins, J., Peck Jr., H. D., Chatelus, C. Y., Choi, E.-S., and Przybyla, A. E., 1990, Cloning and sequencing of a putative *Escherichia coli* hydrogenase-1 operon containing six open reading frames, *J. Bacteriol.* **172:**1969–1977.

Menon, N. K., Robbins, J., Wendt, J. C., Shanmugam, K. T., and Przybyla, A. E., 1991, Mutational analysis and characterization of the *Escherichia coli hya* operon, which encodes [NiFe] hydrogenase 1, *J. Bacteriol.* **173:**4851–4861.

Meyer, J., and Gagnon, J., 1991, Primary structure of hydrogenase I from *Clostridium pasteurianum, Biochemistry* **30:**9697–9704.

Moura, I., Tavares, P., Moura, J. J. G., Ravi, N., Huynh, B.-H., Liu, M.-Y., and LeGall, J., 1992, Direct spectroscopic evidence for the presence of a 6Fe cluster in an iron-sulfur protein isolated from *Desulfovibrio desulfuricans* (ATCC 27774), *J. Biol. Chem.* **267:**4489–4496.

Mus-Veteau, I., Dolla, A., Guerlesquin, F., Payan, F., Czjzek, M., Haser, R., Bianco, P., Haladjian, J., Rapp-Giles, B. J., Wall, J. D., Voordouw, G., and Bruschi, M., 1992, Site-directed mutagenesis of tetraheme cytochrome $c_3$. Modification of oxidoreduction potentials after heme axial ligand replacement, *J. Biol. Chem.* **267:**16851–16858.

Odom, J. M., and Peck Jr., H. D., 1981, Hydrogen cycling as a general mechanism for energy-coupling in the sulfate reducing bacteria *Desulfovibrio* sp., *FEMS Microbiol. Lett.* **12:**47–50.

Odom, J. M., and Peck Jr., H. D., 1984, Hydrogenase, electron-transfer proteins, and energy coupling in the sulfate-reducing bacteria, *Desulfovibrio, Ann. Rev. Biochem.* **38:**551–592.

Peck Jr., H. D., and LeGall, J., 1982, Biochemistry of dissimilatory sulphate reduction, *Phil. Trans. R. Soc. Lond. B.* **298:**443–466.

Pierik, A. J., and Hagen, W. R., 1991, $S = 9/2$ EPR signals are evidence against coupling between siroheme and the Fe/S cluster prosthetic groups in *Desulfovibrio vulgaris* (Hildenborough) dissimilatory sulfite reductase, *Eur. J. Biochem.* **195:**505–516.

Pierik, A. J., Wolbert, R. B. G., Mutsaers, P. H. A., Hagen, W. R., and Veeger, C., 1992a, Purification and characterization of a putative [6Fe-6S] prismane-cluster-containing protein from *Desulfovibrio vulgaris* (Hildenborough), *Eur. J. Biochem.* **206:**697–704.

Pierik, A. J., Hagen, W. R., Dunham, W. R., and Sands, R. H., 1992b, Multi-frequency EPR and high-resolution Mössbauer spectroscopy of a putative [6Fe-6S] prismane-cluster-containing protein from *Desulfovibrio vulgaris* (Hildenborough): characterization of a supercluster and superspin model protein, *Eur. J. Biochem.* **206:**705–719.

Pierik, A. J., Hagen, W. R., Redeker, J. S., Wolbert, R. B. G., Boersma, M., Verhagen, M. F. J. M., Grande, H. J., Veeger, C., Mutsaers, P. H. A., Sands, R. H., and Dunham,

W. R., 1992c, Redox properties of the iron-sulfur clusters in activated Fe-hydrogenase from *Desulfovibrio vulgaris* (Hildenborough), *Eur. J. Biochem.* **209**:63–72.

Pierik, A. J., Wolbert, R. B. G., Portier, G. L., Verhagen, M. F. J. M., and Hagen, W. R., 1993, Nigerythrin and rubrerythrin from *Desulfovibrio vulgaris* each contain two mononuclear iron centers and two dinuclear iron clusters, *Eur. J. Biochem.* **212**:237–245.

Pierrot, M., Haser, R., Frey, M., Bruschi, M., LeGall, J., Sieker, L., and Jensen, L. H., 1976, Some comparisons between two crystallized anaerobic bacterial rubredoxins from *Desulfovibrio gigas* and *D. vulgaris, J. Mol. Biol.* **107**:179–182.

Pierrot, M., Haser, R., Frey, M., Payan, F., and Astier, J. P., 1982, Crystal structure and electron transfer properties of cytochrome $c_3$, *J. Biol. Chem.* **257**:14341–14348.

Pilkington, S. J., and Walker, J. E., 1989, Mitochondrial NADH-ubiquinone reductase: complementary DNA sequences of import precursors of the bovine and human 24 kDa subunits, *Biochemistry* **28**:3257–3264.

Pollock, W. B. R., Chemerika, P. J., Forrest, M. E., Beatty, J. T., and Voordouw, G., 1989, Expression of the gene encoding cytochrome $c_3$ from *Desulfovibrio vulgaris* (Hildenborough) in *Escherichia coli:* export and processing of the apoprotein, *J. Gen. Microbiol.* **135**:2319–2328.

Pollock, W. B. R., Loutfi, M., Bruschi, M., Rapp-Giles, B. J., Wall, J. D., and Voordouw, G., 1991, Cloning, sequencing and expression of the gene encoding the high-molecular-weight cytochrome *c* from *Desulfovibrio vulgaris* Hildenborough, *J. Bacteriol.* **173**:220–228.

Postgate, J. R., Kent, H. M., Robson, R. L., and Chesshyre, J. A., 1984, The genomes of *Desulfovibrio gigas* and *D. vulgaris, J. Gen. Microbiol.* **130**:1597–1601.

Postgate, J. R., Kent, H. M., and Robson, R. L., 1988, Nitrogen fixation by *Desulfovibrio, Symp. of the Society for Gen. Microbiol.* **42**:457–472.

Powell, B. J., 1989, Genetic studies on the sulphate reducing bacteria. Academic thesis, Napier Polytechnic Edinburgh, Scotland and SCK/CEN, Mol, Belgium.

Powell, B., Mergeay, M., and Christofi, N., 1989, Transfer of broad host-range plasmids to sulphate-reducing bacteria, *FEMS Microbiol. Lett.* **59**:269–274.

Priefer, U., Simon, R., and Pühler, A., 1984, Broad host range vectors for gram-negative bacteria, in: *Proc. III$^{rd}$ Eur. Congress on Biotechnology*, volume III, Eur. Fed. of Biotechnology, Verlag Chemie, Weinheim, pp. 207–212.

Prickril, B. C., Czechowski, M. H., Przybyla, A. E., Peck Jr., H. D., and LeGall, J., 1986, Putative signal peptide on the small subunit of the periplasmic hydrogenase from *Desulfovibrio vulgaris, J. Bacteriol.* **167**:722–725.

Prickril, B. C., He, S.-H., Li, C., Menon, N., Choi, E.-S., Przybyla, A. E., DerVartanian, D. V., Peck Jr., H. D., Fauque, G., LeGall, J., Teixeira, M., Moura, I., Moura, J. J. G., Patil, D., and Huynh, B.-H., 1987, Identification of three classes of hydrogenase in the genus, *Desulfovibrio, Biochem. Biophys. Res. Comm.* **149**:369–377.

Przybyla, A. E., Robbins, J., Menon, N., and Peck, Jr., H. D., 1992, Structure-function relationships among the nickel-containing hydrogenases, *FEMS Microbiol. Rev.* **88**:109–136.

Rapp, B. J., and Wall, J. D., 1987, Genetic transfer in *Desulfovibrio desulfuricans, Proc. Natl. Acad. Sci. USA* **84**:9128–9130.

Rieder, R., Cammack, R., and Hall, D. O., 1984, Purification and properties of the soluble hydrogenase from *Desulfovibrio desulfuricans* (strain Norway 4), *Eur. J. Biochem.* **145**:637–643.

Rohde, M., Fürstenau, U., Mayer, F., Przybyla, A. E., Peck Jr., H. D., LeGall, J., Choi, E.-S., and Menon, N. K., 1990, Localization of membrane-associated [NiFe] and [NiFeSe] hydrogenases of *Desulfovibrio vulgaris* using immunoelectron microscopic procedures, *Eur. J. Biochem.* **191**:389–396.

Rossi, M., Pollock, W. B. R., Reij, M. W., Keon, R. G., Fu, R., and Voordouw, G., 1993, The *hmc* operon of *Desulfovibrio vulgaris* subsp. *vulgaris* Hildenborough encodes a potential transmembrane redox protein complex, *J. Bacteriol.* **175**:4699–4711.

Rousset, M., 1991, Approche moléculaire de l'étude du métabolisme de l'hydrogène chez une bactérie sulfato-reductrice: *Desulfovibrio fructosovorans*. Academic thesis, University of Aix-Marseille I, France.

Rousset, M., Dermoun, Z., Chippaux, M., and Bélaich, J. P., 1991, Marker exchange mutagenesis of the *hydN* genes in *Desulfovibrio fructosovorans*, *Mol. Microbiol.* **5**:1735–1740.

Runswick, M. J., Genneis, R. B., Fearnley, I. M., and Walker, J. E., 1989, Mitochondrial NADH-ubiquinone reductase: complementary DNA sequence of the import precursor of the bovine 75-kDa subunit, *Biochemistry* **28**:9432–9439.

Santos, H., Moura, J. J. G., Moura, I., LeGall, J., and Xavier, A. V., 1984, NMR studies of electron transfer mechanisms in a protein with interacting redox centers: *Desulfovibrio gigas* cytochrome $c_3$, *Eur. J. Biochem.* **141**:283–296.

Sauter, M., Böhm, R., and Böck, A., 1992, Mutational analysis of the operon (*hyc*) determining hydrogenase 3 formation in *Escherichia coli*, *Mol. Microbiol.* **6**:1523–1532.

Sawers, G., Heider, J., Zehelein, E., and Böck, A., 1991, Expression and operon structure of the *sel* genes of *Escherichia coli* and identification of a third selenium-containing formate-dehydrogenase isoenzyme, *J. Bacteriol.* **173**:4983–4993.

Sayavedra-Soto, L. A., Powell, G. K., Evans, H. J., and Morris, R. O., 1988, Nucleotide sequence of the genetic loci encoding subunits of *Bradyrhizobium japonicum* uptake hydrogenase, *Proc. Natl. Acad. Sci. USA* **85**:8395–8399.

Schneider, K., Cammack, R., and Schlegel, H. G., 1984, Content and localization of FMN, Fe-S clusters and nickel in the NAD-linked hydrogenase from *Nocardia opaca* 1b, *Eur. J. Biochem.* **142**:75–84.

Seyedirashti, S., Wood, C., and Akagi, J. M., 1991, Induction and partial purification of bacteriophages from *Desulfovibrio vulgaris* (Hildenborough) and *Desulfovibrio desulfuricans* ATCC 13541, *J. Gen. Microbiol.* **137**:1545–1549.

Sjöström, M., Wold, S., Wieslander, Å., and Rilfors, L., 1987, Signal peptide amino acid sequences in *Escherichia coli* contain information related to final protein localization. A multivariate data analysis, *EMBO J.* **6**:823–831.

Skyring, G. W., and Jones, H. E., 1972, Guanine plus cytosine contents of the deoxyribonucleic acids of some sulfate-reducing bacteria: a reassessment, *J. Bacteriol.* **109**:1298–1300.

Stokkermans, J. P. W. G., 1993, Molecular studies on iron-sulfur proteins in *Desulfovibrio*. Academic thesis, Agricultural University, Wageningen, The Netherlands.

Stokkermans, J., van Dongen, W., Kaan, A., van den Berg, W., and Veeger, C., 1989, *HydC*, a gene from *Desulfovibrio vulgaris* (Hildenborough) encodes a polypeptide homologous to the periplasmic hydrogenase, *FEMS Microbiol. Lett.* **58**:217–222.

Stokkermans, J. P. W. G., Pierik, A. J., Wolbert, R. B. G., Hagen, W. R., van Dongen, W. M. A. M., and Veeger, C., 1992a, The primary structure of a protein containing a putative [6Fe-6S] prismane cluster from *Desulfovibrio vulgaris* (Hildenborough), *Eur. J. Biochem.* **208**:435–442.

Stokkermans, J. P. W. G., Houba, P. H. J., Pierik, A. J., Hagen, W. R., van Dongen, W. M. A. M., and Veeger, C., 1992b, Overproduction of prismane protein in *Desulfovibrio vulgaris* (Hildenborough): evidence for a second S=1/2 spin system in the one-electron reduced state, *Eur. J. Biochem.* **210**:983–988.

Stokkermans, J. P. W. G., van den Berg, W. A. M., van Dongen, W. M. A. M., and Veeger, C., 1992c, The primary structure of a protein containing a putative [6Fe-6S] prismane cluster from *Desulfovibrio desulfuricans* (ATCC 27774), *Biochim. Biophys. Acta* **1132**:83–87.

Teixeira, M., Moura, I., Fauque, G., Czechowski, M., Berlier, Y., Lespinat, P. A., LeGall, J., Xavier, A. V., and Moura, J. J. G., 1986, Redox properties and activity studies on a nickel-containing hydrogenase isolated from a halophilic sulfate reducer *Desulfovibrio salexigens, Biochimie* (Paris) **68**:74–84.

Teixeira, M., Moura, I., Xavier, A. V., Moura, J. J. G., LeGall, J., DerVartanian, D. V., Peck Jr., H. D., and Huyhn, B.-H., 1989, Redox intermediates of *Desulfovibrio gigas* hydrogenase generated under hydrogen: Mössbauer and EPR characterization of the Fe-S and Ni centers, *J. Biol. Chem.* **264**:16435–16450.

Tran-Betcke, A., Warnecke, U., Böcker, C., Zaborosch, C., and Friedrich, B., 1990, Cloning and nucleotide sequences of the genes for the subunits of NAD-reducing hydrogenase of *Alcaligenes eutrophus* H16, *J. Bacteriol.* **172**:2920–2929.

Turner, D. L., Salgueiro, C. A., LeGall, J., and Xavier, A. V., 1992, Structural studies of *Desulfovibrio vulgaris* ferrocytochrome $c_3$ by two-dimensional NMR, *Eur. J. Biochem.* **210**:931–936.

Van den Berg, W. A. M., Stokkermans, J. P. W. G., and van Dongen, W. M. A. M., 1989, Development of a plasmid transfer system for the anaerobic sulphate reducer, *Desulfovibrio vulgaris, J. Biotechnol.* **12**:173–184.

Van den Berg, W. A. M., van Dongen, W. M. A. M., and Veeger, C., 1991, Reduction of the amount of periplasmic hydrogenase in *Desulfovibrio vulgaris* (Hildenborough) with antisense RNA: direct evidence for an important role of this hydrogenase in lactate metabolism, *J. Bacteriol.* **173**:3688–3694.

Van den Berg, W. A. M., Stevens, A. A. M., Verhagen, M. F. J. M., van Dongen, W. M. A. M., and Hagen, W. R., 1994, Overproduction of the prismane protein from *Desulfovibrio desulfuricans* ATCC 27774 in *Desulfovibrio vulgaris* (Hildenborough) and EPR spectroscopy of the [6Fe-6S] cluster in different redox states. *Biochim. Biophys. Acta* **1206**:240–246.

Van der Westen, H. M., Mayhew, S. G., and Veeger, C., 1978, Separation of hydrogenase from intact cells of *Desulfovibrio vulgaris* purification and properties, *FEBS Lett.* **86**:122–126.

Van der Zwaan, J. W., Albracht, S. P. J., Fontijn, R. D., and Slater, E. C., 1985, Monovalent nickel in hydrogenase from *Chromatium vinosum:* light sensitivity and evidence for direct interaction with hydrogen, *FEBS Lett.* **179**:271–277.

Van Dongen, W., Hagen, W., van den Berg, W., and Veeger, C., 1988, Evidence for an unusual mechanism of membrane translocation of the periplasmic hydrogenase of *Desulfovibrio vulgaris* (Hildenborough), as derived from expression in *Escherichia coli, FEMS Microbiol. Lett.* **50**:5–9.

Van Dongen, W. M. A. M., Stokkermans, J. P. W. G., and van den Berg, W. A. M., 1994, Genetic manipulation of *Desulfovibrio.* Meth. Enzymol. **243**:319–330.

Verhagen, M. F. J. M., Wolbert, R. B. G., and Hagen, W. R., 1994, Cytochrome $c_{553}$ from *Desulfovibrio vulgaris* (Hildenborough). Electrochemical properties and electron transfer with hydrogenase, *Eur. J. Biochem.* **221**:821–829.

Von Heijne, G., 1985, Signal sequences: the limits of variation, *J. Mol. Biol.* **184**:99–105.

Voordouw, G., 1988, Cloning of genes encoding redox proteins of known aminoacid sequence from a library of the *Desulfovibrio vulgaris* (Hildenborough) genome. *Gene* **69**:75–83.

Voordouw, G., and Brenner, S., 1985, Nucleotide sequence of the gene encoding the hydrogenase from *Desulfovibrio vulgaris* (Hildenborough), *Eur. J. Biochem.* **148:**515–520.

Voordouw, G., and Brenner, S., 1986, Cloning and sequencing of the gene encoding cytochrome $c_3$ from *Desulfovibrio vulgaris* (Hildenborough), *Eur. J. Biochem.* **159:**347–351.

Voordouw, G., Walker, J. E., and Brenner, S., 1985, Cloning of the gene encoding the hydrogenase from *Desulfovibrio vulgaris* (Hildenborough) and determination of the $NH_2$-terminal sequence, *Eur. J. Biochem.* **148:**509–514.

Voordouw, G., Hagen, W. R., Krüse-Wolters, K. M., van Berkel-Arts, A., and Veeger, C., 1987, Purification and characterization of *Desulfovibrio vulgaris* (Hildenborough) hydrogenase expressed in *Escherichia coli*, *Eur. J. Biochem.* **162:**31–36.

Voordouw, G., Menon, N. K., LeGall, J., Choi, E.-S., Peck Jr., H. D., and Przybyla, A. E., 1989a, Analysis and comparison of nucleotide sequences encoding the genes for [NiFe] and [NiFeSe] hydrogenases from *Desulfovibrio gigas* and *Desulfovibrio baculatus*, *J. Bacteriol.* **171:**2894–2899.

Voordouw, G., Strang, J. D., and Wilson, F. R., 1989b, Organization of the genes encoding [Fe] hydrogenase in *Desulfovibrio vulgaris* subsp. *oxamicus* Monticello, *J. Bacteriol.* **171:**3881–3889.

Voordouw, G., Nivière, V., Ferris, F. G., Fedorak, P. M., and Westlake, D. W. S., 1990a, Distribution of hydrogenase genes in *Desulfovibrio* spp. and their use in identification of species from the oil field environment. *Appl. Env. Microbiol.* **56:**3748–3754.

Voordouw, G., Pollock, W. B. R., Bruschi, M., Guerlesquin, B., Rapp-Giles, B. J., and Wall, J. D., 1990b, Functional expression of *Desulfovibrio vulgaris* Hildenborough cytochrome $c_3$ in *Desulfovibrio desulfuricans* G200 after conjugational gene transfer from *Escherichia coli*, *J. Bacteriol.* **172:**6122–6126.

Voordouw, G., Voordouw, J. K., Karkhoff-Schweizer, R. R., Fedorak, P. M., and Westlake, D. W. S., 1991, Reverse Sampling Genome Probing, a new technique for identification of bacteria in environmental samples by DNA hybridization, and its application to the identification of sulfate-reducing bacteria in oil field samples, *Appl. Environ. Microbiol.* **57:**3070–3078.

Wall, J. D., Rapp-Giles, B. J., and Rousset, M., 1993, Characterization of a small plasmid from *Desulfovibrio desulfuricans* and its use for shuttle vector construction, *J. Bacteriol.* **175:**4121–4128.

Woese, C. R., 1987, Bacterial evolution, *Microbiol. Rev.* **51:**221–271.

# Ecology of Sulfate-Reducing Bacteria

# 8

## GUY D. FAUQUE

## 1. INTRODUCTION

Microorganisms play key roles in the assimilation–dissimilation steps and oxidation–reduction processes of the global sulfur cycle. The dissimilatory reduction of sulfur compounds is an essential step in the biological sulfur cycle (LeGall and Fauque, 1988; LeFaou et al., 1990; Fauque et al., 1991; Widdel and Hansen, 1992). This dissimilatory reduction is due mainly to sulfur- and sulfate-reducing bacteria which perform anaerobic oxidative phosphorylation with elemental sulfur, sulfite, or sulfate as terminal electron acceptors (Barton et al., 1972; Fauque et al., 1980, 1991). These bacteria produce large amounts of sulfide, the oxidation of which permits energy generation by phototropic and chemolithotrophic microorganisms (Trüper, 1984).

The taxonomy and the physiology of sulfate-reducing bacteria (SRB) have undergone enormous changes in the last 15 years; 15 genera of sulfate reducers are well characterized (Widdel, 1988; Widdel and Bak, 1992). In this chapter we treat some principal topics of the ecology of SRB; detailed information on the detrimental effects caused by these microorganisms and on the methodology for measurement of sulfate reduction activity is provided in the reviews of Widdel (1988) and Smith (1993).

GUY D. FAUQUE • Centre d'Océanologie de Marseille, Université d'Aix-Marseille II, URA CNRS No. 41 Campus de Luminy, Case 901, F-13288 Marseille Cedex 9, France.

*Sulfate-Reducing Bacteria*, edited by Larry L. Barton. Plenum Press, New York, 1995.

## 2. PROCESSES IN THE ANAEROBIC DEGRADATION OF ORGANIC MATTER

In aerobic ecosystems, heterotrophic bacteria carry out a complete mineralization of organic material to $CO_2$ alone. The anaerobic degradation of organic matter is a more complex process requiring the interaction of different groups of microorganisms (Widdel, 1986). Each microbial group functions in a food chain where the metabolic end product of one group is the substrate for another until complete oxidation has occurred.

Four different trophic groups of bacteria are involved in the active decomposition of organic matter in anaerobic environments (Fig. 1). The first group is the group of fermentative microorganisms that hydrolyse hight molecular weight polymers (proteins, polysaccharides, lipids, nucleic acids) and ferment their respective monomers (amino acids, sugars, fatty acids, nucleotides) to $H_2$, $CO_2$ acetate, other organic acids, and alcohols. Acetate, $H_2$, and $CO_2$ are utilized directly by methanogenic bacteria. The second group consists of acetogenic bacteria that cleave organic acids and alcohols into acetate, $H_2$ and $CO_2$. For thermodynamic reasons, this conversion and its coupling to energy conservation are only feasible in the presence of a second bacterium, such as a methanogen, that consumes molecular hydrogen and thus keeps the $H_2$ concentration very low. The third group of microorganisms is formed by the methanogenic bacteria that utilize the end-products of all previous processes ($H_2$, $CO_2$, acetate, formate) to produce methane. The fourth group consists of sulfate-reducing bacteria which compete with the methanogens and acetogens for available substrates. Methanogenesis and dissimilatory sulfate reduction are the two terminal processes of anaerobic mineralization, and the predominance of one of this process on the other depends mainly on the availability of sulfate (see sections 4 and 7).

## 3. THE BIOLOGICAL SULFUR CYCLE

The reduction of inorganic sulfate to organic or inorganic sulfide and the subsequent oxidation of sulfide back to sulfate is known as the biological sulfur cycle (Fig. 2) (Goldhaber and Kaplan, 1974; Pfennig and Widdel, 1981, 1982; Widdel, 1988; Peck and Lissolo, 1988). The biological sulfur cycle consists of an assimilatory part and a dissimilatory part. The assimilatory part includes sulfide and sulfate assimilation, as well as release of sulfur from dead and living organic substances by decomposition and excretion. The assimilatory reduction of sulfate pro-

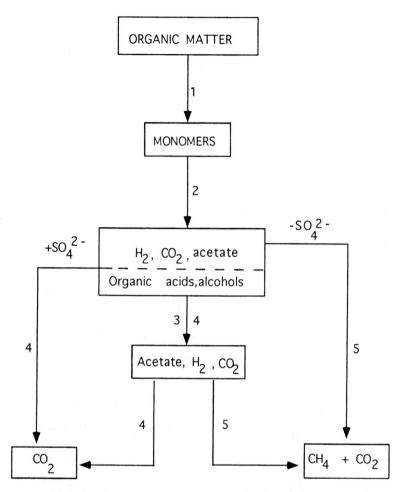

**Figure 1.** Sequence of processes involved in the mineralization of organic matter in anaerobic sediments: (1) hydrolysis; (2) fermentation; (3) acetogenesis; (4) sulfate reduction; (5) methanogenesis. (Modified from Laanbroek and Veldkamp, 1982).

vides bacteria, fungi, algae, and plants' demand for reduced sulfur compounds by synthesizing sulfur-containing amino acids (cysteine, cystine, methionine) and sulfur-containing growth factors (biotin, lipoic acid, thiamin). Animals excrete sulfur in the form of sulfate; however, the bulk of sulfur in living beings is returned to the cycle in the form of sulfide due to death and decomposition by fungi and bacteria. The dissimilatory part of the sulfur cycle includes oxidative processes like

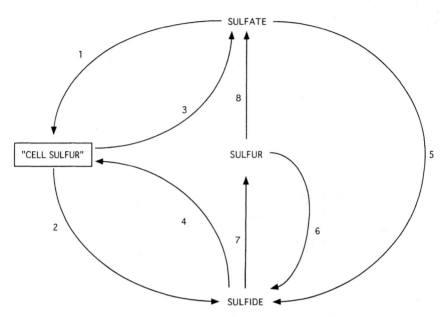

ASSIMILATORY PART                                      DISSIMILATORY PART

**Figure 2.** The biological sulfur cycle (modified from Fischer, 1988). "Cell Sulfur" includes sulfur bound in bacteria, fungi, animals and plants. (1) Assimilatory sulfate reduction by bacteria, plants and fungi; (2) Death and decomposition by bacteria and fungi; (3) Sulfate excretion by animals; (4) Sulfide assimilation by bacteria (and some plants); (5) Dissimilatory sulfate reduction; (6) Dissimilatory elemental sulfur reduction; (7) Chemotrophic and phototrophic sulfide oxidation; (8) Chemotrophic and phototrophic sulfur oxidation.

chemotrophic and phototrophic sulfide, sulfur oxidation, and reductive processes such as microbial sulfate and sulfur reduction.

## 3.1. The Sulfuretum

A sulfuretum is a system in which part or whole of the biological sulfur cycle takes place. The term "sulfuretum," introduced by Baas-Becking (1925) represents ecological communities of sulfate-reducing bacteria and sulfide-oxidizing microorganisms involved together in a biological sulfur cycle. The dissimilatory part of the biological cycle in a sulfuretum is a combination of a "small" sulfur cycle in which sulfur is only cycled between elemental sulfur and sulfide, and a "large" sulfur cycle in which the sulfur is cycled between sulfate and sulfide (Trüper, 1984). The "small" sulfur cycle, occurring mainly in marine sediments

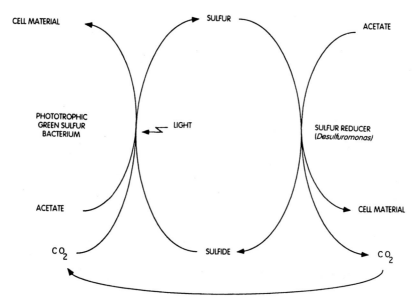

**Figure 3.** "Small" sulfur cycle in a syntrophic coculture of an acetate-oxidizing sulfur reducer (*Desulfuromonas*) and a phototrophic green sulfur bacterium assimilating acetate only together with $CO_2$ and $H_2S$ in the light (modified from Widdel and Pfennig, 1992).

where sufficient elemental sulfur is present near the redoxcline, was first detected by Pfennig and Biebl (1976) as an anaerobic syntrophic community of a phototrophic green sulfur bacterium and a sulfur reducer of the genus *Desulfuromonas* (Fig. 3). The "large" sulfur cycle is more important in balancing sulfur in the environment because it connects the oxic zone of the sulfuretum with the underlying anoxic zone (Trüper, 1984). Many different groups of microorganisms involved in sulfur metabolism in the sulfuretum include sulfate reducers, sulfur reducers, chemotrophic sulfur oxidizers, oxygenic and anoxygenic phototrophic bacteria (see section 3.2). Sulfureta are not restricted to marine sediments and are also found in sulfur springs, cyanobacterial mats, salt-marsh sediments, bogs, peats, meromictic lakes, volcanoes, and other environments where organic matter and sulfate co-exist in anoxic conditions (Postgate, 1984).

### 3.2. Microorganisms Involved in the Sulfur Cycle

Different groups of microorganisms are involved in the different steps of the biological sulfur cycle (Fig. 2). These microorganisms can be classified into five different metabolic types according to the manner they utilize inorganic sulfur compounds as electron donors and acceptors (Trüper, 1984; Fischer, 1988).

### 3.2.1. Dissimilatory Sulfate Reduction

The dissimilatory sulfate-reducing bacteria comprise fourteen eubacterial genera: D. (*Desulfovibrio*); Da. (*Desulfoarculus*); Dba. (*Desulfobacter*); Dbt. (*Desulfobacterium*); Dbo. (*Desulfobotulus*); Dbu. (*Desulfobulbus*); Dc. (*Desulfococcus*); Dh. (*Desulfohalobium*); Dsm. (*Desulfomicrobium*); Dm. (*Desulfomonile*); Dn. (*Desulfonema*); Ds. (*Desulfosarcina*); Dtm. (*Desulfotomaculum*); T. (*Thermodesulfobacterium*), and one archaebacterial genus: A. (*Archaeoglobus*) (Fauque *et al.*, 1991; Widdel, 1992a,b; Widdel and Bak, 1992; Stetter, 1992).

### 3.2.2. Dissimilatory Elemental Sulfur Reduction

Several genera of eubacteria and archaebacteria are able to gain energy for their growth by a dissimilatory reduction of elemental sulfur to hydrogen sulfide in a respiratory type of metabolism (Widdel and Pfennig, 1984, 1992; Widdel, 1988; LeFaou *et al.*, 1990). The eubacterial sulfur reducers are the genera *Desulfuromonas, Campylobacter, Desulfurella, Sulfurospirillum, Thermotoga, Wolinella succinogenes,* and some thiophilic sulfate-reducing bacteria of genera *Desulfovibrio* and *Desulfomicrobium* (Widdel and Pfennig, 1992; Fauque *et al.*, 1979, 1991, 1994; Fauque, 1994). A dissimilatory sulfur reduction metabolism can also be performed only in the dark by phototrophic green and purple sulfur bacteria (Trüper, 1984). The archaebacterial sulfur reducers belong to the methanogenic bacteria and the extremely thermophiles (orders Thermococcales, Thermoplasmales, and Thermoproteales) (LeFaou *et al.*, 1990; Fauque *et al.*, 1991).

### 3.2.3. Phototrophic Anaerobic Oxidation of Sulfur Compounds

Some cyanobacteria and most phototrophic eubacteria such as purple bacteria (families Chromatiaceae, Rhodospirillaceae, Ectothiorhodospiraceae) and green bacteria (families Chloroflexaceae, Chlorobiaceae) use reduced sulfur compounds (e.g., sulfide, sulfur, thiosulfate) as electron donors for their anoxygenic photosynthesis (Trüper and Fischer, 1982).

### 3.2.4. Chemotrophic Aerobic Sulfur Oxidation

Various metabolic microorganisms that grow under aerobic conditions while oxidizing reduced inorganic sulfur compounds to sulfate include chemolithoheterotrophs (*Thiobacillus perometabolis*, *Pseudomonas* species), obligate chemolithotrophs (*Thiomicrospira* and *Thiobacillus* species), facultative chemolithotrophs (*Sulfolobus* and *Thiobacillus* species, *Paracoccus denitrificans*), heterotrophs (*Beggiatoa* and *Pseudomonas* species), and some "morphologically conspicuous sulfur bacteria" (*Thiovulum*, *Thiothrix*, *Thiospira*, *Macromonas*, *Achromatium*) (Kuenen and Beudeker, 1982; Kuenen and Bos, 1989).

### 3.2.5. Chemotrophic Anaerobic Sulfur Oxidation

Besides phototrophic bacteria, only a few other microbial species such as *Thiomicrospira denitrificans* and *Thiobacillus denitrificans* anaerobically oxidize reduced inorganic sulfur compounds to sulfate with nitrate as terminal electron acceptor (Trüper, 1984).

## 4. SULFATE REDUCTION IN NATURAL HABITATS

Dissimilatory sulfate reduction or methanogenesis is the major terminal degradation process in anaerobic environments. Methanogenesis dominates in sulfate-poor habitats such as freshwater environments, whereas sulfate reduction is predominant in marine or other sulfate-rich saline habitats (Widdel, 1988). In freshwater environments, the sulfate concentration is usually low and ranges from about 10 to 200 nmoles per liter (Ingvorsen *et al.*, 1981). In seawater, the sulfate concentration is on the average 28 mmoles per liter (Goldhaber and Kaplan, 1974).

The sulfate-reducing bacteria are widely distributed in anaerobic terrestrial and aquatic environments. The presence of sulfate reducers with a high metabolic activity is easily revealed by the blackening of water and sediment due to the precipitation of iron sulfide, and by the smell of hydrogen sulfide. Marine, estuarine, and saltmarsh sediments, as well as those of saline and hypersaline lakes and ponds are the most permanent and significant habitats of sulfate reducers in nature because of their high sulfate content (Zobell, 1958; Pfennig *et al.*, 1981; Postgate, 1984). Active sulfate reduction has also been reported in nonsaline environments such as soils, and freshwater sediments (Smith and Klug, 1981; Postgate, 1984; Jones and Simon, 1984). Sulfate reducers have been detected in polluted environments such as anaerobic purification plants (Nanninga and Gottschal, 1987), spoiled foods (Campbell *et al.*, 1957), sour whey digesters (Zellner *et al.*, 1989), and sewage plants (Schoberth, 1973). They have also been isolated from ricefields (Ouattara and Jacq, 1992), rumen contents (Huisingh *et al.*, 1974), termite guts (Brauman *et al.*, 1990; Traoré *et al.*, 1990), faeces of man and animals (Moore *et al.*, 1976; Beerens and Romond, 1977; Gibson and Gibson, 1988; Gibson, 1990), and oilfield waters (Antloga and Griffin, 1985; Rosnes *et al.*, 1991).

### 4.1. Gram-Negative Mesophilic Sulfate Reducers

Mesophilic gram-negative nonsporeforming types are the most widespread sulfate-reducing bacteria in nature. They include five genera which oxidize organic compounds incompletely to the level of acetate: *Desulfovibrio* (Table I), *Desulfobotulus*, *Desulfobulbus*, *Desulfohalobium*,

**Table I. Some Characteristic Features of Sulfate-Reducing Bacteria of the Genus *Desulfovibrio***

| Species | Source | Habitat | Growth substrate | N$_2$ fixation |
|---------|--------|---------|------------------|----------------|
| D. africanus | NCIB 8401 | Well water | Lactate | + |
| D. alcoholovorans | DSM 5433 | Pilot fermenter | Glycerol | NR[a] |
| D. carbinolicus | DSM 3852 | Purification plant | Methanol | NR |
| D. desulfuricans | DSM 642 | Soil | Lactate | + |
| D. fructosovorans | DSM 3604 | Estuarine sediment | Fructose | NR |
| D. furfuralis | DSM 2590 | Pulp and paper waste | Furfural | NR |
| D. giganteus | DSM 4123 | Brackish sediment | Glycerol | − |
| D. gigas | NCIB 9332 | Pond water | Lactate | + |
| D. halophilus | DSM 5663 | Benthic microbial mat | Lactate | NR |
| D. longus | DSM 6739 | Oil-producing well | Lactate | NR |
| D. piger | ATCC 29098 | Human faeces | Lactate | NR |
| D. salexigens | NCIB 8403 | "Sling mud" | Lactate | + |
| D. simplex | DSM 4141 | Sour whey digester | Lactate | NR |
| D. sulfodismutans | DSM 3696 | Freswater mud | Thiosulfate dismutation | NR |
| D. termitidis | DSM 5308 | Hindgut of termite | Glucose | NR |
| D. vulgaris | NCIB 8303 | Soil | Lactate | + |

[a]NR: not reported.

**Table II. Some Characteristic Features of Mesophilic Sulfate Reducers with Incomplete Oxidation of Organic Compounds (not Including the Genus *Desulfovibrio*)**

| Species | Source | Habitat | Growth substrate | $N_2$ fixation |
|---------|--------|---------|------------------|-----------------|
| *Dbu. elongatus* | DSM[a] 2908 | Mesophilic digester | Propionate | + |
| *Dbu. marinus* | DSM 2058 | Marine mud | Propionate | + |
| *Dbu. propionicus* | DSM 2032 | Freshwater mud | Propionate | + |
| *Dsm. apsheronum* | AUCCM[b] 1105 | Oil deposits | Lactate | NR[c] |
| *Dsm. baculatum* | AUCCM 1378 | Manganese ore | Lactate | + |
| *Dbo. sapovorans* | DSM 2055 | Freshwater mud | Butyrate | NR |
| *Dh. retbaense* | DSM 5692 | Sediments hyper-saline lake | Lactate | NR |

[a] DSM: Deutsche Sammlung von Mikroorganismen, Braunschweig, F.R.G.
[b] AUCCM: All-Union Culture Collection of Microorganisms, Moscow, Russia.
[c] NR: not reported.

*Desulfomicrobium* (Table II), and seven genera that oxidize organic substrates completely to $CO_2$: *Desulfoarculus, Desulfobacter, Desulfobacterium, Desulfococcus, Desulfomonile, Desulfonema, Desulfosarcina* (Table III). Members of genera *Desulfobulbus* and *Desulfovibrio* are equally common in marine and freshwater environments (Postgate, 1984; Widdel and Bak, 1992). Species of genera *Desulfobacter, Desulfobacterium, Desulfohalobium, Desulfonema,* and *Desulfosarcina* are mainly brackish or marine (section 5.1.). *Desulfoarculus, Desulfobotulus, Desulfomicrobium,* and *Desulfomonile* species have been isolated primarily from freshwater environments (Widdel and Bak, 1992). Fixation of molecular nitrogen has been observed with species of genera *Desulfobacter, Desulfobulbus, Desulfomicrobium,* and *Desulfovibrio* (Lespinat *et al.,* 1987; Postgate *et al.,* 1988; Widdel and Hansen, 1992).

### 4.2. Gram-Positive Sporeforming Sulfate Reducers

Gram-positive sporulated sulfate-reducing bacteria of the genus *Desulfotomaculum* include complete and incomplete oxidizing species (Table IV). Most *Desulfotomaculum* species have been isolated from freshwater environments or other habitats with relatively low salt concentrations. The genus *Desulfotomaculum* contains five moderately thermophilic species and dissimilatory sulfate reduction in environments with temperatures between 50 and 65°C can mainly be due to sporeforming species (Widdel,

**Table III. Some Characteristic Features of Mesophilic Sulfate Reducers with Complete Oxidation of Organic Compounds to $CO_2$**

| Species | Source | Habitat | Growth substrate |
|---|---|---|---|
| Dba. curvatus | DSM 3379 | Marine mud | Acetate |
| Dba. hydrogenophilus | DSM 3380 | Marine mud | Acetate |
| Dba. latus | DSM 3381 | Marine mud | Acetate |
| Dba. postgatei | DSM 2034 | Brackish water | Acetate |
| Dbt. anilini | DSM 4660 | Marine sediment | Aniline |
| Dbt. autotrophicum | DSM 3382 | Marine mud | Succinate |
| Dbt. catecholicum | DSM 3882 | Anoxic mud (sludge) | Catechol |
| Dbt. indolicum | DSM 3383 | Marine mud | Indole |
| Dbt. macestii | AUCCM B-1598 | Sulfide spring | Ethanol |
| Dbt. niacini | DSM 2650 | Marine sediment | Nicotinate |
| Dbt. phenolicum | DSM 3384 | Marine mud | Phenol |
| Dbt. vacuolatum | DSM 3385 | Marine mud | Succinate |
| Dc. biacutus | DSM 5651 | Anaerobic digestor sludge | Acetone |
| Dc multivorans | DSM 2059 | Sewage digestor | Phenylacetate |
| Ds. variabilis | DSM 2060 | Marine mud | Phenylacetate |
| Da. baarsii | DSM 2075 | Freshwater mud | Stearate |
| Dm. tiedjei | ATCC 49306 | Sewage sludge | Benzoate |
| Dn. limicola | DSM 2076 | Marine black mud | Succinate |
| Dn. magnum | DSM 2077 | Marine black mud | Benzoate |

1992a). Diazotrophic growth has been demonstrated in *Dtm. orientis* and *Dtm. ruminis* (Postgate, 1984; Lespinat *et al.*, 1985).

### 4.3. Gram-Negative Thermophilic Eubacterial Sulfate Reducers

The genus *Thermodesulfobacterium* contains two species (Table V) which are nutritionally restricted, incomplete oxidizers and phylogenetically separate from other eubacterial genera (Rozanova and Khudyakova, 1974; Zeikus *et al.*, 1983; Rozanova and Pivovarova, 1988; Widdel, 1992b). *Thermodesulfobacterium* species have been isolated from saline water; however, they do not present a significant salt tolerance. In their natural environment, *Thermodesulfobacterium* strains probably grow with molecular hydrogen produced by geothermal reactions or thermophilic fermentation processes (Widdel, 1992b).

**Table IV. Some Characteristic Features of Sporulated Sulfate-Reducing Bacteria of the Genus *Desulfotomaculum***

| Species | Source | Habitat | Oxidation[a] | Optimum temperature (°C) |
|---------|--------|---------|--------------|--------------------------|
| *Dtm. acetoxidans* | DSM 771 | Muds, animal faeces | C | 36 |
| *Dtm. antarcticum* | IAM[b] 64 | Antarctica muds | I | 20–30 |
| *Dtm. geothermicum* | DSM 3669 | Geothermal groundwater | C | 54 |
| *Dtm. guttoideum* | DSM 4024 | Freshwater muds | I | 31 |
| *Dtm. kuznetsovii* | AUCCM 17 | Thermal mineral waters | C | 60–65 |
| *Dtm. nigrificans* | DSM 574 | Soils | I | 55 |
| *Dtm. orientis* | NCIB[c] 8382 | Soils | I | 30–37 |
| *Dtm. ruminis* | ATCC[d] 23193 | Sheep rumen | I | 37 |
| *Dtm. sapomandens* | DSM 3223 | Aerobic soils | C | 38 |
| *Dtm. thermoace-toxidans* | Strain CAMZ | Thermophilic bioreactor | C | 55–60 |
| *Dtm. thermoben-zoicum* | DSM 6193 | Thermophilic bioreactor | C | 62 |

[a]C: Complete oxidation of organic compounds to $CO_2$; I: incomplete oxidation of organic compounds to acetate as an end product.
[b]IAM: Institute for Applied Microbiology, The University of Tokyo, Japan.
[c]NCIB: National Collection of Industrial Bacteria, Aberdeen, U.K.
[d]ATCC: American Type Culture Collection, Rockville, MD, U.S.A.

## 4.4. Gram-Negative Thermophilic Archaebacterial Sulfate Reducers

Archaebacterial sulfate reducers of the genus *Archaeoglobus,* found only in anaerobic submarine hydrothermal areas, require salt and high temperatures for growth (Table VI). The genus *Archaeoglobus* may represent a biochemical missing link between the methanogenic and sulfur-metabolizing archaebacteria (Stetter, 1992).

## 5. EFFECTS OF ENVIRONMENTAL FACTORS ON GROWTH OF SULFATE-REDUCING BACTERIA

In natural habitats, the capacities to adapt to modifications of biological and physico-chemical factors may be decisive for the growth and activity of a microorganism.

**Table V. Some Characteristic Features of Thermophilic Eubacterial Sulfate Reducers**

| Feature | T. commune | T. mobile |
|---|---|---|
| Source | DSM 2178 | DSM 1276 |
| Habitat | Volcanic thermal vents | Stratal water from a petroleum deposit |
| Motility | − | + (polar flagella) |
| Optimum temperature (°C) | 70 | 65 |
| GC content of DNA (mole %) | 34.4 | 38 |
| $N_2$ fixation | NR[a] | − |

[a] NR: not reported.

## 5.1. Effect of pH, Temperature, and Salts

Sulfate-reducing bacteria grow better under slightly alkaline conditions over a relatively restricted pH range (pH 7.0–7.8) and tolerate pH values ranging from 5.5 to 9.0 (Zobell, 1958; Pfennig et al., 1981). Alkalization or acidification of waters in industrial plants has been utilized as a method of inhibiting bacterial sulfate reduction (Zobell, 1958). However, dissimilatory reduction of sulfate has been observed in highly acidic environments (pH values of about 2.5 to 4.5) in acid mine drainage (Tuttle

**Table VI. Some Characteristic Features of Thermophilic Archaebacterial Sulfate Reducers**

| Feature | A. fulgidus | A. profundus |
|---|---|---|
| Source | DSM 4304 | DSM 5631 |
| Habitat | Marine hydrothermal vents | Deep sea hydrothermal vents |
| Flagellation | Monopolar polytrichous | None |
| Nutrition type | Facultatively chemolithoautotrophic | Obligately chemolithoheterotrophic ($H_2$ + organic carbon sources) |
| Methane production | + (traces) | − |
| GC content of DNA (mole %) | 46 | 41 |

*et al.*, 1969; Herlihy and Mills, 1985; Gyure *et al.*, 1990), and peat from an acidic freshwater wetland (Spratt *et al.*, 1987). The *Desulfovibrio* and *Desulfotomaculum* species isolated from the mixed cultures of acid mine water did not reduce sulfate below pH 5.5 (Tuttle *et al.*, 1969). It was consequently postulated that the sulfate-reducing bacteria in the acid environment occurred in microniches of a higher, more favorable pH.

Seasonal variations in the rates of dissimilatory sulfate reduction have been reported by Jorgensen (1977). Nedwell and Abram (1979) have shown that temperature is the dominant environmental parameter influencing reduction of sulfate in anaerobic saltmarsh sediments. However, Abdollahi and Nedwell (1979) have shown that there was no physiological response or adaptation of the sulfate reducers population to seasonally changing environmental temperature in marine sediments. The temperature dependence for dissimilatory sulfate reduction in marine sediments is variable, nonrandom, with lower rates of activity exhibiting a more pronounced temperature dependence (Westrich and Berner, 1988). Obligate psychrophilic sulfate reducers have not been isolated so far. Mesophilic sulfate reducers grow best between 28 and 38°C and have an upper temperature limit around 45°C (Widdel and Pfennig, 1984; Widdel and Hansen, 1992). Most of the thermophilic sulfate-reducing bacteria have been isolated from geothermal environments and oil field waters; their growth temperatures reflect the habitat from which they were found (Rozanova and Khudyakova, 1974; Zeikus *et al.*, 1983; Daumas *et al.*, 1988; Nazina *et al.*, 1988). The optimum growth temperature for thermophilic eubacterial sulfate reducers of genera *Desulfotomaculum* and *Thermodesulfobacterium* ranges from 54 to 70°C and the maximum growth temperature is between 56 and 85°C. The archaebacterial sulfate reducers of the genus *Archaeoglobus* have a temperature optimum of 83°C and a maximum of 92°C (Stetter, 1992).

Due to their broad metabolic capacities, sulfate reducers are very important in the mineralization of organic matter in anoxic marine sediments. Microbiological sulfate reduction has been observed in various hypersaline ecosystems such as the Great Salt Lake (salinity 24%) (Zeikus, 1983), the Dead Sea (Nissenbaum and Kaplan, 1976), and salterns (Oren, 1988). However, most of the halophilic sulfate-reducing bacteria isolated so far are marine or slightly halophilic microorganisms (with optimum salinity ranging from 1 to 4% NaCl) belonging to different species of the genera *Desulfovibrio*, *Desulfobacterium*, *Desulfobacter*, *Desulfonema*, *Archaeoglobus*, *Ds. variabilis*, and *Dtm. geothermicum* (Ollivier *et al.*, 1994) (Tables VII). Only two moderately halophilic sulfate reducers have been isolated so far: *D. halophilus* and *Dh. retbaense*. *Desulfovibrio halophilus* isolated from a microbial mat of Solar Lake in Sinai, grows in a

Table VII. Salinity Range and Optimum Salinity of Slightly and Moderately Halophilic Species of Sulfate-Reducing Bacteria (Modified from Caumette *et al.*, 1991)

|  | Salinity range (% NaCl) | Optimum salinity (% NaCl) |
|---|---|---|
| Slightly halophilic species |  |  |
| D. salexigens | 0.5–12 | 2–5 |
| D. giganteus | 0.2–5 | 2–2.5 |
| D. desulfuricans subsp. aestuarii | 0.5–6 | 2.5 |
| Desulfobacterium species | NR[a] | 2 |
| Desulfobacter species | NR | 0.7–2 |
| Desulfonema species | NR | 1.5–3 |
| Ds. variabilis | NR | 1.5 |
| Dtm. geothermicum | 0.2–5 | 2.5–3.5 |
| Archaeoglobus species | 0.9–3.6 | 1.5–1.8 |
| Moderately halophilic species |  |  |
| D. halophilus | 3–18 | 6–7 |
| Dh. retbaense | 3–24 | 10 |

[a] NR: Not reported.

salinity range of 3 to 18% NaCl, with optimum growth occurring at 6 to 7% NaCl (Caumette *et al.*, 1991). Another moderately halophilic sulfate-reducing bacterium, *Dh. retbaense,* isolated from Retba Lake, a pink hypersaline lake in Senegal, grows in media containing NaCl at concentrations up to 24%, with an optimum near 10% (Ollivier *et al.*, 1991) (Table VII).

## 5.2. Aerobic Sulfate Reduction

Since their discovery around one century ago, the dissimilatory sulfate-reducing bacteria have been described as obligate anaerobic microorganisms and only the utilization of strict anaerobic techniques permitted the isolation of a broad range of metabolically different sulfate reducers (Widdel and Bak, 1992). However, several recent studies have shown that the sulfate-reducing bacteria are able to survive or even take advantage of the presence of molecular oxygen. It was demonstrated that sulfate reducers remain viable for hours or even days when exposed to molecular oxygen (Hardy and Hamilton, 1981; Cypionka *et al.*, 1985; Fukui and Takii, 1990; Wall *et al.*, 1990). The capability of aerobic respiration has been detected in several genera of sulfate reducers (*De-*

*sulfovibrio, Desulfobulbus, Desulfobacterium,* and *Desulfococcus*) (Abdollahi and Wimpenny, 1990; Dilling and Cypionka, 1990) and a weak aerobic growth has been observed with *D. desulfuricans* Essex 6 and *Dbt. autotrophicum* (Marshall *et al.,* 1993). The presence of protective enzymes against molecular oxygen, such as superoxide dismutase, NADH oxidase and catalase, has also been reported in several *Desulfovibrio* species (Hatchikian *et al.,* 1977). A terminal oxygen reductase has been recently purified and characterized from soluble extracts of *D. gigas;* it is a rubredoxin:oxygen oxidoreductase, a FAD-containing protein, able to couple the reduction of oxygen to water with NADH oxidation (Chen *et al.,* 1993).

Environments from which sulfate reducers can be isolated suggests that a range of aerotolerance exists (Hardy and Hamilton, 1981; Postgate, 1984). Several recent studies have shown that high rates of dissimilatory sulfate reduction activity occur in the oxic zones of microbial mats (Cohen, 1989; Canfield and Des Marais, 1991) and near or within the oxic/anoxic interfaces in sediments and deep waters (Ingvorsen *et al.,* 1981; Laanbroek and Pfennig, 1981; Battersby *et al.,* 1985; Hastings and Emerson, 1988). Jorgensen and Bak (1991) have found that the highest concentrations of sulfate-reducing bacteria ($2 \times 10^6$ cm$^{-3}$) were present in the oxic layer of a marine sediment.

Laminated microbial sediment ecosystems or "microbial mats" are probably the oldest and most widespread biological communities known (Des Marais, 1990). These ecosystems are composed of vertically stratified microbial communities of phototrophic, chemotrophic, and heterotrophic organisms. Microbial mats have been found in a variety of environments, both marine and freshwater, but they develop best in sediments in hypersaline and marine habitats (Stal *et al.,* 1985; Krumbein and Stal, 1991). In most cases, photosynthetic organisms, particularly cyanobacteria, are the mat-building organisms. The colored lamination, clearly visible to the naked eye, can be attributed to the uppermost green layer of cyanobacteria with underneath a red layer of phototrophic purple sulfur bacteria. Underneath the sediment is blackened by ferrous sulfide due to the production of sulfide by sulfate-reducing bacteria (Ward and Olson, 1980; Skyring *et al.,* 1983; Skyring, 1984; Stal *et al.,* 1985; Cohen, 1989). Canfield and Des Marais (1991) have shown that, in hypersaline marine microbial mats, dissimilatory sulfate reduction can occur at rapid rates in the presence of high molecular oxygen tensions. High number of sulfate-reducing bacteria ($1.1 \times 10^8$ cells cm$^{-3}$ sediment), anoxygenic phototrophic microorganisms and chemolithoautotrophic bacteria were found in the uppermost 5 mm of a marine microbial mat in which dissimilatory sulfate reduction took place under oxic conditions (Visscher *et al.,* 1992). Considerable sulfate reduction could be detected in the oxic zone of a cyanobacterial mat and an enhancement of these reduction rates was obtained in the light under exposure to oxygen concentrations as high as 1.2 mM (Fründ and Cohen, 1992).

## 6. GROWTH IN PHOTOTROPHIC ASSOCIATIONS

Special relationships between sulfate-or sulfur-reducing bacteria and photosynthetic microorganisms have been reported as consortia or syntrophic associations. A syntrophic cycling of sulfur does not occur with phototrophic purple sulfur bacteria because these microorganisms store elemental sulfur inside the cells, making sulfur unavailable to the heterotrophs (Van Gemerden and Beeftink, 1981). The first demonstration that mixed cultures of sulfate reducers (*Desulfovibrio* species) and phototrophic green sulfur bacteria (*Chlorobium* species) grown in the light was made by Butlin and Postgate (1954) in order to understand the sulfur deposits in a sulfate rich desert salt lake. This syntrophic association constitutes a miniature sulfur cycle. An interesting example of such a syntrophism is the case of the so-called "*Chloropseudomonas ethylica,*" reported to be the only phototrophic green sulfur bacterium capable of growth on organic substrates such as ethanol and acetate, instead of sulfide (Shaposhnikov *et al.,* 1960). In fact "*Chloropseudomonas ethylica*" was shown to be a consortium of a green sulfur bacterium with one heterotrophic microorganism (Gray *et al.,* 1973). Depending on the source of the "*Chloropseudomonas ethylica*" culture, either sulfur-or sulfate-reducing heterotrophs were isolated (Pfennig and Biebl, 1976; Biebl and Pfennig, 1977). "*Chloropseudomonas ethylica* strain 2 K" was a mixed culture of the green sulfur bacterium *Prosthecochloris aesturaii* and the sulfur reducer *Desulfuromonas acetoxidans* (Pfennig and Biebl, 1976). The phototrophic green sulfur bacterium assimilates in the light organic substrates such as acetate, only in the presence of $CO_2$ and an inorganic electron donor; therefore it depends on hydrogen sulfide produced by *Desulfuromonas acetoxidans*. In the same way, the sulfur excreted by the green sulfur bacterium is immediately reduced back to sulfide by the sulfur reducer at the expense of the organic compound, which is oxidized to $CO_2$ (Fig. 3). In such a syntrophic association, the sulfur plays the role of an electron-carrying catalyst between the two types of microorganisms (Pfennig and Biebl, 1976). "*Chloropseudomonas ethylica* strain $N_2$" contained the green sulfur bacterium *Chlorobium limicola* and a sulfate reducer stain 9974, now classified as *Dsm. baculatum* DSM 1743

(Biebl and Pfennig, 1977). The growth yields of *D. gigas* DSM 496 and *D. desulfuricans* Essex 6 grown on ethanol, that was incompletely oxidized to acetate, with *Chlorobium limicola* have been determined by Biebl and Pfennig (1978) and the stoichiometry of the processes can be described by the following equations:

a. *Desulfovibrio:*

$$34 \; CH_3CH_2OH + 17 \; SO_4^{2-} \rightarrow 34 \; CH_3COO^- + 17 \; H_2S + 34 \; H_2O$$

b. *Chlorobium:*

$$17 \; H_2S + 34 \; CH_3COO^- + 28 \; CO_2 + 16 \; H_2O \overset{light}{\rightarrow} 24 \; (C_4H_7O_3) + 17SO_4^{2-}$$

c. Net Reaction:

$$34 \; CH_3CH_2OH + 28 \; CO_2 \overset{light}{\rightarrow} 24 \; (C_4H_7O_3) + 18 \; H_2O$$

## 7. INTERACTIONS WITH METHANOGENIC BACTERIA

The sulfate-reducing bacteria and the methanogenic bacteria present many ecological and physiological similarities, mainly their common presence in a wide variety of anaerobic ecosystems, such as sediments, and their common utilization of acetate and molecular hydrogen as electron donors (Oremland and Polcin, 1982; Widdel, 1988). Three general relations can occur between sulfate reducers and methanogens (Smith, 1993): (1) coexistence through utilization of separate electron donors; (2) competition between the two groups for the same electron donors; or (3) synergism in which species of one group provide an electron donor needed by the other group. A coexistence between sulfate-reducing bacteria and methanogens can be possible if both microbial groups utilize different electron donors when present together in the same microecosystems. Noncompetitive substrates, such as methylamines and methanol, are used quantitatively by methanogenic bacteria, even in sulfate-rich sediments (Oremland and Polcin, 1982; Winfrey and Ward, 1983).

### 7.1. Synergism: Interspecies Hydrogen Transfer

Interspecies hydrogen transfer can be described as the transfer of molecular hydrogen from a $H_2$-evolving microorganism to a $H_2$-utilizing bacterium in mixed or cocultures with the maintenance of a low

H$_2$ partial pressure (Bryant *et al.*, 1977; Conrad *et al.*, 1985). Bryant *et al.* (1977) have shown that *Desulfovibrio* species cannot ferment lactate or ethanol but may grow with these substrates in the absence of sulfate if co-cultured with H$_2$- scavenging methanogens. In such syntrophic associations, the sulfate-reducing bacteria serve as H$_2$- evolving acetogens. The first demonstration of syntrophic interspecies hydrogen transfer resulted from the discovery that *Methanobacillus omelianskii* was an association of two different bacteria: the nonmethanogenic S organism, which oxidized ethanol to H$_2$ plus acetate, while the second organism, *Methanobacterium bryantii* (formerly called *Methanobacterium* strain MOH), reduced the bicarbonate present in the medium with H$_2$ produced by the S organism (Bryant *et al.*, 1967).

Hydrogenases play a central role in the process of interspecies hydrogen transfer that occurs in the fermentation of organic matter in anaerobic microbial ecosystems (Peck and Odom, 1984; Peck *et al.*, 1987; Fauque *et al.*, 1987, 1988; Moura *et al.*, 1988; Fauque, 1989; Rajagopal *et al.*, 1989).

### 7.2. Competition

Inhibition studies with sediments have clearly demonstrated that sulfate-reducing bacteria are able to outcompete methanogens for the mutual growth substrates acetate and H$_2$ (Oremland, 1988; Widdel, 1988). The success of the sulfate reducers in competing against the methanogens may be attributed to their higher affinities (i.e., lower K$_m$ values) for either H$_2$ (Robinson and Tiedje, 1984) or acetate (Schönheit *et al.*, 1982). The chief competitive result is that dissimilatory sulfate reduction is the predominant terminal process in the anaerobic mineralization in marine and brackish sediments where sulfate is abundant whereas methanogenesis dominates in freshwater environments that are low in sulfate (Ward and Winfrey, 1985; Oremland, 1988; Widdel, 1988). Competitive ineractions for hydrogen between sulfate-reducing bacteria and methanogens also occur in the human large intestine (Gibson *et al.*, 1988, 1990, 1993).

### 8. CONCLUSION

Dissimilatory sulfate-reducing bacteria play a complex role in the sulfur cycle which is the major biogeochemical cycle. As a result of their economic effects and ubiquitous nature, sulfate reducers are currently the subject of considerable scientific attention. The sulfate-reducing bacteria form a phylogenetically diverse and heterogeneous group of microorganisms (Devereux and Stahl, 1993) which share a common physiological and ecological character.

ACKNOWLEDGMENTS. The skillful secretarial assistance of Mrs. Cathy Perrot is gratefully acknowledged.

# REFERENCES

Abdollahi, H., and Nedwell, D. B., 1979, Seasonal temperature as a factor influencing bacterial sulfate reduction in a saltmarsh sediment, *Microbial Ecol.* **5:**73–79.

Abdollahi, H., and Wimpenny, J.W.T., 1990, Effects of oxygen on the growth of *Desulfovibrio desulfuricans, J. Gen. Microbiol.* **136:**1025–1030.

Antloga, K. M., and Griffin, W. M., 1985, Characterization of sulfate-reducing bacteria isolated from oilfield waters, *Dev. Ind. Microbiol.* **26:**597–610.

Baas-Becking, L.G.M., 1925, Studies on the sulphur bacteria, *Ann. Bot.,* **39:**613–650.

Barton, L. L., LeGall, J., and Peck, H. D. Jr., 1972, Oxidative phosphorylation in the obligate anaerobe *Desulfovibrio gigas,* in: *Horizons of Bioenergetics* (A. San Pietro, and H. Gest, eds.), Academic Press, New York, pp. 33–51.

Battersby, N. S., Malcolm, S. J., Brown, C. M., and Stanley, S. O., 1985, Sulphate reduction in oxic and suboxic North East Atlantic sediments, *FEMS Microbiol. Ecol.* **31:**225–228.

Beerens, H ., and Romond, C., 1977, Sulfate-reducing anaerobic bacteria in human feces, *Am. J. Clin. Nutr.* **30:**1770–1776.

Biebl, H., and Pfennig, N., 1977, Growth of sulfate-reducing bacteria with sulfur as electron acceptor, *Arch. Microbiol.* **112:**115–117.

Biebl, H., and Pfennig, N., 1978, Growth yields of green sulfur bacteria in mixed cultures with sulfur and sulfate reducing bacteria, *Arch. Microbiol.* **117:**9–16.

Brauman, A., Koenig, J. F., Dutreix, J., and Garcia, J. L., 1990, Characterization of two sulfate-reducing bacteria from the gut of the soil-feeding termite, *Cubitermes speciosus, Antonie van Leeuwenhoek* **58:**271–275.

Bryant, M. P., Wolin, E. A., Wolin, M. J., and Wolfe, R. S., 1967, *Methanobacillus omelianskii,* a symbiotic association of two species of bacteria, *Arch. Microbiol.* **59:**20–31.

Bryant, M. P., Campbell, L. L., Reddy, C. A., and Crabill, M. R., 1977, Growth of *Desulfovibrio* in lactate or ethanol media low in sulfate in association with H$_2$- utilizing methanogenic bacteria, *Appl. Environ. Microbiol.* **33:**1162–1169.

Butlin, K. R., and Postgate, J. R., 1954, The microbiological formation of sulphur in Cyrenaican lakes, in: *Biology of Deserts* (J. Cloudsley-Thompson, ed.), Institute of Biology, London, pp. 112–122.

Campbell, L. L., Frank, H. A., and Hall, E. R., 1957, Studies on thermophilic sulfate-reducing bacteria. I. Identification of *Sporovibrio desulfuricans* as *Clostridium nigrificans, J. Bacteriol.* **73:**516–521.

Canfield, D. E., and Des Marais, D. J., 1991, Aerobic sulfate reduction in microbial mats, *Science* **251:**1471–1473.

Caumette, P., Cohen, Y., and Matheron, R., 1991, Isolation and characterization of *Desulfovibrio halophilus* sp. nov., a halophilic sulfate-reducing bacterium isolated from Solar Lake (Sinai), *System Appl. Microbiol.* **14:**33–38.

Chen, L., Liu, M.-Y., LeGall, J., Fareleira, P., Santos, H., and Xavier, A. V., 1993, Rubredoxin oxidase, a new flavo-hemo-protein, is the site of oxygen reduction to water by the "strict anaerobe" *Desulfovibrio gigas*, *Biochem. Biophys. Res. Commun.* **193:**100–105.

Cohen, Y., 1989, Photosynthesis in cyanobacterial mats and its relation to the sulfur cycle: A model for microbial sulfur interactions, in: *Microbial Mats: Physiological Ecology of Benthic Microbial Communities*, Chapter 3 (Y. Cohen, and E. Rosenberg, eds.), American Society for Microbiology, Washington, D.C., pp. 22–36.

Conrad, R., Phelps, T. J., and Zeikus, J. G., 1985, Gas metabolism evidence in support of the juxtaposition of hydrogen-producing and methanogenic bacteria in sewage sludge and lakes sediments, *Appl. Environ. Microbiol.* **50:**595–601.

Cypionka, H., Widdel, F., and Pfennig, N., 1985, Survival of sulfate-reducing bacteria after oxygen stress, and growth in sulfate-free oxygen-sulfide gradients, *FEMS Microbiol. Ecol.* **31:**39–45.

Daumas, S., Cord-Ruwisch, R., and Garcia, J. L., 1988, *Desulfotomaculum geothermicum* sp. nov., a thermophilic, fatty acid-degrading, sulfate-reducing bacterium isolated with $H_2$ from geothermal ground water, *Antonie van Leeuwenhoek* **54:**165–178.

Des Marais, D. J., 1990, Microbial mats and the early evolution of life, *Trends Ecol. Evol.* **5:**140–144.

Devereux, R., and Stahl, D. A., 1993, Phylogeny of sulfate-reducing bacteria and a perspective for analyzing their natural communities, in: *The Sulfate-Reducing Bacteria: Contemporary Perspectives*, Chapter 6 (J. M. Odom, and R. Singleton, Jr., eds.), Brock/Springer Series in Contemporary Bioscience, Springer-Verlag, New York, pp. 131–160.

Dilling, W., and Cypionka, H., 1990, Aerobic respiration in sulfate-reducing bacteria, *FEMS Microbiol. Lett.* **71:**123–128.

Fauque, G., 1989, Properties of [NiFe] and [NiFeSe] hydrogenases from methanogenic bacteria, in: *Microbiology of Extreme Environments and its Potential for Biotechnology* (M. S. DaCosta, J. C. Duarte, and R.A.D. Williams, eds.), Elsevier Applied Science, London and New York, pp. 216–236.

Fauque, G. D., 1994, Sulfur reductase from thiophilic sulfate-reducing bacteria, in: *Methods in Enzymology: Inorganic Microbial Sulfur Metabolism*, Volume 243 (H. D. Peck, Jr., and J. LeGall, eds.), Academic Press, Inc., San Diego, pp. 353–367.

Fauque, G., Hervé, D., and LeGall, J., 1979, Structure-function relationship in hemoproteins: The role of cytochrome $c_3$ in the reduction of colloidal sulfur by sulfate reducing bacteria, *Arch. Microbiol.* **121:**261–264.

Fauque, G. D., Barton, L. L., and LeGall, J., 1980, Oxidative phosphorylation linked to the dissimilatory reduction of elemental sulphur by *Desulfovibrio*, in: *Sulphur in Biology*, Ciba Foundation Symposium 72, Excerpta Medica, Amsterdam, pp. 71–86.

Fauque, G. D., Berlier, Y. M., Czechowski, M. H., Dimon, B., Lespinat, P. A., and LeGall, J., 1987, A proton-deuterium exchange study of three types of *Desulfovibrio* hydrogenases, *J. Ind. Microbiol.* **2:**15–29.

Fauque, G., Peck, H. D. Jr., Moura, J.J.G., Huynh, B. H., Berlier, Y., DerVartanian, D. V., Teixeira, M., Przybyla, A. E., Lespinat, P. A., Moura, I., and LeGall, J., 1988, The three classes of hydrogenases from sulfate-reducing bacteria of the genus *Desulfovibrio*, *FEMS Microbiol. Rev.* **54:**299–344.

Fauque, G., LeGall, J., and Barton, L. L., 1991, Sulfate-reducing and sulfur-reducing bacteria, in: *Variations in Autotrophic Life*, Chapter 10 (J. M. Shively, and L. L. Barton, eds.), Academic Press Limited, London, pp. 271–337.

Fauque, G. D., Klimmek, O., and Kröger, A., 1994, Sulfur reductases from spirilloid mesophilic sulfur-reducing eubacteria, in: *Methods in Enzymology: Inorganic Microbial Sulfur Metabolism*, Volume 243 (H. D. Peck, Jr., and J. LeGall, eds.), Academic Press, Inc., San Diego, pp. 367–383.

Fischer, U., 1988, Sulfur in biotechnology, in: *Biotechnology: Special Microbial Processes*, Chapter 15 (H.-J. Rehm, and G. Reeds, eds.), VCH Publishers, Weinheim, pp. 463–496.

Fründ, C., and Cohen, Y., 1992, Diurnal cycles of sulfate reduction under oxic conditions in cyanobacterial mats, *Appl. Environ. Microbiol.* **58:**70–77.

Fukui, M., and Takii, S., 1990, Survival of sulfate-reducing bacteria in oxic surface sediment of a seawater lake, *FEMS Microbiol. Ecol.* **73:**317–322.

Gibson, G. R., 1990, Physiology and ecology of the sulphate-reducing bacteria, *J. Appl. Bacteriol.* **69:**769–797.

Gibson, S.A.W., and Gibson, G. R., 1988, A rapid method for determination of viable sulphate-reducing bacteria in human faeces, *Lett. Appl. Microbiol.* **7:**33–35.

Gibson, G. R., Cummings, J. H., and Macfarlane, G. T., 1988, Competition for hydrogen between sulphate-reducing bacteria and methanogenic bacteria from the human large intestine, *J. Appl. Bacteriol.* **65:**241–247.

Gibson, G. R., Cummings, J. H., and Macfarlane, G. T., 1990, Factors affecting hydrogen uptake by bacteria growing in the human large intestine, in: *Microbiology and Biochemistry of Strict Anaerobes Involved in Interspecies Hydrogen Transfer* (J.-P. Belaich, M. Bruschi, and J. L. Garcia, eds.), Plenum Press, New York and London, pp. 191–202.

Gibson, G. R., Macfarlane, G. T., and Cummings, J. H., 1993, Sulphate-reducing bacteria and hydrogen metabolism in the human large intestine, *Gut* **34:**437–439.

Goldhaber, M. B., and Kaplan, I. R., 1974, The sulphur cycle, in: *The Sea, Marine Chemistry*, Volume 5 (E. D. Goldberg, ed.), Wiley, New York, pp. 569–655.

Gray, B. H., Fowler, C. F., Nugent, N. A., Rigopoulos, N., and Fuller, R. C., 1973, Reevaluation of *Chloropseudomonas ethylica* strain 2-K, *Int. J. Syst. Bacteriol.* **23:**256–264.

Gyure, R. A., Konopka, A., Brooks, A., and Doemel, W., 1990, Microbial sulfate reduction in acidic (pH 3) strip-mine lakes, *FEMS Microbiol. Ecol.* **73:**193–202.

Hardy, J. A., and Hamilton, A., 1981, The oxygen tolerance of sulfate-reducing bacteria isolated from North Sea waters, *Curr. Microbiol.* **6:**259–262.

Hastings, D., and Emerson, S., 1988, Sulfate reduction in the presence of low oxygen levels in the water column of the Cariaco Trench, *Limnol. Oceanogr.* **33:**391–396.

Hatchikian, C. E., LeGall, J., and Bell, G. R., 1977, Significance of superoxide dismutase and catalase activities in the strict anaerobes, sulfate-reducing bacteria, in: *Superoxide and Superoxide Dismutases* (A. M. Michelson, J. M. McCord, and I. Fridovich, eds.), Academic Press, London, pp. 159–172.

Herlihy, A. T., and Mills, A. L., 1985, Sulfate reduction in freshwater sediments receiving acid mine drainage, *Appl. Environ. Microbiol.* **49:**179–186.

Huisingh, J., McNeil, J. J., and Matrone, G., 1974, Sulfate reduction by a *Desulfovibrio* species isolated from sheep rumen, *Appl. Microbiol.* **28:**489–497.

Ingvorsen, K., Zeikus, J. G., and Brock, T. D., 1981, Dynamics of bacterial sulfate reduction in a eutrophic lake, *Appl. Environ. Microbiol.* **42:**1029–1036.

Jones, J. G., and Simon, B. M., 1984, The presence and activity of *Desulfotomaculum* spp. in sulphate-limited freshwater sediments, *FEMS Microbiol. Lett.* **21:**47–50.

Jorgensen, B. B., 1977, The sulfur cycle of a coastal marine sediment (Limfjorden, Denmark), *Limnol. Oceanogr.* **22:**814–832.

Jorgensen, B. B., and Bak, F., 1991, Pathways and microbiology of thiosulfate transformations and sulfate reduction in a marine sediment (Kattegat, Denmark), *Appl. Environ. Microbiol.* **57:**847–856.

Krumbein, W. E., and Stal, L. J., 1991, The geophysiology of marine cyanobacterial mats and biofilms, in: *Distribution and Activity of Microorganisms in the Sea*, Proc. 4th Europ. Mar. Microbiol. Symp., Kieler Meeresforsch., Sonderh. 8, pp. 137–145.

Kuenen, J. G., and Beudeker, R. F., 1982, Microbiology of thiobacilli and other sulphur-oxidizing autotrophs, mixotrophs, and heterotrophs, *Phil. Trans. R. Soc. Lond.* B **298**:473–497.

Kuenen, J. G., and Bos, P., 1989, Habitats and ecological niches of chemolitho(auto)trophic bacteria, in: *Autotrophic Bacteria,* Chapter 4 (H. G. Schlegel, and B. Bowien, eds.), Springer-Verlag, Berlin, pp. 53–80.

Laanbroek, H. J., and Pfennig, N., 1981, Oxidation of short-chain fatty acids by sulfate-reducing bacteria in freshwater and in marine sediments, *Arch. Microbiol.* **128**:330–335.

Laanbroek, H. J., and Veldkamp, H., 1982, Microbial interactions in sediment communities, *Phil. Trans. R. Soc. Lond.* B **298**:533–550.

Le Faou, A., Rajagopal, B. S., Daniels, L., and Fauque, G., 1990, Thiosulfate, polythionates and elemental sulfur assimilation and reduction in the bacterial world, *FEMS Microbiol. Rev.* **75**:351–382.

LeGall, J., and Fauque, G., 1988, Dissimilatory reduction of sulfur compounds, in: *Biology of Anaerobic Microorganisms,* Chapter 11 (A.J.B. Zehnder, ed.), John Wiley and Sons, Inc., New York, pp. 587–639.

Lespinat, P. A., Denariaz, G., Fauque, G., Toci, R., Berlier, Y., and LeGall, J., 1985, Fixation de l'azote atmosphérique et métabolisme de l'hydrogène chez une bactérie sulfato-réductrice sporulante, *Desulfotomaculum orientis, C.R. Acad. Sc. Paris* **301**:707–710.

Lespinat, P. A., Berlier, Y. M., Fauque, G. D., Toci, R., Denariaz, G., and LeGall, J., 1987, The relationship between hydrogen metabolism, sulfate reduction and nitrogen fixation in sulfate reducers, *J. Ind. Microbiol.* **1**:383–388.

Marshall, C., Frenzel, P., and Cypionka, H., 1993, Influence of oxygen on sulfate reduction and growth of sulfate-reducing bacteria, *Arch. Microbiol.* **159**:168–173.

Moore, W.E.C., Johnson, J. L., and Holdeman, L. V., 1976, Emendation of *Bacteroidaceae* and *Butyrivibrio* and descriptions of *Desulfomonas* gen. nov. and ten new species in the genera *Desulfomonas, Butyrivibrio, Eubacterium, Clostridium,* and *Ruminococcus, Int. J. Syst. Bacteriol.* **26**:238–252.

Moura, J.J.G., Moura, I., Teixeira, M., Xavier, A. V., Fauque, G. D., and LeGall, J., 1988, Nickel-containing hydrogenases, in: *Metal Ions in Biological Systems,* Volume 23 (H. Sigel, ed.), Marcel Dekker, Inc., New York and Basel, pp. 285–314.

Nanninga, H. J., and Gottschal, J. C., 1987, Properties of *Desulfovibrio carbinolicus* sp. nov. and other sulfate-reducing bacteria isolated from an anaerobic-purification plant, *Appl. Environ. Microbiol.* **53**:802–809.

Nazina, T. N., Ivanova, A. E., Kanchaveli, L. P., and Rozanova, E. P., 1988, A new spore-forming thermophilic methylotrophic sulfate-reducing bacterium, *Desulfotomaculum kuznetsovii* sp. nov., *Mikrobiologiya* **57**:823–827.

Nedwell, D. B., and Abram, J. W., 1979, Relative influence of temperature and electron donor and electron acceptor concentrations on bacterial sulfate reduction in salt-marsh sediment, *Microbial Ecol.* **5**:67–72.

Nissenbaum, A., and Kaplan, I. R., 1976, Sulfur and carbon isotopic evidence for biogeochemical processes in the Dead Sea ecosystem, in: *Environmental Biogeochemistry,* Volume 1 (J. O. Nriagu, ed.), Ann Arbor Science Publishers, Ann Arbor, Michigan, pp. 309–325.

Ollivier, B., Hatchikian, C. E., Prensier, G., Guezennec, J., and Garcia, J. L., 1991, *Desulfohalobium retbaense* gen. nov., sp. nov., a halophilic sulfate-reducing bacterium from sediments of a hypersaline lake in Senegal, *Int. J. Syst. Bacteriol.* **41**:74–81.

Ollivier, B., Caumette, P., Garcia, J.-L., and Mah, R. A., 1994, Anaerobic bacteria from hypersaline environments, *Microbiol. Rev.* **58**:27–38.

Oremland, R. S., 1988, Biogeochemistry of methanogenic bacteria, in: *Biology of Anaerobic Microorganisms,* Chapter 12 (A.J.B. Zehnder, ed.), John Wiley and Sons, New York, pp. 641–705.

Oremland, R. S., and Polcin, S., 1982, Methanogenesis and sulfate reduction: competitive and noncompetitive substrates in estuarine sediments, *Appl. Environ. Microbiol.* **44:**1270–1276.

Oren, A., 1988, Anaerobic degradation of organic compounds at high salt concentrations, *Antoine van Leeuwenhoek* **54:**267–277.

Ouattara, A. S., and Jacq, V. A., 1992, Characterization of sulfate-reducing bacteria isolated from Senegal ricefields, *FEMS Microbiol. Ecol.* **101:**217–228.

Peck, H. D., Jr., and Lissolo, T., 1988, Assimilatory and dissimilatory sulphate reduction: enzymology and bioenergetics, in: *The Nitrogen and Sulphur Cycles,* (J. A. Cole, and S. J. Ferguson, eds.), 42nd Symposium of the Society for General Microbiology, Cambridge University Press, Cambridge, pp. 99–132.

Peck, H. D. Jr., and Odom, J. M., 1984, Hydrogen cycling in *Desulfovibrio:* A new mechanism for energy coupling in anaerobic microorganisms, in: *Microbial Mats: Stromatolites,* (Y. Cohen, R. W. Castenholz, and H. O. Halverson, eds.), Alan R. Liss, Inc., New York, pp. 215–243.

Peck, H. D. Jr., LeGall, J., Lespinat, P. A., Berlier, Y., and Fauque, G., 1987, A direct demonstration of hydrogen cycling by *Desulfovibrio vulgaris* employing membrane-inlet mass spectrometry, *FEMS Microbiol. Lett.* **40:**295–299.

Pfennig, N., and Biebl, H., 1976, *Desulfuromonas acetoxidans* gen. nov. and sp. nov., a new anaerobic, sulfur-reducing, acetate-oxidizing bacterium, *Arch. Microbiol.* **110:**3–12.

Pfennig, N., and Widdel, F., 1981, Ecology and physiology of some anaerobic bacteria from the microbial sulfur cycle, in: *Biology of Inorganic Nitrogen and Sulfur,* (H. Bothe, and A. Trebst, eds.), Springer-Verlag, Berlin, pp. 169–177.

Pfennig, N., and Widdel, F., 1982, The bacteria of the sulphur cycle, *Phil. Trans. R. Soc. Lond.* B **298:**433–441.

Pfennig, N., Widdel, F., and Trüper, H. G., 1981, The dissimilatory sulfate-reducing bacteria, in: *The Prokaryotes: A Handbook on the Biology of Bacteria: Ecophysiology, Isolation, Identification, Applications,* 2nd ed., Volume 1 (M. P. Starr, H. Stolp, H. G. Trüper, A. Balows, and H. G. Schlegel, eds.), Springer-Verlag, Heidelberg, pp. 926–940.

Postgate, J. R., 1984, *The Sulphate-Reducing Bacteria,* 2nd ed., Cambridge University Press, Cambridge.

Postgate, J. R., Kent, H. M., and Robson, R. L., 1988, Nitrogen fixation by *Desulfovibrio,* in: *The Nitrogen and Sulphur Cycles,* (J. A. Cole, and S. J. Ferguson, eds.), 42nd Symposium of the Society for General Microbiology, Cambridge University Press, Cambridge, pp. 457–471.

Rajagopal, B. S., Lespinat, P. A., Fauque, G., LeGall, J., and Berlier, Y. M., 1989, Mass-spectrometric studies of the interrelations among hydrogenase, carbon monoxide dehydrogenase, and methane-forming activities in pure and mixed cultures of *Desulfovibrio vulgaris, Desulfovibrio desulfuricans,* and *Methanosarcina barkeri,Appl. Environ. Microbiol.* **55:**2123–2129.

Robinson, J. A., and Tiedje, J. M., 1984, Competition between sulfate-reducing and methanogenic bacteria for $H_2$ under resting and growing conditions, *Arch. Microbiol.* **137:**26–32.

Rosnes, J. T., Torsvik, T., and Lien, T., 1991, Spore-forming thermophilic sulfate-reducing bacteria isolated from North Sea oil field waters, *Appl. Environ. Microbiol.* **57:**2302–2307.

Rozanova, E. P., and Khudyakova, A. I., 1974, A new nonspore-forming thermophilic sulfate-reducing organism, *Desulfovibrio thermophilus* nov. sp., *Mikrobiologiya* **43**:908–912.

Rozanova, E. P., and Pivovarova, T. A., 1988, Reclassification of *Desulfovibrio thermophilus* (Rozanova, Khudyakova, 1974), *Mikrobiologiya* **57**:102–106.

Schoberth, S., 1973, A new strain of *Desulfovibrio gigas* isolated from a sewage plant, *Arch. Microbiol.* **92**:365–368.

Schönheit, P., Kristjansson, J. K., and Thauer, R. K., 1982, Kinetic mechanism for the ability of sulfate reducers to out-compete methanogens for acetate, *Arch. Microbiol.* **132**:285–288.

Shaposhnikov, V. V., Kondratieva, E. N., and Fedorov, V. D., 1960, A new species of green sulphur bacteria, *Nature* **187**:167–168.

Skyring, G. W., 1984, Sulfate reduction in marine sediments associated with cyanobacterial mats in Australia, in: *Microbial Mats: Stromatolites* (Y. Cohen, R. W. Castenholz, and H. O. Halvorson, eds.), Alan R. Liss, Inc., New York, pp. 265–275.

Skyring, G. W., Chambers, L. A., and Bauld, J., 1983, Sulfate reduction in sediments colonized by cyanobacteria, Spencer Gulf, South Australia, *Aust. J. Mar. Freshw. Res.* **34**:359–374.

Smith, D. W., 1993, Ecological actions of sulfate-reducing bacteria, in: *The Sulfate-Reducing Bacteria: Contemporary Perspectives,* Chapter 7 (J. M. Odom, and R. Singleton, Jr. eds.), Brock/Springer Series in Contemporary Bioscience, Springer-Verlag, New York, pp. 161–188.

Smith, R. L., and Klug, M. J ., 1981, Electron donors utilized by sulfate-reducing bacteria in eutrophic lake sediments, *Appl. Environ. Microbiol.* **42**:116–121.

Spratt, H. G., Morgan, M. D., and Good, R. E., 1987, Sulfate reduction in Peat from a New Jersey Pinelands cedar swamp, *Appl. Environ. Microbiol.* **53**:1406–1411.

Stal, L. J., Van Gemerden, H., and Krumbein, W. E., 1985, Structure and development of a benthic marine microbial mat, *FEMS Microbiol. Ecol.* **31**:111–125.

Stetter, K. O., 1992, The genus *Archaeoglobus,* in: *The Prokaryotes: A Handbook on the Biology of Bacteria: Ecophysiology, Isolation, Identification, Applications,* 2nd ed., Volume I (A. Balows, H. G. Trüper, M. Dworkin, W. Harder, and K.-H. Schleifer, eds.), Springer-Verlag, New York, pp. 707–711.

Traoré, S. A., Fauque, G., Jacq, V. A., and Belaich, J.-P., 1990, Characterization of a sulfate-reducing bacterium isolated from the gut of a tropical soil termite, in: *Microbiology and Biochemistry of Strict Anaerobes Involved in Interspecies Hydrogen Transfer* (J.-P. Belaich, M. Bruschi, and J. L. Garcia, eds.), Plenum Press, New York and London, pp. 481–483.

Trüper, H. G., 1984, Microorganisms and the sulfur cycle, in: *Sulfur, its Significance for Chemistry, for the Geo-, Bio-, and Cosmosphere and Technology* (A. Müller, and B. Krebs, eds.), *Studies in Inorganic Chemistry,* Volume 5, Elsevier Science Publishers B.V., Amsterdam, pp. 351–365.

Trüper, H. G., and Fisher, U., 1982, Anaerobic oxidation of sulphur compounds as electron donors for bacterial photosynthesis, *Phil. Trans. R. Soc. Lond.* B **298**:529–542.

Tuttle, J. H., Dugan, P. R., MacMillan, C. B., and Randles, C. I., 1969, Microbial dissimilatory sulfur cycle in acid mine water, *J. Bacteriol.* **97**:594–602.

Van Gemerden, H., and Beeftink, H. H., 1981, Coexistence of *Chlorobium* and *Chromatium* in a sulfide-limited continuous culture, *Arch. Microbiol.* **129**:32–34.

Visscher, P. T., Prins, R. A., and Van Gemerden, H., 1992, Rates of sulfate reduction and thiosulfate consumption in a marine microbial mat, *FEMS Microbiol. Ecol.* **86**:283–294.

Wall, J. D., Rapp-Giles, B. J., Brown, M. F., and White, J. A., 1990, Response of *Desulfovibrio desulfuricans* colonies to oxygen stress, *Can. J. Microbiol.* **36**:400–408.

Ward, D. M., and Olson, G. J., 1980, Terminal processes in the anaerobic degradation of an algal-bacterial mat in a high-sulfate hot spring, *Appl. Environ. Microbiol.* **40:**67–74.

Ward, D. M., and Winfrey, M. R., 1985, Interactions between methanogenic and sulfate-reducing bacteria in sediments, in: *Advances in Aquatic Microbiology*, Volume 3 (H. W. Jannasch, and P. J. LeB.Williams, eds.), Academic Press, Orlando, pp. 141–179.

Westrich, J. T., and Berner, R. A., 1988, The effect of temperature on rates of sulfate reduction in marine sediments, *Geomicrobiol. J.* **6:**99–117.

Widdel, F., 1986, Sulphate-reducing bacteria and their ecological niches, in: *Anaerobic Bacteria in Habitats Other than Man*, (E. M. Barnes, and G. C. Mead, eds.), Blackwell Scientific Publications, Oxford, pp. 157–184.

Widdel, F., 1988, Microbiology and ecology of sulfate- and sulfur-reducing bacteria, in: *Biology of Anaerobic Microorganisms*, Chapter 10 (A.J.B. Zehnder, ed.), John Wiley and Sons, Inc., New York, pp. 469–585.

Widdel, F., 1992a, The genus *Desulfotomaculum*, in: *The Prokaryotes: A Handbook on the Biology of Bacteria: Ecophysiology, Isolation, Identification, Applications*, 2nd ed., Volume II (A. Balows, H. G. Trüper, M. Dworkin, W. Harder, and K.-H. Schleifer, eds.), Springer-Verlag, New York, pp. 1792–1799.

Widdel, F., 1992b, The genus *Thermodesulfobacterium*, in: *The Prokaryotes: A Handbook on the Biology of Bacteria: Ecophysiology, Isolation, Identification, Applications*, 2nd ed., Volume IV (A. Balows, H. G. Trüper, M. Dworkin, W. Harder, and K.-H. Schleifer, eds), Springer-Verlag, New York, pp. 3390–3392.

Widdel, F., and Bak, F., 1992, Gram-negative mesophilic sulfate-reducing bacteria, in: *The Prokaryotes: A Handbook on the Biology of Bacteria: Ecophysiology, Isolation, Identification, Applications*, 2nd ed., Volume IV (A. Balows, H. G. Trüper, M. Dworkin, W. Harder, and K.-H. Schleifer, eds.), Springer-Verlag, New York, pp. 3352–3378.

Widdel, F., and Hansen, T. A., 1992, The dissimilatory sulfate- and sulfur-reducing bacteria, in: *The Prokaryotes: A Handbook on the Biology of Bacteria: Ecophysiology, Isolation, Identification, Applications*, 2nd ed., Volume I (A. Balows, H. G. Trüper, M. Dworkin, W. Harder, and K.-H. Schleifer, eds.), Springer-Verlag, New York, pp. 583–624.

Widdel, F., and Pfennig, N., 1984, Dissimilatory sulfate-or sulfur-reducing bacteria, in: *Bergey's Manual of Systematic Bacteriology*, Volume 1 (N.R. Krieg, and J.G. Holt, eds.), The Williams and Wilkins Co., Baltimore, pp. 663–679.

Widdel, F., and Pfennig, N., 1992, The genus *Desulfuromonas* and other Gram-negative sulfur-reducing eubacteria, in: *The Prokaryotes: A Handbook on the Biology of Bacteria: Ecophysiology, Isolation, Identification, Applications*, 2nd ed., Volume IV (A. Balows, H. G. Trüper, M. Dworkin, W. Harder, and K.-H. Schleifer, eds.), Springer-Verlag, New York, pp. 3379–3389.

Winfrey, M. R., and Ward, D. M., 1983, Substrates for sulfate reduction and methane production in intertidal sediments, *Appl. Environ. Microbiol.* **45:**193–199.

Zeikus, J. G., 1983, Metabolic communication between biodegradative populations in nature, in: *Microbes in Their Natural Environments*, (J. H. Slater, R. Whittenbury, and J.W.T. Wimpenny, eds.), 34th Symposium of the Society for General Microbiology, Cambridge University Press, Cambridge, pp. 423–462.

Zeikus, J. G., Dawson, M. A., Thompson, T. E., Ingvorsen, K., and Hatchikian, E. C., 1983, Microbial ecology of volcanic sulphidogenesis: isolation and characterization of *Thermodesulfobacterium commune* gen. nov. and sp. nov., *J. Gen. Microbiol.* **129:**1159–1169.

Zellner, G., Messner, P., Kneifel, H., and Winter, J., 1989, *Desulfovibrio simplex* spec. nov., a new sulfate-reducing bacterium from a sour whey digester, *Arch. Microbiol.* **152:**329–334.

Zobell, C. E., 1958, Ecology of sulfate-reducing bacteria, *Producers Mon. Penn. Oil Prod. Ass.* **22:**12–29.

# Biocorrosion  9

## W. ALLAN HAMILTON and WHONCHEE LEE

## 1. INTRODUCTION

As is obvious from the other chapters in this book, the sulfate-reducing bacteria (SRB) are currently one of the more intensively studied groups of microorganisms. The reasons for this are not difficult to discern. Their obligate anaerobiosis depends on a number of intriguing variations on standard aerobic respiratory metabolism. This property also determines their considerable significance in the carbon, energy, and sulfur turnover in many anoxic microbial ecosystems. Recent insights from molecular analyses have shown the group to be of particular interest in microbial evolution. However, perhaps the single most important factor stimulating the upsurge of interest in the SRB in recent years has been their considerable, albeit largely negative, ecological and economic impact. In the offshore oil and gas industries, for example, the SRB are implicated in sulfide build-up in enclosed working environments and in seabed pollution under deposits of organically rich drill cuttings (Sanders and Tibbetts, 1987); they are the principal causative organism in microbially influenced corrosion of platform structures, transmission lines, and general equipment (Hamilton, 1985; Cord-Ruwisch et al., 1987); they are the likely cause of major reservoir damage including souring of the produced oil and gas (high sulfide content), and plugging of the geological formation (Herbert, 1987). In applied microbiology, therefore, these various problems are manifestly of major importance to the industries affected by them. The study of the cellular mechanisms involved, however, can also offer rich rewards both to our understanding of microbiological processes themselves, and to our appreciation of

W. ALLAN HAMILTON • Molecular & Cell Biology, Marischal College, University, Aberdeen AB9 1AS, Scotland, United Kingdom    WHONCHEE LEE • Center for Biofilm Engineering, Montana State University, Bozeman, MT 59717-0398.

*Sulfate-Reducing Bacteria*, edited by Larry L. Barton. Plenum Press, New York, 1995.

**243**

how microbial ecosystems interact with and depend on physicochemical components of their immediate environment.

In the study of the role of SRB in biocorrosion there is the additional challenge of marrying the experimental approaches and technical language of the corrosion engineer and materials scientist to those of the microbiologist. This is by no means a trivial matter, and whereas the major advances in our understanding have come from such marriages, it is equally true to say that many of the outstanding contradictions and confusions, whether real or apparent, have their origin in one or other discipline seeking to address the problem entirely through its own view of the world. Consideration of this aspect of the problem will be an important feature of this exposition on microbial biocorrosion.

## 2. CORROSION

### 2.1. Abiotic Corrosion

Corrosion is an electrochemical phenomenon in which the electrical potentials of two adjacent areas of a metallic surface within a conductive medium differ relative to each other such that one area is anodic and the other cathodic. Abiotically, such an electrochemical cell can arise from two dissimilar metals in contact, or from inclusions or other surface heterogeneities within a single metal or alloy. Metal dissolution takes place at the anode:

$$M \rightarrow M^{2+} + 2e$$

This half reaction needs to be balanced by an equivalent cathodic reaction involving the reduction of an electron acceptor. Oxygen is the classical cathodic electron acceptor:

$$2e + 1/2O_2 + H_2O \rightarrow 2OH^-$$

with the combination of the products of the anodic and cathodic reactions giving rise to the metal hydroxides and oxides characteristic of rust. The accumulation at the metal–electrolyte interface of products of these reactions tends to slow the rate of corrosive metal loss. This process is termed polarization, and extensive corrosion damage over a protracted time scale is associated with secondary reactions leading to depolarization of the electrochemical cell and consequent continuing metal dissolution.

### 2.1.1. Analysis by Electrochemical Techniques

The nature of the corrosion occurring in any given situation can be monitored by visual or microscopic observation, and by chemical analysis of the corrosion products. Quantitative measures of the rate and extent of corrosion, however, are gained either by determination of weight loss after removal of the corrosion products, or on a more continuous basis by one or more of a number of electrochemical measurements. Whereas visual observation can indicate whether any metal loss is generalized, or localized in the form of pitting corrosion, analysis of corrosion products and electrochemical measurements can give more extensive information on the possible mechanism of the observed corrosion.

Examples of electrochemical techniques are direct current polarization, AC impedance, and polarization resistance (Dexter *et al.*, 1991). In direct current polarization, for example, the open circuit, or corrosion, potential is first measured against a standard reference electrode using a potentiostat. The specimen can then be polarized, either in the so-called active (cathodic) direction, or in the noble (anodic) direction. These polarizations may be large (several hundred mV) or small (10 to 20 mV). They may be induced either by controlling the applied potential and measuring the resultant current, or by applying a current and measuring the resultant potential. The polarization can be applied in discrete steps (e.g., 25 mV every 5 min) or dynamically at a constant scan rate. The data are then recorded in the form of a plot of potential (E, in mV) against current density (i, in $mA.cm^{-2}$), or more correctly as E against log i. Although changes in the open circuit corrosion potential against time indicate the onset of a corrosive process, a great deal more information can be obtained from analysis of the potential against current density plots. It can be deduced, for example, whether the corrosion processes involve stimulation of the anodic or the cathodic reaction, such stimulation generally being referred to as depolarization. This, in turn, can suggest the mechanism underlying the corrosive weight loss.

In theory, both the anodic and cathodic curves should be linear when the applied potential is plotted against the log of the current density. In practice, however, both curves deviate from linearity, particularly in the region of the open circuit corrosion potential, although they do normally contain significant linear segments which are referred to as Tafel regions. Extrapolation of these Tafel regions to the point of intersection of the anodic and cathodic curves, gives values for the corrosion

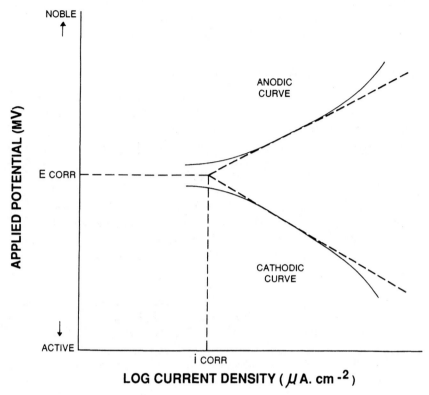

**Figure 1.** Tafel plot obtained by direct current polarization.

potential and for the corrosion current density (Fig. 1). From these values the corrosion rate can be determined from the relationship:

$$R = 0.13 \ i_{corr} \ k/\rho$$

where R is the corrosion rate in $mm.y^{-1}$, $i_{corr}$ is the corrosion current density in $\mu A.cm^{-2}$, k is the equivalent weight of the metal, and $\rho$ is the metal density in $g.cm^{-3}$. It should be stressed, however, that the value for the corrosion rate determined in this manner gives an average figure on the assumption of metal loss in a generalized corrosion process across the metal surface. This relationship is not, therefore, strictly applicable in cases of pitting corrosion.

In fact, corrosion often involves a series of reaction steps with, for example, an initial phase of corrosive weight loss being followed by a period of reduced corrosion arising from the protective character of the primary deposition of corrosion products on the metal surface. Thereafter, rupture or chemical modification of this protective layer can lead to a second active stage of corrosion, often more rapid and prolonged than the initial phase and in many instances associated with a localized pitting. This pattern is characteristic of microbial corrosion in general, and of corrosion resulting from sulfide in particular (Salvarezza *et al.*, 1983). The use of potentiodynamic sweep techniques (Dexter *et al.*, 1991) allows the characterization of this phenomenon, with identification of the so-called pitting potential at which there is a rapid increase in the corrosion current associated with the transition from the passive protected phase to the second active phase. In microbial corrosion it is significant that the initial phase of generalized corrosion may be directly influenced by microbial activity, but that the second and major phase of active and prolonged corrosion is much more likely to be dependent on variations in environmental parameters such as local free iron concentrations, hydrodynamics of the system, and the presence or absence of oxygen.

A further important aspect of abiotic corrosion processes is that, both in theory and in practice, the application of electrochemical techniques assumes that the metal surface is in direct contact with the electrolyte, and that the anodic and cathodic reactions taking place at the metal–solution interface are not subject to diffusional limitations or interference from competing reactions. Whereas modifications to the theory, and differing interpretations of the experimental data obtained have been made in this regard with respect to inorganic passivating films and deposits of corrosion products, the potentially much more pronounced effects of microbial biofilms have largely been ignored in the application of electrochemical techniques to the study of microbially influenced corrosion (Dexter *et al.*, 1991).

### 2.2. Microbially Influenced Corrosion

This brief statement of the fundamental nature of corrosion processes and the techniques for their analyses, allows us to identify a number of key questions that require answers in any attempt to describe the mechanisms underlying microbial corrosion.

1. Are microorganisms involved in the creation of an electrochemical cell, or in its continued functioning over a prolonged period?
2. Do they affect principally the anodic or the cathodic reaction?
3. Is the mechanism direct, or indirect through the action of products of the cells' metabolism?
4. What is the identity of the cathodic electron acceptor, particularly under anoxic conditions?
5. Is the influence of biofilm growth primarily metabolic through the combined action of the organisms present as a consortium, or

physical through the development of diffusional gradients and microenvironments?

6. Can pitting corrosion be the consequence of colonial growth or a patchy biofilm?

7. Do the corrosion products themselves have an influence on the nature and extent of any further corrosion?

Before considering these questions in greater detail with particular reference to SRB and the anaerobic corrosion of mild steel, a general statement can be made. Corrosion is a generic term covering a number of electrochemical processes (e.g., pitting, generalized weight loss, graphitization, stress corrosion cracking, hydrogen embrittlement) which may affect iron, mild steel, stainless steels, and various aluminium, copper, nickel and cobalt alloys. Microbial corrosion can involve a plethora of organisms and mechanisms; effects may be specific, as with $Fe^{2+}$ to $Fe^{3+}$ oxidation by *Gallionella* or the production of organic acids by *Cladosporium,* or more general, as with differential aeration cells arising from colonial growth or biofilm patchiness. Although the nature of the corrosion process can generally be defined within narrow limits in respect to its metallurgy, the microbiological component is seldom simple or easily identified as a single organism or a unique mechanism (Hamilton, 1991; Tatnall, 1991).

An example of this last point is the oxygen concentration or differential aeration cell which can arise from localized growth of aerobic organisms. Here the region immediately below the area of active growth becomes anoxic and hence anodic relative to the surrounding metallic surface exposed to the air, and therefore acts as a site of metal dissolution. It is likely that this constitutes the most common and widespread mechanism of microbial corrosion associated with generalized microbial growth, either as a discontinuous biofilm or in discrete colonies. The further microbial oxidation of iron from the ferrous to the ferric state can occasionally increase the extent of corrosion and lead to the formation of tubercles. Also, the anoxic regions created by such microbial activities can serve as ideal microenvironments for the growth of SRB, with the consequent development of anaerobic corrosion. Thus three organisms and three separate mechanisms of corrosion may be operating simultaneously, and to some extent in concert (Hamilton, 1985).

### 2.2.1. Sulfate-Reducing Bacteria and Biofilms

Although the SRB are obligate anaerobes, they can readily be isolated from aerobic environments. Also, it is widely documented that the worst cases of corrosion involving SRB are often associated with an ingress of oxygen. Both of these features arise from the common occurrence of SRB in natural environments as component members of mixed microbial consortia, within which aerobic and facultative organisms both supply the nutrients for the SRB from the products of their own partial metabolism of the primary nutrients, and generate the necessary reducing conditions to allow growth of SRB within the consortium (Hamilton, 1985). Where such a consortium takes the form of a biofilm on a metal or other substratum, anoxic conditions can readily be generated in the base region at biofilm thicknesses in excess of $20\mu m$. It is only relatively recently, however, that the direct role of oxygen in stimulating SRB-induced corrosion per se has been clearly identified and subjected to careful experimental study. Whereas earlier investigations tended to use pure cultures or selected enrichments of SRB under controlled conditions of anaerobiosis, more recently both environmentally based analyses and laboratory simulations of naturally occurring corrosion processes have expanded their approach to examine the effects of mixed microbial biofilms formed within oxygen-containing environments.

There is thus something of a hierarchy in the methodologies and their underlying theoretical bases that have been applied, at different times and by different scientific disciplines, to the study of microbial corrosion: from electrochemical analyses in the absence of physical interference from biofilms; to homogeneous suspensions of SRB under conditions of anoxia; to structured communities displaying physicochemical heterogeneities. While this progression has been neither absolute nor linear, it does indicate the development in our understanding of an extremely complex system, in which the cardinal feature is now seen to be the interdependence between the biological and the physical components of a given environment.

### 2.2.2. Hypotheses for the Mechanism of Anaerobic Microbial Corrosion

1. Probably the most widely quoted model for the mechanism of SRB-induced biocorrosion is the classical *cathodic depolarization* hypothesis (von Wolzogen Kuhr and van der Vlught, 1934). In this scheme it is envisaged that in the absence of oxygen and under acid conditions, protons act as the cathodic electron acceptor with the formation of, first, atomic, then molecular hydrogen.

$$2H^+ + 2e \rightarrow 2H \rightarrow H_2$$

This reaction sequence, and consequently the overall corrosion process, is inhibited by the polarizing effect of the accumulation of a film of molecular hydrogen at the metal surface. The oxidation of this hydrogen by hydrogenase-positive SRB is then considered to result in cathodic depolarization with a consequent marked stimulation of metal dissolution at the anode.

2. Alternatively, Costello (1974) has proposed that at neutral pH values hydrogen sulfide is the electron acceptor, again with hydrogen as the key cathodic product.

$$2H_2S + 2e \rightarrow 2HS^- + H_2$$

These two proposals have in common cathodic stimulation, and the production of metal sulfide as a corrosion product.

$$4H_2 + SO_4^{2-} \rightarrow 4H_2O + S^{2-}$$

$$M^{2+} + S^{2-} \rightarrow MS$$

The stoichiometry of this last reaction is critically influenced by the extent of sulfide production, independent of cathodic depolarization and due to sulfate reduction from the growth of SRB on various organic substrates. Equally, the concentration of soluble metal ions can determine the nature of the final product. In this regard, it is highly significant that the iron sulfide generated by the corrosion of cast iron or mild steel can exist in a number of forms, differing in their iron to sulfur stoichiometries, and in their physical and chemical properties: for example, mackinawite (tetragonal, $FeS_{1-x}$), ferrous sulfide (cubic, $FeS$), troilite (hexagonal, $FeS$), pyrrhotite (hexagonal, $Fe_{1-x}S$), pyrite (cubic, $FeS_2$).

3. Taking a contrary view, other workers have concluded that SRB stimulate corrosion by a mechanism of *anodic depolarization,* essentially dependent upon sulfide production (Wanklyn and Spruit, 1952; Salvarezza and Videla, 1980; Salvarezza *et al.,* 1983; Crolet, 1992). In an extensive series of papers, Crolet has argued that the basic tenet of cathodic depolarization is incompatible with our present day understanding of the irreversible nature of the electrochemical reactions of corrosion. He has further pointed out that the equations used to describe the metabolism of the causative organisms in microbial corrosion are seldom correctly balanced, in that the dissociation of the weak acids, including $H_2S$, are generally ignored. The theoretical calculation of the pH dependence of $H^+$ production with SRB metabolizing lactate, and

the experimental verification of the $H^+/HS^-$ stoichiometry, demonstrate that SRB are capable of regulating their immediate environmental pH to that value at which $H^+/HS^-$ is zero (Crolet *et al.*, 1993; Daumas *et al.*, 1993). These are the theoretical and experimental bases for a model of the mechanism of SRB-induced corrosion that incorporates the two principal characteristics unique to this process: the central role of sulphide production, and localized pitting.

It is proposed that an anode is first created by local $H^+$ production at a focus of SRB metabolic activity, with ensuing metal dissolution. This initial phase of corrosive activity may be generalized across the exposed metal surface with random nucleation being by statistical fluctuation in the anodic reaction. If kinetic conditions become favorable, however, a chain reaction can be established leading to both a stable galvanic current associated with a particular anode, and the initiation of pit growth at that site. A key element in generating the necessary kinetic conditions is the increased localized acidification at the anode resulting from the formation of iron sulfide corrosion products (Crolet, 1992):

$$Fe^{2+} + HS^- \rightarrow FeS \downarrow + H^+$$

This reaction also has the secondary effect of removing $HS^-$ and so reducing the effectiveness of the $H_2S/HS^-$ buffer system, which would otherwise reduce the extent of acidification. Crolet (1993) has also discussed the effects of the relative availabilities of sulfide and free iron on the nature of the corrosion products formed. Where the local sulfide concentration is low, as would be the case with the above reaction in the presence of high soluble iron concentration, the product formed is most likely to be mackinawite ($FeS_{1-x}$) which is considered to be nonprotective. Where sulfide is in excess, however, the product will be the more protective pyrite ($FeS_2$). Similarly, the recognized stimulatory action of oxygen is predicted on the basis of, firstly, the formation of thiosulfate, followed by a dismutation to sulfate and sulfide with further acidification.

$$2HS^- + 2O_2 \rightarrow S_2O_3^{2-} + H_2O \rightarrow SO_4^{2-} + HS^- + H^+$$

It is claimed that by this means continuous access of oxygen "constitutes an infinite source of acidity" (Crolet *et al.*, 1991). Furthermore, in their analysis of $H^+/HS^-$ stoichiometries with different electron acceptors, Crolet *et al.*, (1993) predict that the severity of the attendant corrosion should be $S^o > S_2O_3^{2-} \gg SO_4^{2-} > SO_3^{2-}$.

It is characteristic of pitting corrosion that a small anode should be surrounded by a much larger cathodic area, and Campaignolle *et al.* (1993) have generated such a situation experimentally, using two concentric electrodes separated by a Teflon insulator, with the ratio of their surface areas being of the order of 150 to 1. Under conditions of sterile deaerated seawater it is possible to initiate a galvanic coupling between the anode and cathode through the application of a preconditioning current. This coupling is unstable and it disappears within a few hours after removal of the applied current. In the presence of *Desulfovibrio vulgaris,* however, the galvanic coupling is stabilized. This effect requires the SRB to grow within the biofilm that forms on both electrodes. Where growth was only present on the cathode, no stabilization of the galvanic current was found.

These data constitute a partial verification of Crolet's model of anodic stimulation arising from local acidification.

4. Although not itself a direct product of SRB metabolism, *elemental sulfur (S°)* can be derived from the oxidation of sulfide consequent to oxygen exposure, and its presence has been noted around the periphery of corrosion pits in several instances of SRB microbial corrosion. In a number of studies, it has been directly demonstrated that sulfur is extremely corrosive to mild steel, with evidence of shallow pitting and calculated corrosion rates of $2mm.y^{-1}$. There are, however, at least three models proposed for the mechanism of sulfur-induced corrosion. Maldonado-Zagal and Boden (1982) suggest that the high local acidity generated on particles of solid sulfur reacting with water could be responsible for the observed high rates of corrosion. Schaschl (1980), on the other hand, favors a concentration cell mechanism, directly analogous to a differential aeration cell, with the principal role of the bacteria being to promote the concentration cell by shielding the underlying metal (anode) from the higher concentration of dissolved sulfur in the surrounding medium. More recently, the corrosive interaction of wet elemental sulfur with mild steel has been reviewed by Schmitt (1991). Five reaction steps have been recognized:

1. sulfur disproportion in water

$$4S° + 4H_2O \rightarrow 3H_2S + H_2SO_4$$

2. formation of iron sulfide film on steel

$$Fe + H_2S \rightarrow FeS + H_2$$

3. sulfur as cathodic electron acceptor

$$S° + H_2O + 2e \rightarrow HS^- + OH^-$$

4. anodic metal dissolution

$$Fe \rightarrow Fe^{2+} + 2e$$

5. chemical formation of iron sulfides

$$Fe^{2+} + HS^- + OH^- \rightarrow FeS + H_2O$$

It has been suggested that the reaction sequence might proceed in two stages. First, the iron sulfide on the steel surface may be a protective film of mackinawite exerting anodic control on the metal dissolution by limiting ferrous ion diffusion through the sulfide film. Subsequently, breakdown of this protective film will create an electrochemical cell (unreacted steel as anode, iron sulfide as cathode) with sulfur as the electron acceptor stimulating cathodic depolarization.

5. *Metal sulfide corrosion products* appear, therefore, to play a central role in determining the nature and extent of further corrosion events, whatever is conceived to be the primary anodic or cathodic reaction. The pattern of events outlined above is repeated in a wide range of experimental studies of SRB microbial corrosion where, broadly speaking, a thin, adherent, continuous sulfide film is found to be protective. This exactly parallels the naturally occurring oxide film which is the basis of the corrosion resistance of stainless steel in aerobic environments. Where, however, the sulphide film is bulky and loosely adherent, or where the adherent film ruptures with exposure of the underlying unreacted metal, extensive and prolonged corrosion will take place, independent of microbial activity. The study of microbial corrosion, therefore, as a real event of practical importance in a range of industrial and natural environments, requires elucidation of the chemical and physical nature of the sulfide corrosion products, and of the factors influencing their formation and activity. Principal among these factors are the relative concentrations of soluble iron and of produced sulfide, and the continuous or intermittent availability of oxygen.

### 2.2.3. Soluble Iron Concentration and Anaerobic Microbial Corrosion

During the 1960s there was a very active period of research into the mechanisms of microbial corrosion, which has been fully documented in the review by Tiller (1982). The predominant focus of this research was the attempt to obtain data in support of the classical cathodic depolarization hypothesis, using polarization and potentiostatic techniques. Initially studies were carried out with single species batch cultures under controlled conditions of anaerobiosis. Although corrosion weight loss was demonstrated, hydrogenase-negative strains were largely inactive in this system. When in later studies continuous culture was used over extended time periods, the picture became less clear-cut. Corrosion rates were generally low, and a thin ferrous sulfide film formed on test specimens. After a period of some months, however, this film fractured and there was a considerable increase in the rate of corrosion, with very little difference between hydrogenase-positive and -negative strains. In the presence of higher soluble iron concentration (5 $g.l^{-1}$) the primary corrosion product was in the form of a bulky black precipitate rather than as an adherent film, and with both hydrogenase-positive and -negative strains, high rates of corrosion were found (1.1 $mm.y^{-1}$), comparable with the rates observed in naturally corrosive environments such as clay soils. Mara and Williams (1972) came to the same general conclusion and suggested that the time-dependent film breakdown in low-iron media resulted from sulfidation of the primary corrosion product mackinawite ($FeS_{1-x}$) to greigite ($Fe_3S_4$). Furthermore, whereas the initial low rate of corrosion was related to bacterial growth rate, after film rupture the much higher rate of corrosion was independent of microbial activity. In electrochemical potential, ferrous sulfide is cathodic to unreacted iron or steel so that where there is contact between corrosion product and exposed metal an electrochemical corrosion cell is established.

These studies were greatly extended by the work of King and Miller and their colleagues (King and Miller, 1971; King, et al. 1973, 1976). They noted that under all conditions the initial sulfide film formed during corrosion is in the form of a continuous and adherent layer of mackinawite, which is protective. However, as noted above, this film can show a degree of physical disruption in a time-dependent modification to greigite. In the presence of higher soluble iron concentrations there is a similar loss of the protective mackinawite film with the conversion to smythite ($Fe_3S_4$) and pyrrhotite ($Fe_{1-x}S$). In each case this loss of the uniform protective mackinawite film generates active electrochemical cells between areas of unreacted, and unprotected, steel and deposits of the various ferrous sulfides, with the resultant corrosion occurring by cathodic stimulation. The reduced ferrous sulfides, mackinawite, greigite, smythite, and pyrrhotite are not thermodynamically stable in aqueous solution where they tend to revert to iron oxides or hydroxides by reaction with water and/or oxygen. Continued sulfide production from SRB is therefore required to maintain their chemical integrity and electrochemical activity (Newman et al., 1991).

By the mid 1970's, therefore, the consensus view was that microbial corrosion by the SRB proceeded under reduced conditions through cathodic stimulation of electrochemical cells established between areas of unreacted steel (anode) and deposits of various reduced ferrous sulfide corrosion products (cathode). The role of $H^+$ as cathodic electron acceptor and $H_2$ oxidation by SRB were implicit in most formulations of the mechanistic model, but not quantified, and the major role of SRB metabolic activity was seen as, first, generating the sulfide to foster the creation of the electrochemical cells and, second, thereafter maintaining their long-term reactivity. The key issue of the part played by the various chemical forms of the ferrous sulfides was seen to relate to their physical structure, and in particular to their respective abilities to form homogeneous adherent and therefore protective layers, or to give rise to surface heterogeneities leading to anode/cathode interactions.

In more recent years there have been a number of studies that have confirmed and extended this general model. For example, Hardy (1983), Pankhania et al. (1986), and Cord-Ruwisch et al. (1987) have all demonstrated that SRB can oxidize cathodically produced $H_2$ and utilize it as a legitimate source of metabolic energy. Significantly, however, the resultant corrosion in these experimental systems was shown to be only transient. Even Bryant et al. (1991), in their attempt to relate hydrogenase activity quantitatively to the observed corrosion, noted that the highest rates of corrosion occurred after the period of microbial activity and, as in the earlier studies, they also suggested that ferrous sulfide corrosion products may have a direct role to play in corrosion mechanisms over and above any effects of hydrogenase.

An interesting commentary on the oxidation of cathodic $H_2$ by SRB is provided in the report by Guezennec et al., (1991) that the application of an impressed current to confer protection against corrosion weight loss, also had the effect of simulating sulfate reducer growth and activity in direct response to the increased tendency to produce $H_2$ at the cathode. This, therefore, parallels the findings of Moosavi et al. (1991) and of Lee and Characklis (1993) that there is, in fact, no direct or quantitative link between SRB numbers or activity in a given environmental or laboratory system, and the rate of corrosion measured in that same system.

The most recent statement of the effects of ferrous sulfide corrosion products on the rate and extent of SRB microbial corrosion under strictly anaerobic conditions is contained in the paper by Lee and Characklis (1993) and in the review by Lee et al. (1995). Using *Desulfovibrio desulfuricans* under nonaseptic conditions in a continuous flow biofilm reactor, Lee and Characklis (1993) have shown that in an iron-free medium where biofilm thickness and SRB activity are controlled by substrate (lactate) loading rate, there is no discernible corrosive weight loss of a mild steel substratum nor any evidence of iron content within the bio-film, even where sulfate reduction activity increases more than tenfold at a high substrate loading rate. When, however, the steel surface is precoated with an iron sulfide film prior to development of the biofilm at the high substrate-loading rate, localized corrosion is observed at sites of inclusions and grain boundaries. This corrosion is associated with the pitting potential derived from the anodic polarization curve shifting to a more active value ($-300$ mV to $-450$ mV) with only little change in the cathodic polarization curve. Aggregates of bacteria and iron sulfide crystals are also associated with these corrosion events. In a further experiment designed to measure the effect of increasing the soluble iron content of the medium, it has been found that significant corrosion only occurs above 60 mg.l$^{-1}$ ferrous ion. Although iron sulfide particles increase within the biofilm at lower ferrous ion concentrations, it is only at 60 mg.l$^{-1}$ that all biogenic sulfide is precipitated and direct physical contact is established between iron sulfide deposits in the biofilm and the metal surface, with consequent dramatic increases in both anodic and cathodic currents. From these findings the authors conclude: "The loose iron sulfide particles apparently play a more important role than the bacteria in the anaerobic corrosion process. The role of the mixed SRB biofilm is to continuously supply hydrogen sulfide to keep loose iron sulfide particles cathodically active" (Lee and Characklis, 1993).

In their review, Lee *et al.* (1995) also consider the rather vexing question of which iron sulfides are protective, which corrosive, and why. The nature of sulfide film formed abiotically from H$_2$S, and particularly the formation of pyrite, is influenced by ferrous ion concentration, temperature, and the fluid dynamics of the system under study. At low H$_2$S concentrations the iron sulfide film formed is relatively protective and is mainly composed of troilite and pyrite. At higher H$_2$S, however, the film is less protective and is mainly mackinawite. Although mackinawite is generally protective as a thin adherent film (or tarnish), when it becomes thicker and less continuous (as a scale) it is also less protective due to its increasing electronic and ionic conductivity. The factors determining the corrosivity of sulfide films have been identified as good electron conductivity, low overvoltage for hydrogen evolution, noble electrode potential, and defect structure. There remains some controversy, however, as to which individual iron sulfide corrosion products are, in fact, the most corrosive, and specific conclusions appear to differ somewhat depending on the system under study. In this regard, one should note in particular Crolet's contention that mackinawite is nonprotective and is formed under conditions of high iron-to-sulfide ratios, with the more protective pyrite being characteristic of the high sulfide concentrations.

Also, it is important to restate Crolet's thesis that the fundamental action of microbial corrosion by SRB is the stimulation of the anodic and not the cathodic reaction.

The only clear generalization that can be made with any degree of confidence at the present state of our knowledge is that thin adherent films tend to be protective, whereas loose, bulky or disrupted surface layers actively stimulate corrosion processes. Furthermore, it is the prevailing environmental conditions and not the growth and activity of SRB that determine the critically important chemical structure and physical form of the iron sulfide corrosion products that, in turn, determine the rate and extent of corrosion.

### 2.2.4. The Central Role of Oxygen in Anaerobic Microbial Corrosion

Reference has already been made to the fact that in the field the most dramatic instances of SRB corrosion are generally associated with access to oxygen. An explanation of this apparently paradoxical finding has consequently been the principal objective of a number of recent studies. Hardy and Bown (1984) exposed a sheet of mild steel to a growing culture of a marine *Desulfovibrio* sp. under controlled conditions of anoxia and noted the formation of a nonhomogeneous black gelatinous biofilm with a number of small tubercles. They recorded a low rate of corrosion, as measured by weight loss and electrical resistance probe. After completion of the growth phase, the vessel was subjected to short pulses of air and during these it was noted that the corrosion rate showed an instantaneous 90-fold increase to a value of 650 $\mu$m.y$^{-1}$. The nature of the corrosion was pitting, directly associated with the tubercles. The design of this experiment clearly excludes the possibility of oxygen having its effect on the growth of the SRB, or on the initial stages of biofilm development and the accumulation of ferrous sulfide corrosion products. It is the further modification of these sulfide products that is crucially affected by the sudden ingress of oxygen.

These findings have been confirmed and extended in a series of experiments with corroding steel coupons exposed either in seabed sediments associated with an oil production platform, or in a laboratory simulation of a stratified marine ecosystem (Moosavi *et al.*, 1991; McKenzie and Hamilton, 1992). Maximal rates of corrosion, associated with pitting, were noted in those bulk environments where conditions were aerobic. Three factors appeared to be of particular significance under these circumstances. First, there was no correlation between SRB numbers or activity and the rate or extent of corrosion. Second, visual observation revealed three distinct layers of corrosion product: a thin black adherent layer, surmounted by first a bulky layer of looser material, again black, and finally an overlying layer of brown, oxidised products. Preliminary attempts at chemical analysis served only to establish an extremely complex mixture of sulfides, oxides, and carbonates. Third, it was noted that in the more oxidizing regions associated with the higher corrosion rates, up to 92% of the sulfides were in a form that did not give rise to $H_2S$ on treatment with cold acid. The principal forms of so-called nonacid volatile sulfur compounds have been identified as pyrite and elemental sulfur, the formation of each of which is favored by oxidizing conditions. It remains to be established, however, whether the marked relationship between the nonacid volatile sulfur compounds and the higher rates of corrosion is causal, or merely casual.

This facet has been further explored by Lee and colleagues in a series of laboratory studies designed to obtain data on the effects of increasing oxygen concentrations on the corrosion resulting from the growth of SRB in a defined mixed culture biofilm (Lee *et al.*, 1993a,b; Nielsen *et al.*, 1993). The experimental system employed was an open flow channel reactor that could be operated in either a batch or continuous culture mode, with or without the incorporation of a recycling loop. Growth was established with a mixed inoculum of *Pseudomonas aeruginosa*, *Klebsiella aerogenes*, and *Desulfovibrio desulfuricans*, but no attempt was made to operate the reactor aseptically. The growth medium was buffered artificial seawater, supplemented with yeast extract and a carbon source, generally lactate, and the liquid phase dissolved oxygen concentration was controlled at levels between zero and 230 μM (saturation). Biofilm growth and corrosion occurred on mild steel coupons mounted in the base of the polycarbonate reactor, and corrosion was monitored by direct current polarization, AC impedance, and polarization resistance. Flush-mounted probes in the base of the reactor and microprobes penetrating the biofilm perpendicular to the metal surface allowed continuous monitoring of dissolved oxygen (DO), sulfide, and pH within the biofilm. During the course of the experiment, selected coupons were sacrificed and the microbial flora of the biofilm determined by viable counting after growth in selective media. At the end of the experimental run, the corrosion coupons were examined both microscopically and by Auger spectroscopy.

Under conditions of low DO (1.5 mg.l$^{-1}$), it was noted that during the first fifteen days the measured corrosion declined, accompanied by a decrease in the cathodic current density (polarization). At the same time the DO concentration in the bulk phase decreased, and levels of 0.6 to 1.0 mg.l$^{-1}$ were recorded at the metal surface. These data were interpreted as decreasing aerobic corrosion resulting from the lowered oxygen levels consequent to microbial activity within the developing biofilm. It was also noticeable that during this period there was a considerable build-up of SRB in the biofilm, which, in view of the oxygen levels recorded by microelectrode, is a clear indication of heterogeneities within the biofilm, with regard to both the prevailing physicochemical conditions and the organisms present. During the remaining three weeks of the experiment, DO of the bulk phase dropped further to 0.4 mg.l$^{-1}$, zero oxygen was recorded at the base of the biofilm, and corrosion increased with rises in both anodic and cathodic current densities. This phase of corrosion was due to SRB activity and was characterized by a high incidence of pitting.

After removal of the corrosion products, the steel surface was examined with Auger electron spectroscopy incorporating argon ion sputter etching, which gave information on surface morphology and on elemental analysis and mapping throughout the depth of any pits that formed. After the initial two weeks exposure in the reactor there was no evidence of pitting, and Auger spectroscopy indicated that the overlying film consisted of iron oxides. During the final three week period when pitting became evident, sulfur, iron, and oxygen signals were detected from the steel surface. There was a close physical association of the sulfur and oxygen signals, with the former being coincident with pits at depths down to 3600 angstroms, and the oxygen signal coming from the areas immediately surrounding the pits.

In a further study (Lee *et al.*, 1993b) the dissolved oxygen concentration was raised to 230 μM (saturation). Also chemical determinations of acid-volatile sulfide (e.g., mackinawite), chromium-reducible sulfide (principally pyrite), and elemental sulfur were carried out, in addition to SEM and EDAX analyses of the surface of corroded coupons. Biofilm thickness up to 3 mm was noted by the end of the experimental run with total dissolved sulfide varying from 10 mg.l$^{-1}$ at the metal surface to zero at a distance of 300 μM out from the metal. A constant pH of 6.9 was recorded throughout the thickness of the biofilm. A layering of corrosion products, similar to that noted earlier by Moosavi *et al.* (1991), was observed , with pitting occurring directly beneath orange colored deposits. Pyrite was observed throughout the experimental run, with sulfur only becoming evident in the final stages. It was concluded that significant SRB-induced pitting corrosion depended upon the establishment of relatively large areas of conductive iron sulfides surrounding and generating an active galvanic cell with the unreacted metal anode at the base of the pit or crevice.

Under alternating periods (12 h) of oxic and anoxic conditions for

35 days, high rates of corrosion up to 4 mm.y$^{-1}$ were recorded (Nielsen *et al.*, 1993). These were accompanied by significant increases in the cathodic current density as measured by DC polarization, with little alteration in the anodic current density. During a further incubation for one month with constant saturating dissolved oxygen but no added carbon source, the high corrosion rate was maintained without significant alteration in either cathodic or anodic current densities.

The biofilm thickness was about 2 mm, and at 1 mm depth the recorded oxygen concentration went from saturation to zero over a distance of less than 100 μM, thus giving a very steep oxygen concentration gradient and a clearly defined oxic/anoxic interface within the biofilm. The limited oxygen penetration was caused by a very high oxygen consumption rate due to oxygenation of reduced chemical species originating from the metal dissolution in the corrosion process. High *in situ* levels of sulfate reduction were noted in the anoxic layer of the biofilm, and the biofilm contained 157 μmol.cm$^{-3}$ total sulfur in the form of acid-volatile sulfides, chromium-reducible sulfides, and elemental sulfur. Under these conditions, therefore, a stable structured biofilm has become established, with a marked differentiation in the vertical dimension but most likely relatively homogeneous in the horizontal dimension. It is highly significant that under these circumstances although the system shows a high rate of general corrosion, it does not display the characteristic pitting of natural environmental SRB corrosion.

On the basis of these studies, the authors (Nielsen *et al.*, 1993) put forward the interesting hypothesis that SRB and their sulfide and sulfur corrosion products stimulate corrosion cathodically by acting as electron carriers between the metal surface and the oxic interface within the biofilm. This proposal closely parallels the suggestion of Newman *et al.* (1991) that the major component of the cathodic current may be supplied by a remote oxygen or sulfur cathode, with the action of H$_2$S being to stimulate the localized metal dissolution at an anaerobic anodic site.

## 3. CONCLUSIONS

It is now possible to identify with some degree of certainty many aspects of the mechanism of SRB-induced biocorrosion.

1. SRB-induced biocorrosion is an electrochemical event.
2. The crucial property is the production of sulfide which reacts with the released metal ions to produce a range of iron sulfides. The chemical and physical nature of these sulfides is critically influenced by environmental parameters.

a. The relative concentrations of free sulfide anion and of soluble iron will determine both the chemical nature (e.g., mackinawite, $FeS_{1-x}$, or pyrite, $FeS_2$) and physical form (adherent film, or bulky precipitate) of the sulfide corrosion products. Whereas thin adherent films or tarnishes are found to be protective, bulky precipitates or ruptured films simulate corrosion processes.

b. Exposure to oxygen both markedly accelerates corrosion and leads to the conversion of the primary sulfide products to pyrite and elemental sulfur.

3. Naturally occurring SRB-induced biocorrosion is invariably characterized by localized metal dissolution and the formation of pits. This is not always the case in various laboratory simulations of corrosive processes. Equally, weight loss measurements and certain electrochemical techniques do not, of themselves, give a direct measure of the incidence of pitting.

There remain, however, a significant number of questions that still require an unequivocal answer.

1. What is the quantitative significance of $H_2$ oxidation by SRB? Does it influence cathodic depolarization, or merely act as an additional energy substrate leading to further sulfide production?

2. Recognizing that the primary event in corrosion is metal dissolution at the anode, do SRB and their corrosion products accelerate that corrosion by stimulating anodic or cathodic reactions? Does such a distinction have any real theoretical or practical validity?

3. Can pitting corrosion be fully explained by colonial or patchy biofilm growth? Or is pit formation a stochastic event triggered by, for example, rupture of a protective film?

4. Is there any evidence of species specificity with respect to SRB and metal corrosion, or is it a property of the group as a whole?

5. Can protective/corrosive character be unambiguously assigned to particular iron sulfides?

6. What is the exact role of oxygen in stimulating SRB corrosion? Are the relationships between oxygen and pyrite and sulfur, and between pyrite and sulfur and increased corrosion causal or casual?

Undoubtedly many of these points remain unresolved as a consequence of the inherent complexity of the process of biologically influenced corrosion. To a certain degree, however, this situation is exacerbated by the extensive range of experimental models that have been used in an effort both to simplify the complexity, and to obtain answers to specific points. In many cases these models neither mirror natural corroding systems, nor are capable of supplying the unequivocal answers sought. It is to be hoped that the six questions raised above may help to formulate further experimental analyses of this intriguing phenomenon.

## REFERENCES

Bryant, R. D., Jansen, W., Boivin, J., Laishley, E. J., and Costerton, J. W., 1991, Effects of hydrogenase and mixed sulfate-reducing bacterial populations on the corrosion of steel, *Appl. Environ. Microbiol.* **57**:2804–2809.

Campaignolle, X., Luo, J. S., Bullen, J., White, D. C., Guezennec, J., and Crolet, J-L., 1993, Stabilization of localized corrosion of carbon steel by sulfate-reducing bacteria, in: *Corrosion 93*, National Association of Corrosion Engineers, Houston, Texas, paper no. 302.

Cord-Ruwisch, R., Kleinitz, W., and Widdel, F., 1987, Sulfate-reducing bacteria and their activities in oil production, *J. Pet. Technol.* **Jan:**97–106.

Costello, J. A., 1974, Cathodic depolarization by sulphate-reducing bacteria, *S. Afr. J. Sci.* **70**:202–204.

Crolet, J.-L., 1992, From biology and corrosion to biocorrosion, *Oceanologica Acta* **15(1)**:87–94.

Crolet, J.-L., 1993, Mechanism of uniform corrosion under corrosion deposits, *J. Material Sci.* **28**:2589–2606.

Crolet, J.-L., Daumas, S., and Magot, M., 1993, pH regulation by sulfate-reducing bacteria, in: *Corrosion 93*, National Association of Corrosion Engineers, Houston, Texas, paper no. 303.

Crolet, J.-L., Pourbaix, M., and Pourbaix, A., 1991, The role of trace amounts of oxygen on the corrosivity of $H_2S$ media, in: *Corrosion 91* NACE, Houston, Texas, paper no. 22.

Daumas, S., Magot, M., and Crolet, J.-L., 1993, Measurement of the net production of acidity by a sulphate-reducing bacterium: experimental checking of theoretical models of microbially influenced corrosion, *Res. Microbiol.* **144**:327–332.

Dexter, S. C., Duquette, D. J., Siebert, O. W., and Videla, H. A., 1991, Use and limitations of electrochemical techniques for investigating microbiological corrosion, *Corrosion* **47**:308–318.

Guezennec, J., Dowling, N. J., Conte, M., Antoine, E., and Fiksdal, L., 1991, Cathodic protection in marine sediments and the aerated seawater column, in: *Microbially Influenced Corrosion and Biodeterioration* (N. J. Dowling, M. W. Mittleman, and J. C. Danko, eds.), National Association of Corrosion Engineers, Washington, pp. 643–650.

Hamilton, W. A., 1985, Sulphate-reducing bacteria and anaerobic corrosion, *Ann. Rev. Microbiol.* **39**:195–217.

Hamilton, W. A., 1991, Sulphate-reducing bacteria and their role in biocorrosion, in: *Biofouling and Biocorrosion in Industrial Water Systems*, (H-C. Flemming and G. G. Geesey, eds.), Springer-Verlag, Berlin, pp. 187–193.

Hardy, J. A., and Bown, J., 1984, The corrosion of mild steel by biogenic sulphide films exposed to air, *Corrosion* **40**:650–654.

Hardy, J. A., 1983, Utilization of cathodic hydrogen by sulphate-reducing bacteria, *Brit. Corr. J.* **18**:190–193.

Herbert, B. N., 1987, Reservoir souring, in: *Microbial Problems in the Offshore Oil Industry*, (E. C. Hill, J. L. Shennan, and R. J. Watkinson, eds.) Wiley, Chichester, pp. 63–73.

King, R. A., and Miller, J.D.A., 1971, Corrosion by sulphate-reducing bacteria, *Nature* **233**:491–492.

King, R. A., Dittmer, C. K., and Miller, J.D.A., 1976, Effect of ferrous iron concentration on the corrosion of iron in semicontinuous cultures of sulphate-reducing bacteria, *Brit. Corr. J.* **11**:105–107.

King, R. A. Miller, J.D.A., and Wakerley, D. S., 1973, Corrosion of mild steel in cultures of sulphate-reducing bacteria: effect of changing the soluble iron concentration during growth, *Brit. Corr. J.* **8**:89–93.

Lee, W., and Characklis, W. G., 1993, Corrosion of mild steel under anaerobic biofilm, *Corrosion* **49**:186–199.

Lee, W., Lewandowski, Z., Okabe, S., Characklis, W. G., and Avci, R., 1993a, Corrosion of mild steel underneath aerobic biofilms containing sulfate-reducing bacteria. Part 1: at low dissolved oxygen concentration, *Biofouling* **7**:197–216.

Lee, W., Lewandowski, Z., Morrison, M., Characklis, W. G., Avci, R., and Nielsen, P. H., 1993b, Corrosion of mild steel underneath aerobic biofilms containing sulfate-reducing bacteria. Part II: at high bulk oxygen concentration, *Biofouling* **7**:217–239.

Lee, W., Lewandowski, Z., Nielsen, P. H., and Hamilton, W. A., 1995, Role of sulfate-reducing bacteria in corrosion of mild steel: a review, *Biofouling*, **8**:165–194.

Maldonado-Zagal, S. B., and Boden, P. J., 1982, Hydrolysis of elemental sulfur in water and its effect on the corrosion of mild steel, *Brit, Corr. J.* **17**:116–120.

Mara, D. D., and Williams, D.J.A., 1972, The mechanism of sulphide corrosion by sulphate-reducing bacteria, in: *Biodeterioration of Materials*, Vol. 2 (A. M. Walters and E. H. Hueck van der Plas, eds.), Applied Science Publishers, London, pp. 103–113.

McKenzie, J., and Hamilton, W. A., 1992, The assay of *in-situ* activities of sulphate-reducing bacteria in a laboratory marine corrosion model, *Internat. Biodeterior. Biodegrad.* **29**:285–297.

Moosavi, A. N., Pirrie, R. S., and Hamilton, W. A., 1991, Effect of sulphate-reducing bacteria activity on performance of sacrificial anodes, in: *Microbially Influenced Corrosion and Biodeterioration* (N. J. Dowling, M. W. Mittleman, and J. C. Danko, eds.), National Association of Corrosion Engineers, Washington, pp. 3.13–3.27

Newman, R. C., Webster, B. J., and Kelly, R. G., 1991, The electrochemistry of SRB corrosion and related inorganic phenomena, *ISIJ International*, **31**:201–209.

Nielsen, P. H., Lee, W., Lewandowski, Z., Morrison, M., and Characklis, W. G., 1993, Corrosion of mild steel in an alternating oxic and anoxic biofilm system, *Biofouling* **7**:267–284.

Pankhania, I. P., Moosavi, A. N., and Hamilton, W. A. 1986, Utilization of cathodic hydrogen by *Desulfovibrio vulgaris* (Hildenborough), *J. Gen. Microbiol.* **132**:3357–3365.

Salvarezza, R. C., and Videla, H. A., 1980, Passivity breakdown of mild steel in sea water in the presence of sulfate reducing bacteria, *Corrosion* **36(10)**:550–554.

Salvarezza, R. C., Videla, H. A., and Arvia, A. J., 1983, The electrochemical behaviour of mild steel in phosphate-borate-sulphide solutions, *Corrosion Sci.* **23(7)**:717–732.

Sanders, P. F., and Tibbetts, P. J., 1987, Effects of discarded drill muds on microbial populations, *Phil. Trans. Roy. Soc. Lond. Ser. B*, **316**:567–585.

Schaschl, E., 1980, Elemental sulfur as a corrodent in deaerated, neutral aqueous solutions, *Materials Performance* **19**:9–12.

Schmitt, G., 1991, Effect of elemental sulfur on corrosion in sour gas systems, *Corrosion* **47**:285–308.

Tatnall, R. E., 1991, Case histories: biocorrosion, in: *Biofouling and Biocorrosion in Industrial Water Systems*, (H.-C. Flemming and G. G. Geesey, eds.), Springer-Verlag, Berlin, pp. 165–185.

Tiller, A. K., 1982, Aspects of microbial corrosion, in: *Corrosion Processes* (R. N. Parkins, ed.), Applied Science Publishers, London, pp. 115–159.

von Wolzogen Kuhr, C.A.M., and van der Vlught, I. S., 1934, The graphitisation of cast iron as an electrobiochemical process in anaerobic soils, *Water* **18**:147–165.

Wanklyn, J. N., and Spruit, J.C.P., 1952, *Nature* **169**:928–929.

# Control in Industrial Settings  10

## T. R. JACK and D. W. S. WESTLAKE

## 1. INTRODUCTION

The main concern of the oil and gas industry is the protection from corrosion of carbon steel in extensive production, transportation, and processing facilities. Hundreds of millions of dollars are spent annually to minimize economic and environmental effects attributed to corrosion. The importance of microorganisms in corrosion has been recognized for more than half a century. In the 1930s pioneering work by Von Wolzogen Kuhr and Van der Klugt (1934) identified the cause of acceleration of anaerobic external corrosion on unprotected pipe in wet soil as the action of sulfate-reducing bacteria (SRB). This and subsequent observations confirming this relationship formed the basis for the oil and gas industry's interest in developing techniques for the detection, enumeration, and control of SRB. The wide range of detrimental microbial activities occurring in oil-field and water handling facilities in the 1940s are summarized in a pamphlet published by the National Association of Corrosion Engineers (TPC Publication 3, 1976). The control of undesirable microbial activity in industrial processes continues to this day.

### 1.1. *In Situ* Microbial Activity

The importance of microorganisms in the diagenesis of crude oil was first described in the 1940s when the American Petroleum Institute funded research by Zobell (see Davis, 1967). By the 1970s, the continuing influence of microbes in the maturation of oil was recognized (Evans *et al.*, 1971; Milner *et al.*, 1977; Connan, 1984) and the role of adventitious

T. R. JACK • Novacor Research & Technology Corporation, 2928 - 16 St. N.E., Calgary, Alberta, Canada T2E 7K7.    D. W. S. WESTLAKE • Westec Microbes Ltd., 3375 Anchorage Ave., Victoria, British Columbia, Canada, V9C 1X4.

*Sulfate-Reducing Bacteria*, edited by Larry L. Barton. Plenum Press, New York, 1995.

oxygen in promoting the biodegradation of lighter alkane and aromatic components at air or water contacts in the reservoir was documented (Orr, 1979). Kuznetsova and coworkers (1963, 1964) showed that *in situ* sulfate reduction could be stimulated by the injection of freshwater containing sulfate and a mixture of microorganisms. The microbial hydrogen sulfide produces can cause economic consequences by souring fluid production from waterflood operations in which water is intentionally injected into the reservoir to displace oil. Kuznetsova *et al.* (1963, 1964) concluded that souring was a complicated process requiring the activity of aerobic hydrocarbon utilizing pseudomonads as well as SRB. Laboratory support came from the work of Bailey *et al.* (1973) and Jobson *et al.* (1979) who developed the hypothesis that microbial activity in a reservoir depends on the interaction of a mixed population of organisms breaking down hydrocarbon through the use of adventitious oxygen. Evidence, old (Davis, 1967) and new (Aeckersberg *et al.*, 1991), supports the direct anaerobic degradation of selected simple hydrocarbons by sulfate reducers, but metabolism of bulk oil under anoxic conditions appears to be limited (Jack *et al.*, 1985). Microbial activity in the reservoir can have practical consequences beyond souring. Plugging, loss of lighter hydrocarbon components, and destruction of chemicals injected to enhance oil recovery can also occur (Iverson and Olson, 1984).

## 1.2. Microbial Activities in Surface Facilities

In pipelines and water handling facilities, biofilms and not planktonic organisms are responsible for the undesirable effects. The importance of sessile populations in nature was recognized by only a few workers in the first half of the century (e.g., ZoBell, 1943). By the 1970s, serious studies were appearing on mechanisms of adhesion, growth rates, and metabolism of such organisms (Marshall *et al.*, 1971; Bott, 1975; Geesey *et al.*, 1978) including the effect of the surface itself on the development of the initial microbial community (Marszalek *et al.*, 1979). In the biofilm environment, bacteria can grow on complex carbon substances not normally utilized by individual free-floating cells. The behavior of bacteria in biofilms has been recently reviewed (Costerton and Lappin-Scott, 1989) and a general overview has been published by Marshall (1992).

## 2. MANAGEMENT

### 2.1. Cathodic Protection and Coatings to Reduce Corrosion

Management of microbial corrosion problems relies on a combination of prevention and control. Table I summarizes methods applicable

**Table I. Management Methods by Type of Facility**

| Facility type | Materials selection | Cathodic protection | Coatings | Water removal | Mechanical cleaning | Corrosion control chemicals (biocide) |
|---|---|---|---|---|---|---|
| | | | | Approach | | |
| **Field production** | | | | | | |
| wells | X[a] | X | | | | X |
| pipe | | (X)[b] | | | | X |
| tanks | | X | X | | (X)[c] | X |
| **Pipelines** | | | | | | |
| external | | X | X | | | |
| internal | | | | | | |
| gas | | | | X | X | |
| oil | | | | X | X | X |
| **Refined products** | | | | | | |
| tanks | | X | | X | | |
| fuel systems | X | | | X | (X)[c] | (X)[d] |
| **Cooling towers and heat exchangers** | X | (X)[b] | | | (X)[c] | X |

[a] Approach used for this type of facility.
[b] Partial application on outside of pipes, on selected surfaces, or through use of galvanizing in cooling towers.
[c] Occasional cleaning, e.g. periodic cleaning of tank at turnaround.
[d] Use of fuel soluble biocides in some applications.

to various industrial facilities. Microbially influenced corrosion problems can be precluded by materials selection; however, economic and process constraints are often limiting factors. Prevention of biocorrosion, therefore, focuses heavily on use of cathodic protection and coatings.

In cathodic protection, an impressed electrical current or sacrificial anode of a more active metal such as magnesium or zinc is used to render the potential at the steel surface thermodynamically stable with respect to the dissolution of iron (Peabody, 1970). While many facilities can be protected in this way, intricately shaped surfaces defy adequate potential distribution and the need for unrestricted flow precludes installation of anodes inside most heat exchange systems. A potential of $-950$ mV ($Cu/CuSO_4$) is sufficient to protect steel from the worst bacterial corrosion scenarios.

Impermeable coatings (Hamner, 1970) are also an effective means to prevent microorganisms from directly colonizing metal surfaces. These coatings can be purchased already applied to steel components or can be applied during facility construction. Retrofits are obviously difficult where surfaces are inaccessible after installation but, to extend the life of existing systems, a number of plastic liner products are commercially available which can be inserted into low-pressure piping suffering pitting corrosion. When coatings are durable, nonbiodegradable, impermeable, and bonded to the steel surface, they are effective at stopping bacterial corrosion.

## 2.2. Changes in the Environment

Microorganisms require an energy source, nutrients, favorable temperature, pH, and salinity to flourish. In most cases, it is impractical to alter the pH or salinity conditions associated with a process or operation to prevent growth, but instances do occur where temperature shifts can be used to discourage sustained microbial activity. Similarly, eliminating nutrients in source water through expensive purification technologies is rarely attractive economically, but selection of alternate water sources is often a practical option. Figure 1 shows a case where one unit in an oil field waterflooding operation was experiencing corrosion failures far in excess of surrounding units. Removal of a highly variable make-up water source received from oil field operations in other areas resulted in a dramatic performance improvement.

In hydrocarbon handling facilities removal of water is an effective strategy. Fuel systems are particularly susceptible to microbial infestations if a standing water phase is present (Davis, 1967). Spectacular problems have occurred when seawater has been used to displace hydro-

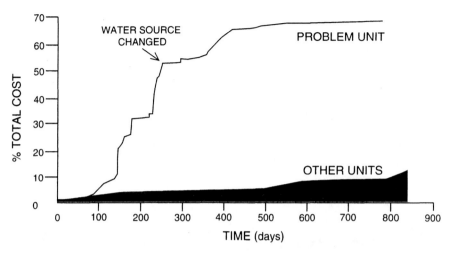

**Figure 1.** Effect of altering the make-up water source on corrosion costs in a produced water facility in an oil field waterflood relative to neighboring units.

carbon directly from oil production platform storage legs or warship fuel tanks in an effort to provide ballast. Early fuel filter plugging and wing tank corrosion in aircraft clearly demonstrated the risk of leaving water in jet fuels. Internal corrosion problems do not exist if the gas pipeline has been properly dehydrated after construction and if water is removed from incoming gas. Proper design and good housekeeping are usually the most cost-effective approaches to manage microbial problems.

Algae in cooling towers or trace hydrocarbons left in water streams are *in situ* sources for carbon substrates. Similarly, essential salts present in cooling tower water or brines from the oil reservoir are excellent sources of inorganic nutrients for growth. Advanced oxidation pretreatments of source water by ultraviolet radiation or enhanced ozonation can reduce organics and create biocidal oxidized species of sufficient lifetime to discourage downstream growth of bacteria. Nutrient elimination is not routinely done, presumably for reasons of cost (Paillard *et al.*, 1987).

Assessing nutrient limitation is inherently difficult. When biofilms harvest essential nutrients like sulfate or organic acids from a flowing stream, adequate amounts to sustain growth are more dependent on flow rate than on nutrient concentration. For example, in high-volume-oil-field-brine-handling systems where excess barium restricts soluble sulfate, limitation of the sessile SRB is not evident even at 2 ppm sulfate.

The water flow in these systems is sufficient to bring in enough sulfate each day to meet the needs of those biofilm organisms.

## 2.3. Control with Biocides

Use of biocides to kill microorganisms has become an institutional approach to microbial control in fluid handling systems. Biocides are often part of an overall chemical control strategy that includes corrosion inhibitors, antiscaling additives, oxygen scavengers, dispersants, chelators, and surface conditioners. Obviously chemical compatibility is a serious challenge. Comprehensive packages are offered by commercial companies but much remains to be done in this area to improve the efficiency of treatment.

Biocide products vary according to the nature of the active ingredient and the requirements of the target application. Table II illustrates some of the components that might be used in a biocide formulation. The active ingredient is a chemical which has a general biocidal effect on a wide range of SRB and other organisms. Biocides can be classed into oxidizing and nonoxidizing products. Strong oxidants such as chlorine or ozone kill organisms by aggressive chemical oxidation, which leads to hydrolysis and dispersal of cells and polymer slimes of biofilms. Nonoxidizing biocides act by cross-linking cellular components (aldehydes) or by altering cell membrane integrity (quaternary ammonium salts). Other nonoxidizing agents react with functional groups commonly found in biological systems or otherwise interfere with vital processes.

Some active ingredients, chlorine or acrolein for example, can be applied directly as pure compounds. In other cases, the active ingredient is sold in solution as part of a formulation. Water is the most common solvent employed because most biocides are targeted to control water-phase corrosion problems. In special applications, hydrocarbons (Angrykovich and Neohof, 1987) or mildly biocidal solvents are used. Solubility limitations and the need to prevent freezing in cold climates often requires inclusion of an alcohol. This can constitute 40% of an aqueous product and in some cases can be used without water. An addition of glutaraldehyde in methanol, for example, can disinfect and dry out a pipeline system in a single treatment.

Figure 2 illustrates the performance of two biocides in an oil-field water-handling system. Product A appeared to slowly reduce bacterial counts, in absolute number of viable SRB present in *biofilms*, over the first year it was introduced downstream of biocide injection point (Fig. 2A). However, a plot of the increase in SRB from upstream of the biocide injection point to the downstream sampling point (i.e., the dif-

**Table II. Components for Biocide Products**

| Active Ingredient (1–95%) | | |
|---|---|---|
| *Strong oxidants* | *Aldehydes* | *Quaternary ammonium salts and amines* | *Others* |
| Chlorine | Formaldehyde | Cetylpyridinium chloride | Triazine |
| Hypochlorite | Glutaraldehyde | Cetytrimethylammonium chloride | Methylene(bis)cyanate |
| Chlorine dioxide | Acrolein | Cocodiamine | Dibromonitrilopropionamide |
| Ozone | | | Isothiazolone |

| Solvents | Solubilizers/antifreeze (20–40%) | Dispersants/penetrants |
|---|---|---|
| Water | Methanol | Various polyethoxylates |
| Tall oil | Ethanol | Fatty acid amides |
| Petroleum distillates | Isopropanol | |

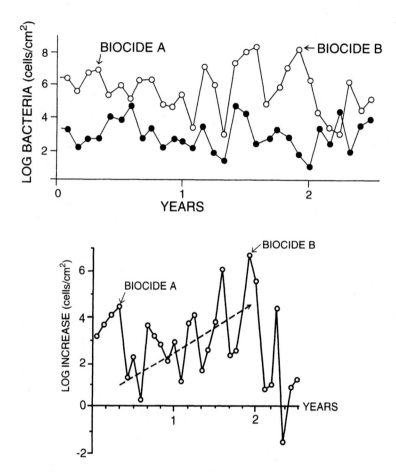

**Figure 2.** Relative performance of two cocodiamine-based biocide products in an oil field produced water handling system. (A) Sessile viable counts for SRB upstream (--●--) and downstream (--○--) of the biocide injection point. (B) As above, but plotting the difference between the above plots to reflect increase in SRB numbers downstream of the biocide injection point.

ference in the plots in Fig. 2A) revealed that biocide A had begun to lose control of growth only two or three months after its introduction (Fig. 2B). When replaced by product B, an impressive kill was achieved and further growth was suppressed. Both product A and B contained the same active ingredient applied at comparable dosages.

Biocide products in general are complex formulations in which a variety of complementary agents are blended to achieve superior kills, cleaning, shelf life, handling characteristics, and corrosion mitigation. While there are hundreds of products on the market, only certain ones will fit a given application. For example, chlorine may do well in cooling tower applications but is not suitable for oil water systems with a high chlorine demand. Hydrogen sulfide present in many oil field waters is chemically incompatible with many biocide products. Methanol used as an antifreeze in cold temperature formulations can damage certain kinds of seals and gaskets. Chromates, mercurials, tin derivatives, chlorinated products, and copper salts are no longer used because of occupational health concerns and fear of environmental liability.

The amount of a biocide product required can be substantial. An oil field waterflood operation in which produced water is separated and recycled through the oil reservoir to displace more oil may handle millions of cubic meters of water per year. Given a continuous treatment of 100 ppm, such a field would require 100,000 liters of biocide or more per year. Clearly even small price differences become significant at these volumes. A further element of cost is the labor and travel needed to apply a product. A waterflood operation might be spread over a wide geographic area with a dozen or more injection points to be supplied and serviced year round.

## 3. BIOCIDE APPLICATION

### 3.1. Bacterial Detection Systems

The reduction in bacterial numbers, and in particular the SRB, is used for measuring the effectiveness of biocides applied in the oil and gas industries. Traditionally, this has involved the use of the Most Probable Number technique, which can require a four-week incubation time before usable information is obtained. Assays for the identification and assessment of bacterial populations in industrial systems are under very active research and commercial development. At least one new method has appeared in the market every year over the last decade. Following the work of Costerton and others (Ruseska *et al.*, 1982), attention is now focused almost exclusively on the sessile organisms in biofilms on the metal surface.

### 3.1.1. Commercial Kits

Commercially available kits for use by operators in directly assaying samples scraped from steel surfaces are based either on traditional growth methods or on new enzyme activity tests. Growth tests are able to provide viable counts for selected organisms present before or after treatment. SRB have been the traditional indicator organisms (API RP-38, 1965). In recent years, a new medium developed for the Gas Research Institute has come into use for organisms able to ferment a simple sugar to acid. These are referred to as "acid producing bacteria" and reflect a more general population than the SRB (GRI Field Guide I, 1990; GRI Field Guide II, 1992). Enzyme assays are faster than growth tests. Commercial products are available for hydrogenase and APS-reductase activity. Hydrogenase relates to a specific corrosion mechanism, cathodic depolarization, in which hydrogen is removed from anaerobic metal surfaces in the corrosion cell, a process which accelerates corrosion (Booth *et al.*, 1968; Pankhania, 1988).

Several of these commercial kits have been compared in recent tests (Scott and Davies, 1992). Because microbial numbers can vary over eight-to-ten orders of magnitude, assays need only have an order of magnitude accuracy. Speed, ease, cost, precision, and a sensitivity range appropriate to the target system are the factors dictating selection. Biocide application or other action is indicated where high numbers of SRB in the presence of black iron sulfide corrosion products are found (API RP-38, 1965; TPC Publication 3, 1976). Other criteria are set out in newer industry guides (GRI Field Guide I, 1990; GRI Field Guide II, 1992).

### 3.1.2. Other Assays

Other tests include oligonucleotide probes (Devereux *et al.*, 1992) specific for certain SRB genera and groups, use of radioisotopically labelled substrates to measure metabolic processes such as sulfate reduction (Maxwell and Hamilton, 1986), fluorescent stains to count both general populations and specific bacteria by fluorescence microscopy and photometric assays (Jones *et al.*, 1988) for transient biochemicals such as adenosine triphosphate can be related to viable cell density. These techniques are largely used by research groups and involve extraordinary equipment and expertise not commonly found in operating groups. Indirect monitoring methods such as changes in heat exchange efficiencies or the electrochemical noise being generated at the metal surface are available (Strauss, 1992) but require expert interpretation to separate bacterial effects from other surface phenomena.

### 3.2. Screening of Biocide Products

Proving product efficacy in the target system is essential. Even for suitable products, performance varies widely in a given application. It is

**Table III. Effectiveness of Various Biocides on Sessile Microbial Populations in Side Stream Test Loops in an Oil Handling Facility**[a]

| Biocide | Concentration (ppm) | Acid-producing bacteria | Sulfate-reducing bacteria |
|---|---|---|---|
| Quarternary ammonium salt | 1000 | − | − |
| Cocodiamine | 1000 | − | − |
| Triazine | 1000 | − | − |
| Isothiazolone | 1000 | − | + |
| Formaldehyde + isothiazolone | 1000 | + | + |
| Glutaraldehyde | 1000 | + + | + |

[a] Peformance ranges from (+ +) very effective to (−) no effect. Adapted from (Pope *et al.*, 1989)

not uncommon for most recommended products to fail to provide a minimal kill (99.9%). Table III illustrates some typical results (Pope *et al.,* 1989). A quick, reliable screening test, therefore, is required before products are applied in full scale operations. Such a test can also provide preliminary dosage requirements allowing cost estimates to be made in the selection process.

Recommended practice is to screen products under the most realistic conditions possible. In one case for example, sessile populations on bull plugs pulled from the target facility were placed in fluid taken from the system and transported quickly to the laboratory under anaerobic conditions. These populations were then challenged in still-jar tests under anaerobic conditions at the temperature of the operating facility with various products at dosages recommended by the suppliers. Sulfate reduction rates using radioisotopically labeled sulfate and kills assayed by growth of SRB on Postgate's B medium (Postgate, 1979) were measured against untreated controls. Only three of eight products tested gave satisfactory results. These were subsequently tested in instrumented side streams diverted from the main flow at the target facility. Results from the still-jar tests proved to be good indicators of field performance. Still-jar tests in the lab based on sulfate reduction had a turn around time of three days, results based on growth tests took three weeks and field trials in the side stream facility took three months. Implementation in the field required many months to optimize dosages. Figures 2 and 3 can be used to compare performance in side stream tests with full scale field performance. Side streams provide a relatively controlled environment in which sampling is done within a few meters of the biocide injection point. Turbulent flow, clean initial conditions, and a

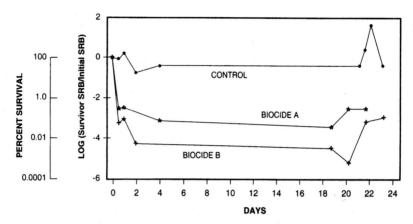

**Figure 3.** Performance trials for two cocodiamine-based biocide products in a side stream test facility in an oil field produced water handling system.

controlled environment allow relatively clear distinction of biocide effects.

In more complex systems involving tanks, headers, pumps, flow lines, elbows, etc. the exposure of biofilms to biocide will be highly variable. Use of a fluorescein dye tracer in an oil-field water-handling system showed that travel time for an injected chemical was actually less than one-quarter that predicted from calculation, assuming perfect mixing. Obviously, such a system has surfaces which are not seeing the dosage of treatment chemicals intended. Biofilms in dead spots and on poorly exposed surfaces under preexisting deposits will not receive effective dosages and will act to reseed the system with viable organisms. System response to treatment will consequently be slow, variable, and difficult to maintain, as is evident in Fig. 2. It is notable that biocide product B showed superior performance to product A in the side stream trials (Fig. 3) a prediction borne out in full scale application (Fig. 2B).

### 3.3. Biocide Dosages

The approach to introducing a biocide and setting injection protocols varies with the target. A highly fouled system or one full of deposits should ideally be cleaned prior to application. Biocides cannot replace good housekeeping and good operating practice. Failure to remove deposits will interfere with biocide performance, consume excessive amounts of chemical, and permit under-deposit corrosion to proceed even if good kills are obtained. A high concentration (several thousand ppm glutaraldehyde for example) is sometimes used to disinfect a system prior to establishing a lower routine dosage. Continuous injection of biocide can be very expensive and not very effective where costs dictate the allowable concentration for the application. Continuous injection of nonoxidizing biocides, especially at low levels, can lead to growth of "resistant" populations (Pope *et al.*, 1989). For the same cost higher concentration batch injections can be more effective over longer periods.

The general biocidal effects of the active agents cited in Table II make it hard to accept that truly resistant organisms develop. Given the nature of the biofilm, however, this may be a question of exposure rather than resistance. Deeply buried or otherwise protected organisms may begin to flourish because they fail to experience deadly doses of the active agent. This is consistent with the success of a different formulation of the same active ingredient seen in Fig. 2 and with the observation that strong oxidizing agents capable of breaking up the biofilm matrix do not experience this "resistance" effect. Alternating use of several products or periodic mechanical or chemical disruption of the biofilm is necessary to maintain control over the long haul.

### 3.4. Factors Affecting Biocide Activity

### 3.4.1. Water Chemistry

Water chemistry is an active and influential variable. Figure 2A shows the behavior of sessile populations upstream and downstream of a biocide injection point. Wide variations are evident, often affecting both populations. This indicates that changes in water chemistry in the system can have as dramatic an impact on the biofilm as biocide injection. Investigation of water chemistry effects showed that the SRB in the biofilm are affected by small changes in pH (over a range of 0.5 units), hardness (especially magnesium concentration), nitrate concentration, and other parameters of water chemistry. The growth response in SRB in this biofilm due to increasing the concentration of lactate from $\leq 1$ to 16 ppm and acetate from $\leq 4$ to 45 ppm in a side-stream experiment supports the idea that in these waters growth is limited by available energy material. The corrosive activity of other bacteria isolated from produced oil field waters has been shown to require the presence of an available energy source (Obuekwe *et al.*, 1987).

### 3.4.2. Treatment Chemicals

The corrosion control chemicals used today, while constituting more or less successful packages, are not efficient, often warring with

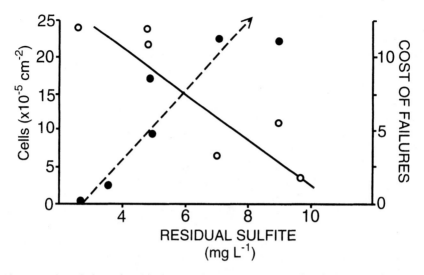

**Figure 4.** Correlation of residual oxygen scavenger (ammonium bisulfite) with sessile viable cell counts and the relative cost of corrosion failures for various units in an oil field waterflood operation.

each other in their effects. Figure 4 shows that the cost of corrosion failures in a particular field correlates inversely with the residual oxygen scavenger present a clearly desirable effect. Figure 4, however, also shows that the same oxygen scavenger (ammonium bisulfate) simultaneously promotes growth of bacteria in the system. Similar observations can be made for the scale inhibitor (phosphate based) and the corrosion inhibitor (a filming amine). It is also likely that the diluted methanol from the biocide formulation itself is a food source in the system. Evidently the treatment chemicals, while superficially compatible, are promoting opposing effects, making the balance of components in the treatment package a delicate matter. Optimization of multicomponent systems in which components provide not simply additive but competitive effects is a formidable proposition.

### 3.4.3. Organisms

Sessile SRB invariably function in a mixed population. The other organisms present are dynamic members of the sessile community and may be quite important in determining the overall aggressivity of the biofilm. Figure 5 shows a correlation between the general population present and the increase in SRB through the produced water handling

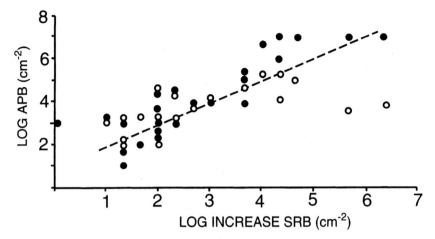

**Figure 5.** Relationship between the general sessile microbial population assayed as "acid producing bacteria" and the increase in sulfate-reducing bacteria in biofilm through an oil field water handling facility. Viable counts for acid-producing bacteria present upstream (○) and downstream (●) are identified separately.

system in an oil field (Jack *et al.*, 1993). The correlation between acid producing bacteria and SRB at a given sampling point is not pronounced; however, the role of other bacteria in support of the growth of biofilm SRB may be significant.

### 3.4.4. Flow Regimes

The behavior of fluid flowing through a pipe depends on the flow rate, pipe size, and surface as well as the properties of the fluid itself. These parameters are captured in a dimensionless constant known as the Reynold's number. At high Reynold's number the fluid in the pipe will be in the turbulent flow regime. This means that the fluid is being constantly mixed during flow and concentration gradients will not exist for dissolved species, including essential nutrients and biocides. At low Reynold's number, water traveling next to the pipe wall continues to do so, slowed with respect to the bulk flow by the drag of the pipe surface. Biofilms are particularly good at increasing the friction factor responsible. A concentration gradient (diffusion limited) will be set up from the bulk fluid in the center of the pipe to the biofilm at the pipe wall. Residence times for water traveling next to the biofilm will be longer than the average for the bulk flow. It may be expected, therefore, that laminar flow will demonstrate more complex response to biocide appli-

cation than turbulent flow systems. Most oil field water handling systems
are in laminar flow.

Flow regime effects on the biofilm itself are poorly understood.
McCoy showed that the characteristics of a biofilm are markedly different in the laminar and turbulent flow regimes (McCoy *et al.*, 1981).
Experiments in an oil-field side-stream test facility showed that taking a
sessile population from turbulent flow to laminar flow resulted not in an
immediate kill but in a fall-off of activity (as measured by sulfate reduction potential, Fig. 6) accompanied by a loss of diversity.

### 3.5. Field Expectations

Figure 2 shows performance of nonoxidizing biocide formulations
applied at the prevailing economic limit to a high salinity of oil-field
brine-handling system. Even at higher dosages preliminary work in experimental side streams showed that reducing the biofilm population in
this system below $10^2$–$10^3$ cm$^{-2}$ is very difficult. Corrosion rates by
weight loss coupon have been related to the increase of SRB between the
sampling points plotted in Fig. 2 (Jack *et al.*, 1993). On this basis risk can
be related to the difference between the plots in Fig. 2. Because the aim
is to control corrosion risk effectively, not to sterilize the system, it appears that risk is being successfully curtailed in this system despite the
absolute number of organisms present (Fig. 2B). Establishing the relationship between corrosion risk and bacterial parameters remains the

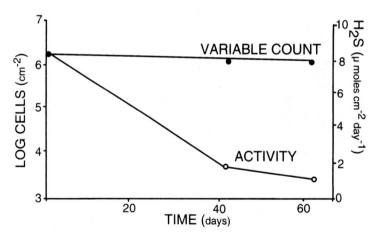

**Figure 6.** Effect on viable cell counts and ability to reduce sulfate in lab assays for a biofilm
population in an oil field water handling system shifted from turbulent to laminar flow.

greatest challenge facing the systematic application of biocides under field conditions. The object is to control the corrosive effects of bacteria, not sterilize the system.

## 3.6. Improving Biocide Performance

### 3.6.1. Mechanical Cleaning of Surfaces

Much more effective control of sessile populations can be achieved through the use of biocides with mechanical cleaning. In oil pipelines where huge capital investments are at risk and leaks can have significant environmental consequences, use of mechanical cleaning devices or "pigs" is normal practice. These devices are pushed through the pipe by fluid flow, scraping deposits off the pipe walls and displacing any water which may have settled out in low spots to a "pig trap," where water and debris can be removed. Pigs are often run with slugs of corrosion control chemicals between them or following them. This practice disrupts the biofilm, exposing organisms to the active ingredients in the biocide formulation. This results in much more effective kills than with application of biocide formulations alone. Unfortunately, pigging is only possible where the facility allows. Small diameter or irregular piping, and complex internal structures do not allow this approach and special pig launching and receiving structures must be permanently installed, adding a significant capital cost to this approach.

### 3.6.2. Use of Ultraviolet Radiation

Use of ultraviolet radiation to disinfect clear water streams is a routine water treatment process used to improve water quality. Use of ultraviolet radiation as a pretreatment for water entering an industrial system has little effect. If other conditions favor growth, reducing incoming organisms even one thousand fold will not prevent survivors colonizing and eventually populating surfaces downstream of the UV unit. For seriously fouled systems recruitment from an incoming planktonic phase is unnecessary to sustain the existing biofilm or colonize small areas of fresh metal surface such as biofilm sampling plugs. Advanced oxidation techniques may have some downstream benefit but have yet to be demonstrated in this context.

### 3.6.3. Bioelectric Effect

A new discovery, the "bioelectric effect," may provide another way to improve biocide performance. Costerton and co-workers have found that biocide kills in the presence of modest electric fields is very much improved over use of chemicals alone (Blenkinsopp *et al.*, 1992). If sufficient fields can be induced in structures like heat exchangers, improved protection should result. This would be very welcome. While cathodic protection relies on a similar creation of an electric field gradient, it must establish a set potential uniformly across the entire metal surface to be effective. Field gradients may be easier to implement and may be sufficient for the bioelectric enhancement of biocide performance, even if uneven potentials are attained.

## 4. MISCELLANEOUS CONTROL OPTIONS

### 4.1. Heat Shock

Short-term heat shocks of a few minutes duration have been demonstrated to control surface populations from offshore production facilities in side stream test loops (Pope, 1992). Exposures at 65°C for five minutes have reduced levels of viable microorganisms by 99%. Field observations are consistent with this suggestion. A change from 20°C to 60°C is sufficient to cause a shift from one set of organisms to another within an oil field system (Cord-Ruwisch *et al.*, 1987). Application of slugs of heated fluids or heating lines periodically may prevent development of a stable biofilm and provide an alternative or complementary treatment to biocide application.

### 4.2. Ice Nucleation

Temperature reduction to induce freezing has been suggested (Costerton, 1983) as a means to remove biofilms from metal surfaces in fluid handling systems. In chilled piping, freezing begins with the nucleation of ice crystals on the metal surface. Subsequent ice crystal growth can effectively detach otherwise tightly bound biofilm deposits allowing them to be swept away in the fluid flow to leave a clean metal surface. While demonstrating in the laboratory, practical applications have yet to be tried.

### 4.3. Manipulation of Microbial Growth Patterns

An alternative control strategy to suppressing microbial growth by biocide application is to manipulate it. Simple alteration of nutrient or physical parameters in an industrial system may promote a different, potentially more benign biofilm community. This concept has been tried in the field in a side stream test facility handling oil field brines. The presence of low levels of nitrate has long been known to control hydrogen sulfide production in mixed microbial communities (Jack *et al.*, 1985). Nitrate addition has even been developed to suppress hydrogen sulfide formation in microbially enhanced oil recovery schemes (Mc-Inerney *et al.*, 1992). Addition of 100 ppm nitrate to oil field brine in a field test facility resulted in a significant change in the biofilm and surface deposits inside the piping. Over 30 days, SRB counts increased up to one-hundredfold, but no iron sulfide could be detected in surface deposits on the piping at the end of the run. This was in contrast to the normal observation of rich iron sulfide content in a control line without nitrate addition and in previous surface samples from the test line itself. The disappearance of sulfide is consistent with observations in oil field test cores (McInerney *et al.*, 1992), where oxidation by nitrate respiring organisms was cited as the mechanism of sulfide destruction.

Scanning electron microscopy revealed a much higher density of organisms in the biofilm under nitrate treatment. This population included increased numbers of selected lactate and acetate utilizing SRB as identified by nucleic acid probe techniques. This presents the intriguing possibility that small amounts of nitrate may have shifted some of the SRB to use of nitrate and cathodic hydrogen under lactate limitation as suggested in the work of Rajagopal *et al.* (1988). When marked changes were induced in the biofilm population and in the associated corrosion products, this shift unfortunately led to higher not lower corrosion rates. Electrochemical measurements and weight loss coupons showed that the nitrate-treated line had a corrosion rate four-to-six times that of the control.

While this experiment failed to yield a better corrosion control treatment, it did establish the ease with which profound changes can be induced in the biofilm community by subtle manipulation of nutrients. It also emphasizes the need to monitor corrosion rates as well as bacterial numbers. The composition and activity of biofilm communities in industrial systems is open to manipulation at a level of cost and difficulty no more and probably less than that needed for traditional biocide applications. Understanding more about biofilm communities and how they work should open up new opportunities for corrosion control strategies based on manipulation rather than suppression of indigenous organisms.

## 5. OIL RESERVOIR SOURING

One source for microbial populations in oil field operations is from the reservoir itself. SRB are probably indigenous to oil reservoirs or adjacent aquifiers, but in any case are common inhabitants of the reservoir during oil and gas recovery operations (Davis, 1967). In some cases, this leads to unwanted hydrogen sulfide production, especially in water-flooding operations. If severe, microbial hydrogen sulfide production can force complete retrofits of field, transportation, and processing equipment for sour gas service. This involves different metallurgy, materials, and operating procedures as well as further capital investment in sulfur recovery facilities. Reservoir souring has also shown up in depleted gas fields, which are used to store natural gas or dewatered crude oil awaiting market distribution or processing.

A complete discussion of reservoir souring lies beyond the scope of this article. It is a controversial area of research driven by the early recognition of serious problems developing in major reservoirs around the world. The likely success of biocide treatment depends on the location of the biological activity responsible for sulfide generation. Saturation of a reservoir with biocidal solutions is economically impractical while limited injections only reach the near-wellbore region. Back-flowing wells has shown that oil wells are surrounded in the oil reservoir by complex microbial communities, including hydrocarbon degrading and methanogenic bacteria (Ivanov and Belyaev, 1983). The sequence with which these organisms leave the formation suggests that this community is focused around the well itself as a source of trace oxygen and extraordinary nutrients. If the SRB are members of this community, then biocide "squeezes" that inject a modest volume of treatment chemicals into the formation in the near-wellbore region should effectively suppress microbial hydrogen sulfide production. A field trial using bleach soaks on 15 injection wells in a waterflood operation resulted in marginally reduced hydrogen sulfide at related production wells in only six cases (Lizama and Sankey, 1993). Apparently, the sulfate reducing bacterial activity responsible for souring pervaded the reservoir beyond the wellbore region and was not accessible to economic levels of biocide injection.

Alternate treatment suggestions for the near-wellbore region include the injection of nitrate or even a mixture of nitrate with sulfide resistant *Thiobacillus denitrificans* to oxidize sulfide to sulfate (McInerney *et al.*, 1992). This is another example of manipulation rather than suppression since nutrients and organisms are added to the microbial community to shift its overall effects. Use of a glutaraldehyde resistant *T. denitrificans* strain creates the opportunity to use both control and manipulation together. Major waterflooding operations can push microbial activity deep into the reservoir. In the North Sea, Ligthelm *et al.* (1992) suggest that hydrogen sulfide production is focused in the region in which injected seawater mixes with connate water deep in the reservoir. This is a moving zone unlikely to be accessible to biocides injected late in the waterflood operation. In storage fields, use of very high salinity brines (140 g/L) is a proposed method of suppressing hydrogen sulfide formation during storage of dewatered petroleum (Gonik *et al.*, 1992). It is noted that storage of hydrocarbons in salt domes avoids the souring problem. Research activities focused on reservoir souring may well have implications for control of the mixed communities in associated surface treatment facilities. In effect many of the same organisms are likely to be involved. Reservoir souring in a recycled waterflood occurs through the action of SRB as the water passes through the reservoir, while corrosion problems involving SRB occur in the surface facilities in the other half of the cycle.

## 6. FUTURE NEEDS

### 6.1. Correlation of Biological Parameters to Corrosion Risk

The most outstanding need is to be able to correlate measurable biological parameter(s) with corrosion rate or risk in a given system. Without this correlation it is impossible to estimate the cost benefit or even cost-effectiveness of proposed microbial control treatment. It is also impossible to predict field performance under treatment in anything but a general way. The current situation of trying to sterilize poorly characterized systems to the limit of allowable economics is not a satisfactory basis of operation, at least from the biocide user's point of view. More efficient chemical packages are needed in which the chemicals used to control scaling or oxygen or corrosion do not simultaneously promote the growth of undesirable microorganisms.

### 6.2. Relating Corrosion Mechanisms and Bacterial Populations

The mechanisms and interactions responsible for corrosion need to be much better understood, as does the microbial ecology in biofilm communities. These populations are unique to each facility and are surprisingly stable over time (Jack *et al.*, 1992). There is preliminary evidence that the population changes in a rational way in going from production well to injection well through surface facilities in waterflood operations (Jack *et al.*, 1994). These systems may behave as flow reactors with immobilized bacteria playing the role of sequential catalysts in nutrient breakdown. More needs to be done. Without an understanding of how communities evolve through a target system, point measurement of biological parameters is of limited use.

## 7. RECOMMENDATIONS

### 7.1. Treatment Chemicals

To improve the performance of chemical control packages, which include biocides, more appropriate agents must be found. Phosphate-free scale inhibitors and better methods to remove oxygen are required. Unfortunately the interest of biocide supplies in further investment in research in this area has been stifled in part by increased regulation in recent years. In some jurisdictions extensive testing of each variation of formulation in a family of products and shifts in occupational health specifications for component chemicals has added costs and uncertainty to new product development. Improved tailoring of chemical packages to specific needs can also result in reduced market size for the final product. These are potent disincentives for development. More formal partnerships between user groups, regulators, and supplies may be needed to encourage action.

### 7.2. Understanding Biofilm Communities

The understanding of the role of microbes in accelerating corrosion awaits our resolution of the structure of communities of organisms involved. This involves not only the identification of organisms present but also identifying their spatial relationships with others in the biofilm environment. Present challenges include a need for methods to characterize such populations *in situ* and the recovery of key organisms many of which may not be easily isolated using present techniques.

Information on community composition and spatial structure has to be obtained from field samples without involving sample growth. Biomarkers and nucleic acid hybridization techniques offer an opportunity to obtain such information. With biomarkers, specific chemical compounds can be analyzed directly from soil or water samples and can be used to assay general biomass or specific groups of organisms. Parkes (1987) concluded that while there are challenges to interpretation this is the best existing technology for obtaining information on community composition. For example, the cellular fatty acid composition of *Desulfovibrio* spp. was recently evaluated and found to be in good agreement with existing SRB phylogenetic relationships (Vainshtein *et al.*,

1992). Fatty acid markers have also been used to identify sulfate-reducing bacteria in substrate stimulated *in situ* marine studies (Dowling *et al.*, 1986). Nucleic acid hybridization technologies (Sayler and Layton, 1990) have applied to SRB populations. The phylogenetic relationships of SRB (Devereux *et al.*, 1990), and *Desulfovibrio* genus (Devereux *et al.*, 1989) have been assessed by comparing 16S and 23R rRNA sequences. The determined relationships, as with the fatty acid biomarkers, were generally consistent with the existing classification based on classical microbiological characteristics. Six oligonucleotide probes representing the phylogenetically defined groups of SRB have been used for determinative and environmental purposes (Devereux *et al.*, 1992). A fluorescent SRB probe based on nucleic acid hybridization has been used to visualize specific SRB populations within developing and established biofilms (Amman *et al.*, 1992). A similar technique was used to monitor the enrichment and isolation of an SRB from a multispecies anaerobic bioreactor (Kane *et al.*, 1993). These probes allow the identification of specific groups without involving growth and when combined with fluorescent microscopy can illuminate the spatial arrangements of bacteria within a biofilm.

Application of nucleic hybridization techniques to oil field populations has begun. Voordouw *et al.* (1990) have used a DNA gene probe for the [NiFe] form of the enzyme hydrogenase for identification of *Desulfovibrio* spp. in oil field fluids. The development and use of a novel DNA hybridization technique, Reverse Sample Genome Probing, indicated at least twenty genotypically different SRB in Alberta oil field fluids (Voordouw *et al.*, 1991). An analysis of bacteria in production waters from 56 sites in seven locations revealed the presence of distinct SRB communities in fresh water and saline operations (Voordouw *et al.*, 1992). The comparison of planktonic and related biofilm populations. (Voordouw *et al.*, unpublished data) showed that only a few of the SRB genotypes present in the planktonic population dominate the related biofilm and that these sessile populations are stable and unique to each operation even within a single oil field. Further effort is required to characterize these bacterial communities both in composition and corrosion consequences.

### 7.3.3. Biofilm Processes

Corrosion is a kinetic phenomenon. The difference between a problem site and a benign one is ultimately a question of process rates yet remarkably little has been done on the investigation of biofilm processes and factors controlling their kinetics. Nutrient flux may be a controlling parameter in establishing a corrosive biofilm. Information obtained in the last decade indicates that the SRB are a diverse group of bacteria, which depend on the activity of associated organisms for their specific nutritional requirements. The growth and activity of the SRB in industrial systems is thus likely to depend on their association with other bacteria able to provide appropriate carbon sources. More needs to be understood about these other organisms and the processes involved in specific associations.

Transformation of inorganic species and the flux of reducing power in these communities must be important. These are the constituent processes involved in corrosion and bacteria can control them. For example, facultative bacteria able to reduce ferric ion can accelerate removal of passivating oxides from the steel surface under microbial colonies (Obuekwe et al., 1981). Sulfite released by SRB under slow continuous growth conditions (Jobson, 1975) may promote a host of other organisms in the "sulfide cascade" to generate corrosive sulfides (Obuekwe et al., 1983). Iron sulfides, long recognized as having the potential to carry reducing power in the form of electrons from the corrosion process through the biofilm (Miller and King, 1975), have been largely ignored to date in terms of community structure. Polysaccharide slimes often associated with biofilms will bind ions and may even order entrapped water. The consequences of this on the behavior of the biofilm as the electrolyte for a corrosion cell have yet to receive attention. More information is needed on the kinetics and flux of nutrients, reducing power, and inorganic species within the biofilm structure.

## 8. SUMMARY

Understanding the composition and action of biofilm communities is a prerequisite to advancing management strategies beyond the present state of the art. For more than 50 years application of biocides monitored by general viable counts to establish kill has been the only approach to microbially influenced corrosion in many systems. It has not been a totally successful approach and without a basis of correlation between biocide application and corrosion performance, lacks proper management and even assessment tools. With increasing regulation and environmental and occupational health concerns, this approach faces further restriction in terms of product selection. Sophisticated new technology is emerging which will enable the development of a better picture of the biofilm and its constituent community. This in turn offers the hope of more subtle and cost effective ways to manipulate the corrosivity of indigenous populations. The challenge is to establish the integrated and interdisciplinary industrial and academic effort to take this next major step.

## REFERENCES

Aeckersberg, F., Bak, F., and Widdel, F., 1991, Anaerobic oxidation of saturated hydrocarbons to $CO_2$ by a new type of sulfate-reducing bacterium, *Arch. Microbiol.* **156**:5–14.

Amman, R. I., Stromley, J., Devereux, R., Key, R., and Stahl, D. A., 1992, Molecular and microscopic identification of sulfate-reducing bacteria in multispecies biofilms, *Appl. Environ. Microbiol.* **58**:614–623.

Angrykovitch, G., and Neohof, R. A., 1987, Fuel-soluble biocides for control of *Cladosporium resinae* in hydrocarbon fuels, *J. of Ind. Microbiol.* **2**:35–40.

American Petroleum Institute Research Publication 38, 1965, Recommended practice for biological analysis of subsurface injection waters, Second Edition, American Petroleum Institute, Washington.

Bailey, N.S.L., Jobson, A. M., and Rogers, M. A., 1973, Bacterial degradation of crude oil: comparison of laboratory and field data, *Am. Assoc. Pet. Geol. Bull.* **57**: 1276–1290.

Blenkinsopp, S. A., Khoury, A. E., and Costerton, J. W., 1992, Electrical enhancement of biocide efficacy against *Pseudomonas aeruginosa* biofilms, *Appl. Environ. Microbiol.* **58(11)**:3770–3773.

Booth, G. H., Elford, L., and Wakerley, D. S., 1968, Corrosion of mild steel by sulphate-reducing bacteria: an alternative mechanism, *Br. Corros, J.***3**:242–245.

Bott, T. L., 1975, Bacterial growth rates and temperature optima in a stream with fluctuating thermal regime, *Liminol. Oceanogr.* **20**:191–197.

Connan, J., 1984, Biodegradation of crude oils in reservoirs, *Adv. Pet. Geochem.* **1**:299–335.

Cord-Ruwisch, R., Kleinitz, W., and Widdel, F., 1987, Sulfate-reducing bacteria and their activities in oil production, *J. of Petroleum Technology,* **3**:97–106.

Costerton, J. W., 1983, Biofilm Removal, U.S. Patent No. 4,419,248.

Costerton, J. W., and Lappin-Scott, H. M., 1989, Behavior of bacteria in biofilms, *Amer. Soc. Microbiol. News* **55**:650–654.

Davis, J. B., 1967, *Petroleum Microbiology*, Elsevier, New York.

Devereux, R., Delaney, M., Widdel, F., and Stahl, D. A., 1989, Natural relationships among sulfate-reducing bacteria, *J. Bacteriol.* **171**:6689–6695.

Devereux, R., He, S., Doyle, C., Orland, S., and Stahl, D., 1990, Diversity and origin of *Desulfovibrio* species: phylogenetic definition of a family, *J. Bacteriol.* **172**:3609–3619.

Devereux, R., Kane, M. D., Winfrey, J., and Stahl, D. A., 1992, Genus- and group-specific hybridization probes for determinative and environmental studies of sulfate-reducing bacteria, *System. Appl. Microbiol.* **15**:601–609.

Dowling, N.J.E., Widdel, F., and White, D., 1986, Phospholipid ester-linked fatty acid biomarkers of acetate-oxidizing sulfate-reducers and other sulfide forming bacteria, *J. of Gen. Microbiol.* **132**:1812–1825.

Evans, C. R., Rogers, M. A., and Bailey, N.J.L., 1971, Evolution and alteration of petroleum in western Canada, *Chem. Geol.* **8**:147–170.

Geesey, G. G., Mutch, R., Costerton, J. W., and Green, R. B., 1978, Sessile bacteria: an important component of the microbial population in small mountain streams, *Liminol. Oceanogr.* **23**:1214–1223.

Gonik, A. A., Sabirova, A. Kh., and Yudina, E. G., 1992. Suppression of the formation of hydrogen sulfide in stored crude oil by placing concentrated brine in reservoir layer and maintaining its concentration. Soviet Union 1714097. Derwent NO. 93-025503/03 H01 Q49.

Gas Research Institute Field Guide I, 1990, Microbiologically Influenced Corrosion (MIC): Methods of detection in the field, Gas Research Institute, Chicago.

Gas Research Institute Field Guide II, 1992, Microbiologically Influenced Corrosion (MIC): Investigation of internal MIC and testing mitigation measures, Gas Research Institute, Chicago.

Hamner, N. E., 1970, Coatings for corrosion protection, in: *NACE Basic Corrosion Course.* National Association of Corrosion Engineers, Houston, pp. 11.1–11.20.

Ivanov, M. V., and Belyaev, S. S., 1983, Microbial activity in waterflooded oil fields and its possible regulation, in: *Proceedings of the International Conference on Microbial Enhancement of Oil Recovery,* Shanghri-la, Oklahoma, May 16–21, 1982, (E. C. Donaldson and J. B. Clark, eds.), U.S. Department of Energy, Bartlesville, Oklahoma, pp. 48–57.

Iverson, W. P., and Olson, G. J., 1984, Problems related to sulfate-reducing bacteria in the petroleum industry, in: *Petroleum Microbiology* (R. M. Atlas, ed.) Macmillan Publishing Co., New York, pp. 619–641.

Jack, T. R., Lee, E., and Mueller, J., 1985, Anaerobic gas production from crude oil, in: *Microbes and Oil Recovery: Proceedings of the Second International Conference on Microbially Enhanced Oil Recovery, Fountainhead, Oklahoma, May 20–25, 1984,* (J. E. Zajic and E. C. Donaldson, eds.), Bioresource Publications, El Paso, Texas, pp. 167–180.

Jack, T. R., Bramhill, B. J., and Ferris, F. G., 1992, Field studies of biocorrosion and its control in water piping systems, *in: BIOMINET Proceedings, Calgary, Alberta, October 31, 1991* (W. D. Gould and S. Lord, eds.) CANMET-Mineral Sciences Laboratory, Energy, Mines and Resources, Ottawa.

Jack, T. R., Rogoz, E., Bramhill, B. J., and Roberge, P. R., 1994, The characterization of sulfate-reducing bacteria in heavy oil waterfloods operations, in *Microbially Influenced Corrosion Testing,* ASTM STP1232 (J. R. Kearns and B. J. Little, eds.), American Society for Testing and Materials, Philadelphia, pp. 108–117.

Jobson, A. M., 1975, Physiological characterization of *Desulfovibrio* sp. isolated from crude oil, Ph.D. Thesis, University of Alberta, Edmonton, Alberta, Canada.

Jobson, A. M., Cook, F. D., and Westlake, D.W.S., 1979. Interaction of aerobic and anaerobic bacteria in petroleum degradation, *Chem. Geol.* **24:**355–365.

Jones, C. A., Leidlein, J. H., and Grierson, J. G., 1988, Methods for evaluating the efficacy of biocides against sessile bacteria, *Journal of the Cooling Tower Institute* **9(1):**28–33.

Kane, M. D., Poulsen, L. K., and Stahl, D. A., 1993, Monitoring the enrichment and isolation of sulfate-reducing bacteria by using oligonucleotide hybridization probes designed from environmentally derived 16S rRNA sequences, *Appl. Environ. Microbiol.* **59:**682–686.

Kuznetsova, V. A., Li, A. D., and Tifora, N. N., 1963, A determination of source contamination of oil-bearing Devonian strata of Roamshinko oil field, *Mikrobiologyia* **32:**581–585.

Kuznetsova, V. A., and Li, A. D., 1964, Developments of sulfate-reducing bacteria in the flooded oil-bearing Devonian strata of the Roamshinko oil field, *Mikrobiologiya* **33:**276–280.

Ligthelm, D. J., de Boer, R. B., and Brint, J. F., 1992, Reservoir souring: An analytical model for $H_2S$ generation and transportation in an oil reservoir owing to bacterial activity, SPE 23141, Society of Petroleum Engineers, Richardson, Texas.

Lizama, H. M., and Sankey, B. M., 1993, On the use of bleach soaks to control bacteria-mediated formation souring, *Journal of Petroleum Science and Engineering* **9**:145–153.

Marshall, K. C., 1992, Biofilms: an overview of bacterial adhesion, activity and control at surfaces, *Amer. Soc. Microbiol. News* **58**:201–207.

Marshall, K.C.R., Stout, R., and Mitchell, R., 1971, Selective sorption of bacteria from seawater, *Can. J. Microbiol.* **17**:1413–1416.

Marszalek, D. S., Gerchakov, S. M., and Udey, L. R., 1979, Influence of substrate composition on marine microfouling, *Appl. Environ. Microbiol.* **38**:987–995.

Maxwell, S., and Hamilton, W. A., 1986, Modified radio respirometric assay for determining the sulfate reduction activity of biofilms on metal surfaces, *J. Microbiol. Methods* **5(2)**:83–91.

McCoy, W. F., Byers, J. D., Robbins, J., and Costerton, J.W.C., 1981, Observations of biofilm formation, *Can. J. Microbiol.* **27(9)**:910–917.

McInerney, M. J., Bhupathiraju, V. K., Vishvesh, K., and Sublette, K. L., 1992, Evaluation of a microbial method to reduce hydrogen sulfide levels in a porous rock biofilm, *Journal of Industrial Microbiology,* **11(1)**:53–58.

Miller, J.D.A., and King, R. A., 1975, Biodeterioration of metals, in: *Microbial Aspects of the Deterioration of Material* (D. W. Lovelock and R. J. Gilbert, eds.), Academic Press, London, pp. 83–105.

Milner, C.W.D., Roger, M.A., and Evans, C. R., 1977, Petroleum transformation in reservoirs, *J. of Geochem. Explor.* **7**:101–153.

Obuekwe, C. O., Westlake, D.W.S., Cook, F. D., and Costerton, J. W., 1981, Surface changes in mild steel coupons from the action of corrosion causing bacteria, *Appl. Environ. Microbiol.* **41**:766–774.

Obuekwe, C. O., Westlake, D.W.S., and Cook, F. D., 1983, Corrosion of Pembina crude oil pipeline: the origin and mode of formation of hydrogen sulfide, *Appl. Microbiol. Biotech.* **17**:173–177.

Obuekwe, C. O., Westlake, D.W.S., and Plambeck, J.A., 1987, Evidence that available energy is a limiting factor in the bacterial corrosion of mild steel by a *Pseudomonas* sp., *Can. J. Microbiol.* **33**:272–275.

Orr, N. L., 1979, Sulfur in heavy oils, oil sands and oilshales, in: *Oil Sand and Oil Shale Chemistry.* (O. P. Strausz and E. M. Lown, editors), Verlag Chemie, New York, pp. 223–243.

Paillard, H ., Brunet, R., and Dore, M., 1987, Application of oxidation by a combined ozone/ultraviolet radiation system to the treatment of natural water, *Ozone Science and Engineering* **9**:391–418.

Pankhania, I. P., 1988, Hydrogen metabolism in sulfate-reducing bacteria and its role in anaerobic corrosion, *Biofouling,* **1**:27–47.

Parkes, R. J., 1987, Analysis of microbial communities within sediments using biomarkers, in: *Ecology of Microbial Communities* (M. Fletcher, T.R.G. Gray and J. G. Jones, eds.), Cambridge University Press, New York, pp. 147–177.

Peabody, A. W., 1970, Principles of cathodic protection, in: *NACE Basic Corrosion Course*, National Association of Corrosion Engineers, Houston, pp. 5.1–5.37.

Pope, D. H., 1992, *State-of-the-Art Report on Monitoring, Prevention and Mitigation of Microbiologically Influenced Corrosion in the Natural Gas Industry*, Gas Research Institute, Chicago.

Pope, D. H., Zintel, T. P., Cookingham, B. A., Morris, R. G., Howard, D., Day, R. A., Frand, J. R., and Pogemiller, G. E., 1989, Mitigation strategies for microbiologically influenced corrosion in gas industry facilities. Paper 192, *Corrosion '89*, National Association of Corrosion Engineers, Houston.

Postgate, J. R., 1979, *The Sulfate Reducing Bacteria,* Cambridge University Press, London.

Rajagopal, B. S., Libert, M. F., and LeGall, J., 1988, Utilization of cathodic hydrogen by *Desulfovibrio* species with sulfate or nitrate as terminal electron acceptor, in: *Microbial Corrosion 1* (C.A.C. Sequina and A. K. Tiller, eds.), Elsevier Applied Sciences, New York, pp. 66–78.

Ruseska, I., Robbins, J., Costerton, J. W., and Lashen, E. S., 1982, Biocide testing against corrosion-causing oil field bacteria helps control plugging, *Oil and Gas Journal* **8:**253–264.

Sayler, G. S., and Layton, A. C., 1990, Environmental application of nucleic acid hybridization, *Ann. Rev. Microbiol.* **44:**625–648.

Scott, P.J.B., and Davies, M., 1992, Survey of field kits for sulfate-reducing bacteria, *Materials Performance* **31:**64–68.

Strauss, S. D., 1992, Instrumentation advances improve fouling, corrosion monitoring, Power, **September:**17–19.

Technical Publication Committee Publication 3, 1976, *The Role of Bacteria in the Corrosion of Oil Field Equipment,* National Association of Corrosion Engineers, Houston.

Vainshtein, M., Hippe, H., and Kroppenstedt, R. M., 1992, Cellular fatty acid composition of *Desulfovibrio* species and its use in the classification of sulfate-reducing bacteria, System. Appl. Microbiol. **15:**554–566.

Von Wolzogen Kuhr, C.A.H., and Van der Klugt, I. S., 1934, De grafiteering van gietzer als electrobiochemisch proces in anerobe groden, Water (The Hague), **18:**147–165.

Voordouw, G., Niviere, V., Ferris, F. G., Fedorak, P. M., and Westlake, D.W.S., 1990, Distribution of hydrogenase genes in *Desulfovibrio* spp. and their use in the identification of species from the oil field environment, *Appl. Environ. Microbiol.* **56:**3748–3753.

Voordouw, G., Voordouw, J. K., Karkhoff-Schweizer, R. R., Fedorak, P. M., and Westlake, D.W.S., 1991, Reverse sampling genome probing, a new technique for identification of bacteria in environmental samples by DNA hybridization, and its application to the identification of sulfate-reducing bacteria in oil field samples, *Appl. Environ. Microbiol.* **57:**3070–3078.

Voordouw, G., Voordouw, J. K., Jack, T. R., Foght, J., Fedorak, P. M., and Westlake, D.W.S., 1992, Identification of distinct communities of sulfate-reducing bacteria in oil fields by reverse sample genome probing, *Appl. Environ. Microbiol.* **58:**3542–3552.

ZoBell, C. E., 1943, The effect of solid surfaces upon bacterial activity, *J. Bacteriol.* **46:**39–56.

# Metabolism of Environmental Contaminants by Mixed and Pure Cultures of Sulfate-Reducing Bacteria

# 11

## BURT D. ENSLEY and JOSEPH M. SUFLITA

## 1. INTRODUCTION

Many environments are anoxic or rapidly become so after contamination with carbon-rich compounds such as gasoline, crude oil, or a myriad of other pollutants. Removal mechanisms for recalcitrant compounds become very important if these potentially hazardous molecules make their way into drinking or ground water supplies. The anaerobic metabolism of organic compounds is an environmental sink for many of these molecules. There are multiple routes for the anaerobic metabolism of organic chemicals. When external electron acceptors are absent, the oxidation of organic matter can be coupled with the reduction of other atoms of the energy source in anaerobic fermentation processes. Alternate metabolic routes for energy generation include anaerobic respiration with nitrate, iron, manganese, sulfate, or carbon dioxide as electron acceptors. Organic electron acceptors are also available for energy conservation in anaerobes, as some organisms are capable of using phenylmethylethers or even haloorganic substances in this capacity with simple organic molecules or hydrogen acting as electron donors.

BURT D. ENSLEY • 7 Colts Neck Drive, Newton, PA 18940.    JOSEPH M. SUFLITA • Department of Botany and Microbiology, The University of Oklahoma, Norman, OK 73019.

*Sulfate-Reducing Bacteria*, edited by Larry L. Barton. Plenum Press, New York, 1995.

In recent years there has been a burgeoning of information on the anaerobic biodegradation of environmentally important chemical compounds. This knowledge surge has been paralleled by the growth in information on the basic ecology and physiology of sulfate- and sulfur-reducing bacteria. A little over a decade ago, only two genera of dissimilatory sulfate-reducing bacteria (SRB) were known (*Desulfovibrio* and *Desulfomaculum*), and these organisms were considered nutritionally very limited. This situation has changed dramatically; many genera are currently recognized, some of which are listed in Table I. A study of these genera reveals that this group of organisms is far more nutritionally diverse than originally perceived. Electron donors used to support sulfate-reduction include hydrogen, a variety of alcohols, fatty acids, other monocarboxylic acids, dicarboxylic acids, some amino acids, sugars, phenylsubstituted acids, homocyclic aromatic compounds, and long-chain saturated alkanes (See Widdel and Bak, 1992, for a review). In addition, there are many compounds whose anaerobic metabolism can be linked with the consumption of sulfate as a terminal electron acceptor. That is, these compounds can be metabolized under sulfate-reducing conditions (Table II).

The recognition of the almost ubiquitous distribution of SRB and the proliferation of information regarding the anaerobic biotransformation of environmental contaminants under sulfate-reducing conditions prompts this review. Such information is critically important for a consideration of the transport and fate characteristics associated with pollutants. As already noted, many environments are anaerobic or become so upon the introduction of labile forms of organic matter. Many of these environments will also have sulfate available as a terminal electron acceptor. For several reasons, biodegradation in these environments is important to understand. First, it is important to determine the total pollutant burden that a habitat can withstand, and this is at least in part a function of the turnover of contaminants. Second, the partial metabolism of some compounds can lead to the formation of degradative intermediates that are generally more polar, water soluble, and more mobile in the environment than the parent material. Metabolic intermediates of pollutants can have their own impact on both humans and recipient ecosystems. Before the risks of contamination can be fully appreciated, knowledge of which compounds are persistent and which are likely to degrade, and how they are degraded is essential. Conceivably, such information will also provide a fundamental knowledge base upon which bioremedial approaches to environmental contamination can be built.

**Table I. Examples of Chemicals Transformed by Pure Cultures of Sulfate-Reducing Bacteria**

| Organism | | | | |
|---|---|---|---|---|
| Genus | Species | Strain | Substrate | Reference notes |
| *Desulfovibrio* | | | | |
| | *gigas* | | 2,4,6-Trinitrotoluene | Boopathy et al., 1993 |
| | | | Furfural | Bruen et al., 1983 |
| | *vulgaris* | Marburg | 2,4,6-Trinitrotoluene | Boopathy et al., 1993 |
| | *vulgaris* | ATCC43938 | 2-Methoxyethanol | Tanaka, 1992 |
| | *desulfuricans* | ATCC7757 | 2,4,6-Trinitrotoluene | Boopathy et al., 1993 |
| | *desulfuricans* | | Kraft lignin | Ziomek and Williams, 1989 |
| | | | Lignosulfonate | |
| | *desulfuricans* | M6 | Phenylsulfide | Kim et al., 1990a |
| | | | Benzylsulfide | |
| | | | Benzyldisulfide | |
| | | | Benzothiophene | |
| | | | Dibenzothiophene | |
| | sp. | | Crude oil | Kim et al., 1990b |
| | | | Dibenzothiophene | Köhler et al., 1984 |
| | | | Dibenzylsulfide | |
| | | | Benzothiophene | |
| | | | Dibenzyldisulfide | |
| | | | Butylsulfide | |
| | | | Octysulfide | |
| | | B | Furfural | Boopathy and Daniels, 1991 |
| | | | 2,4,6-Trinitrotoluene | Boopathy et al., 1993 |
| | | | 2,4-Dinitrophenol | Boopathy and Kulpa, 1993 |
| | | | 2,4-Dinitrotoluene | |
| | | | 2,6-Dinitrotoluene | |
| | | | Aniline | |
| | *furfuralis* | F-1 | Furfural | Brune et al., 1983 |
| | | | Furfural | Folkerts et al., 1989 |

*(continued)*

**Table I.** (*Continued*)

| Organism | | Strain | Substrate | Reference notes |
|---|---|---|---|---|
| Genus | Species | | | |
| *Desulfobacterium* | | | | |
| | *phenolicum* | | Benzoate | Bak and Widdel, 1986b |
| | | | 2-Hydroxybenzoate | |
| | | | 4-Hydroxybenzoate | |
| | | | Phenol | |
| | | | p-Cresol | |
| | | | Phenylacetate | |
| | | | 4-Hydroxyphenylacetate | |
| | | | Indole | |
| | *catecholicum* | | Toluene | Rabus et al., 1993 |
| | | | Benzoate | Szewzyk and Pfennig, 1987 |
| | | | 4-Hydroxybenzoate | |
| | | | Protocatechuate | |
| | | | 2-Aminobenzoate | |
| | | | 3,4-Dihydroxybenzoate | |
| | | | Catechol | |
| | | | Resorcinol | |
| | | | Hydroquinone | |
| | | | Pyrogallol | |
| | | | Phloroglucinol | |
| | *anilini* | | Benzoate | Schnell et al., 1989 |
| | | | 3,4-Dihydroxybenzoate | |
| | | | 2,5-Dihydroxybenzoate | |
| | | | 2,4-Dihydroxybenzoate | |
| | | | Quinoline | |
| | | | Aniline | |
| | | | 2-Aminobenzoate | |
| | | | 4-Aminobenzoate | |
| | | | Phenol | |

| | | | |
|---|---|---|---|
| sp. | Cat 2 | Catechol | Schnell et al., 1989 |
| | | p-Cresol | |
| | | Pyrogallol | |
| | | Phenylacetate | |
| | | Phenylpropionate | |
| | | 2-Hydroxybenzoate | |
| | | 3-Hydroxybenzoate | |
| | | 4-Hydroxybenzoate | |
| | | 2-Hydroxybenzoate | |
| | | 3-Hydroxybenzoate | |
| | | 4-Hydroxybenzoate | |
| | | 3,4-Dihydroxybenzoate | |
| | | 3,5-Dihydroxybenzoate | |
| | | m-Cresol | |
| | | p-Cresol | |
| | | Phenol | |
| | | Catechol | |
| | | Pyrogallol | |
| | | Benzoate | |
| | | Phenylacetate | |
| | | Phenylpropionate | |
| | | 2-aminobenzoate | |
| | | Indole | |
| | | Quinoline | |
| | | Hippurate | |
| oleovorans | Hxd3 | n-Dodencane | Aeckersberg et al., 1991 |
| | | n-Tetradecane | |
| | | n-Pentadecane | |
| | | n-Heptadecane | |
| | | n-Octadecane | |
| | | n-Eicosane | |
| | | 1-Hexadecene | |
| vacuolatum | | Phenylpropionate | Widdel, 1988 |

(continued)

**Table I.** (*Continued*)

| Organism | | Strain | Substrate | Reference notes |
| Genus | Species | | | |
| --- | --- | --- | --- | --- |
| | *cetonicum* | | Benzoate | Galushko and Rozanova, 1991 |
| | *indolicum* | | 2,4,6-Trinitrotoluene | Boopathy and Wilson, 1993 |
| | | | 2-Aminobenzoate | Bak and Widdel, 1986a |
| | | | Indole | |
| | | | Quinoline | |
| *Desulfotomaculum* | *sapomandens* | | Benzoate | Widdel, 1988; Cord-Ruwisch and Garcia, 1985 |
| | | | Phenylacetate | |
| | | | 3-Phenylpropionate | |
| | | | 4-Hydroxybenzoate | |
| | *orientis* | TEP | 3,4,5-Trimethoxybenzoate | Klemps *et al.*, 1985 |
| | | TWC | 3,4,5-Trimethoxybenzoate | |
| | | TWP | 3,4,5-Trimethoxybenzoate | |
| | | Groll | Benzoate | |
| | sp. | | Benzoate | Kuever *et al.*, 1993 |
| | | | 3-Hydroxybenzoate | |
| | | | 4-Hydroxybenzoate | |
| | | | 2-Amimobenzoate | |
| | | | Phenol | |
| | | | Catechol | |
| | | | 3,4-Dihydroxybenzoate | |
| | | | Phenylacetate | |
| | | | Phenylpropionate | |
| | | | Vanillate | |
| | | | Protocatechuate | |

| Genus | Species | Strain | Substrate | Reference |
|---|---|---|---|---|
| *Desulfococcus* | *thermobenzoicum* | | m-Cresol | Tasaki, M et al., 1991 |
| | | | p-Cresol | |
| | | | Ferulate | |
| | | | Cinnamate | |
| | | | Syringate | |
| | | | 2,6-Dimethoxyphenol | |
| | | | 2-Methoxybenzoate | |
| | | | Benzoate | |
| | *australicum* | | Benzoate | Love et al., 1993 |
| | *multivorans* | | Benzoate | Widdel, 1988 |
| | | | Phenylacetate | |
| | | | Phenylpropionate | |
| | | | 2-Hydroxybenzoate | |
| | *multivorans* | Hy5 | Benzoate | Schnell et al., 1989 |
| | | | Phenylacetate | |
| | | | Phenylpropionate | |
| | | | Hydroquinone | |
| | | | 2,5-Dihydroxybenzoate | |
| | *niacini* | | Nicotinic acid | Imhoff-Stuckle and Pfennig, 1983 |
| | | | Phenylpropionate | |
| *Desulfosarcina* | *variabilis* | | Benzoate | Widdel, 1988 |
| | | | Phenylacetate | |
| | | | 3-Phenylpropionate | Widdel and Pfennig, 1984 |
| | | | 3-Hydroxybenzoate | |
| | | | 4-Hydroxybenzoate | |
| | | | Hippurate | |
| *Desulfomonile* | *tiedjei* | | 3-Anisate | DeWeerd et al., 1990 |
| | | | 4-Anisate | |
| | | | Benzoate | |

*(continued)*

**Table I.** (*Continued*)

| Organism | | | Substrate | Reference notes |
|---|---|---|---|---|
| Genus | Species | Strain | | |
| *Desulfonema* | | | | |
| | *magnum* | | Benzoate | Widdel *et al.*, 1983 |
| | | | Phenylacetate | |
| | | | 3-Phenylpropionate | |
| | | | Hippurate | |
| | | | 4-Hydroxybenzoate | |
| *Desulfoarculus* | | | | |
| | sp. | SAX | Benzoate | Drzyzga *et al.*, 1993 |
| | | | 4-Hydroxybenzoate | |
| | | | Phenol | |
| | | | Phenylacetate | |
| *Desulfobacula* | | | | |
| | *toluolica* | | Toluene | Rabus *et al.*, 1993 |
| | | | p-Cresol | |
| | | | Benzaldehyde | |
| | | | Benzoate | |
| | | | Phenylacetate | |
| | | | p-Hydroxybenzaldehyde | |
| | | | 4-Hydroxybenzoate | |
| *Other Organisms* | | | | |
| | *Isolate* | Re10 | Resorcinol | Schnell *et al.*, 1989 |
| | *Isolate* | | 2,4-Dihydroxybenzoate | |
| | | | 2,4,6-Trinitrotoluene | Preuss and Dickert, 1993 |

**Table II. Examples of Substrates Amenable to Anaerobic Biotransformation under Sulfate-Reducing Conditions with Different Inoculum Sources**

| Inoculum source | Compound transformed | Reference |
|---|---|---|
| Aquifer sediment | Benzene | Edwards and Grbić-Galić, 1992 |
| | Toluene | Edwards et al., 1992; Haag et al., 1991 |
| | o-Xylene | |
| | m-Xylene | |
| | p-Xylene | |
| | o-Cresol | Suflita et al., 1989; Smolenski and Suflita, 1987 |
| | m-Cresol | |
| | p-Cresol | |
| | 2-Hydroxybiphenyl | Liu et al., 1991 |
| | Acridine | Knezovich et al., 1990 |
| | Benzoate | Gibson and Suflita, 1986 |
| | Phenol | |
| | 2-Aminobenzoate | Kuhn and Suflita, 1989a |
| | 3-Aminobenzoate | |
| | 4-Aminobenzoate | |
| | Benzamide | |
| | N-Methylbenzamide | |
| | p-Toluamide | |
| | m-Toluidine | |
| | 2-Hydroxybenzoate | Kuhn et al., 1989 |
| | 3-Hydroxybenzoate | |
| | 4-Hydroxybenzoate | |
| | 2-Picoline | Kuhn and Suflita, 1989b |
| | 3-Picoline | |
| | 4-Picoline | |
| | Nicotinic acid | |
| | Pyridine | |
| | 2-Methylfuran | |
| | 2-Furoic Acid | |
| | Indole | Shanker et al., 1991 |
| | Pyridine | |
| Sewage sludge | Pyridine | Kaiser and Bollag, 1991 |
| | 3-Hydroxypyridine | |
| | Indole | Shanker and Bollag, 1990; Bak and Widdel, 1986a |

(*continued*)

**Table II.** *Continued*

| Inoculum source | Compound transformed | Reference |
|---|---|---|
| Marine sediment | n-Hexadecane | Aeckersberg *et al.*, 1991 |
| | n-Dodecane | |
| | 2-Hydroxybenzoate | Bak and Widdel, 1986b |
| | 4-Hydroxyphenylacetate | |
| | 4-Hydroxybenzoate | |
| | o-Cresol | |
| | p-Cresol | Bak and Widdel, 1986b; Schnell *et al.*, 1989 |
| | Quinoline | Bak and Widdel, 1986a |
| | Pyridine | |
| | Benzoate | Widdel *et al.*, 1983; Schnell *et al.*, 1989 |
| | Catechol | |
| | Phenol | Bak and Widdel, 1986b; King, 1988; Schnell *et al.*, 1989 |
| | Nicotinic Acid | Imhoff-Stuckle and Pfennig, 1983; Schnell *et al.*, 1989 |
| | Resorcinol | Schnell *et al.*, 1989 |
| | Hydroquinone | |
| | 2,5-Dihydroxybenzoate | |
| | Indole | Bak and Widdel, 1986a,b |
| | Skatole | |
| | Aniline | Schnell *et al.*, 1989 |
| Freshwater sediment | Benzoate | Genthner *et al.*, 1989 |
| | Toluene | Beller *et al.*, 1992 |
| | p-Cresol | Haggblom *et al.*, 1990 |
| | Phenol | Genthner *et al.*, 1989a |
| | Catechol | Schnell *et al.*, 1989 |
| | Resorcinol | |
| Fresh water | Benzoate | Szewzyk and Pfennig, 1987 |
| | Catechol | |
| | Resorcinol | Schnell *et al.*, 1989 |
| | 4-Hydroxybenzoate | Szewzyk and Pfennig, 1987 |
| Surface soil | Benzoate | Cord-Ruwisch and Garcia, 1985 |
| | 3-Hydroxypyridine | Kaiser and Bollag, 1992 |
| | 2-Picoline | |
| | 3-Picoline | |
| | 4-Picoline | |
| | 2,6-Lutidine | |
| | 2,4-Lutidine | |
| | 3,4-Lutidine | |

(*continued*)

**Table II.** *Continued*

| Inoculum source | Compound transformed | Reference |
|---|---|---|
| Subsurface soil | 3-Picoline | Kaiser *et al.*, 1993 |
| | 2-Picoline | |
| | 4-Picoline | |
| | 2,6-Lutidine | |
| | 2,4-Lutidine | |
| | 3,4-Lutidine | |
| | 2-Hydroxypyridine | |
| | 3-Hydroxypyridine | |
| | 4-Hydroxypyridine | |
| | 4-Picoline | |
| Oil-Water separator | Hexadecane | Aeckersberg *et al.*, 1991 |
| precipitates oily | Thiophene | Kurita *et al.*, 1971 |
| sludges/tank | Benzothiophene | Kurita *et al.*, 1971 |
| bottoms | Crude Oil | Eckart *et al.*, 1986 |
| Unknown | Benzoate | Balba and Evans, 1980a |

## 2. METABOLISM OF NONHALOGENATED COMPOUNDS BY MIXED CULTURES

The number of substrates amenable to biodegradation under sulfate-reducing conditions has increased dramatically in recent years (Table II). Sometimes the biodegradation pathways of these compounds can be deduced from the detection of metabolic intermediates. More commonly, the metabolic routes employed during the composition of these materials are not known. However, it is generally assumed that only a few similar metabolic pathways are employed among anaerobic microorganisms. With this perspective, it is not astonishing that microbial degradation information gleaned from the study of one type of anaerobic condition can at least provide a prospectus of the types of transformations that might reasonably be expected under sulfate-reducing conditions.

Until recently, the common environmental contaminant benzene was believed to be completely recalcitrant to microbial attack under anaerobic conditions. However, unambiguous evidence for its destruction under methanogenic conditions has been obtained by Grbić-Galić and Vogel (Vogel and Grbić-Galić, 1986; Grbić-Galić and Vogel, 1987). Comparable evidence for the biodecay of benzene under sulfate-reducing conditions does not exist. Edwards and Grbić-Galić (1992) demonstrated that benzene could be mineralized to $CO_2$ by aquifer-derived microorganisms inoculated into an anaerobic medium containing sulfate as a potential electron acceptor. While the authors demonstrated that 90% of [14]C-labeled benzene was converted to [14]$CO_2$, they could not confirm that sulfate served as the electron acceptor for this metabolism.

In contrast, the anaerobic destruction of toluene and xylene isomers under sulfate-reducing conditions has been noted with several types of inocula (Table II). Column microcosms established with aquifer material obtained from the vicinity of a gasoline spill at Vero Beach have shown that toluene and p-xylene were metabolized. The addition of sulfate, the only electron acceptor that was detected in the microcosms, stimulated the transformation of toluene (Haag *et al.*, 1991). Later studies (Edwards *et al.*, 1992; Beller *et al.*, 1992) confirmed that sulfate was the terminal electron acceptor during toluene and xylene metabolism. Studies using [14]C-labeled substrates demonstrated the complete mineralization of both toluene and o-xylene to $CO_2$. This metabolism resulted in very little assimilation of the pollutant carbon into biomass, with essentially all of the label recovered as [14]$CO_2$. These studies also indicated that the product of sulfate-reduction, sulfide, inhibited the transformation of the aromatic substrates. This contention was strengthened by the observation that added sodium sulfide strongly inhibited the degradation rate of both toluene and xylene. Flushing the cultures with $H_2$-$CO_2$ or the addition of $FeSO_4$ increased biodegradation rates.

Similar studies were been carried out with another inoculum from subsurface sediments taken from a fuel storage facility (Beller *et al.*, 1992). Toluene was also metabolized by the microorganism in this material and the stoichiometric ratio of sulfate consumed per mole of toluene oxidized was consistent with the complete oxidation of the parent molecule of $CO_2$. Toluene degradation ceased when sulfate was depleted in the experiments and toluene was required for sulfate-reduction. The use of ring [14]C-labeled toluene confirmed that this aromatic compound was mineralized to [14]$CO_2$. This study also showed that the addition of amorphous $Fe(OH)_3$ strongly enhanced the rate and extent of toluene degradation after an incubation period of approximately 30 days.

Toluene degradation was also noted in sulfate-reducing enrichments inoculated with marine sediments (Rabus *et al.*, 1993). Sulfide levels in toluene-amended incubations were greater than in comparable toluene-free controls. Increased toluene levels were toxic to the requisite microflora such that the levels of sulfide in these incubations were less than that produced in toluene-free controls.

The anaerobic biodegradation of cresol isomer has been detected in a variety of field and laboratory studies. Some of these studies have focused specifically on the fate of cresol isomers under sulfate-reducing conditions (Table II). A selective pattern of cresol biodegradation under these conditions was noted by Smolenski and Suflita (1987) in a laboratory study using inocula obtained from a shallow aquifer polluted by municipal landfill leachate. In sulfate-reducing incubations, $p$-cresol persisted for less than 10 days before significant substrate disappearance was noted, while $o$-cresol metabolism took over 90 days to commence. The time necessary for the onset of $m$-cresol decomposition under sulfate-reducing conditions was intermediate between these extremes.

Stable enrichments of $p$-cresol-degrading anaerobes under sulfate-reducing conditions could be obtained from aquifer sediments. Cresol metabolism was dependent on the addition of sulfate, sulfite, or nitrate (Suflita et al., 1988). No degradation was observed when these electron acceptors were omitted. Uniformly [14]C-labeled cresol was completely mineralized by these cells and the detection of metabolites in the culture allowed for a degradation pathway to be suggested (Fig. 1).

Not only do the cresol isomers vary in their susceptibility to anaerobic decay, but also in the suggested pathways for their decomposition. The metabolism of $p$-cresol likely proceeds through the initial oxidation of the aryl methyl group under a variety of anaerobic conditions, including sulfate-reducing conditions (Smolenski and Suflita, 1987; Suflita et al., 1988). In similar fashion, $o$-cresol may be anaerobically oxidized to $o$-hydroxybenzoate under these conditions (Suflita et al., 1988). The reaction mechanism in these cases presumably involves the incorporation of water via a methyl hydroxylase, but firm evidence to this effect is not available. The pathway for $m$-cresol metabolism under sulfate-reducing conditions is distinctly different (Roberts et al., 1990; Ramanand and Suflita, 1991, 1993). These groups demonstrated that the metabolism of $m$-cresol by either methanogenic or sulfate-reducing enrichment cultures proceeded through the initial carboxylation of the parent substrate. The subsequent steps in the transformation of the carboxylated intermediate (4-hydroxy-2-methylbenzoic acid) have yet to be fully elucidated, but both groups report that $m$-cresol is completely mineralized.

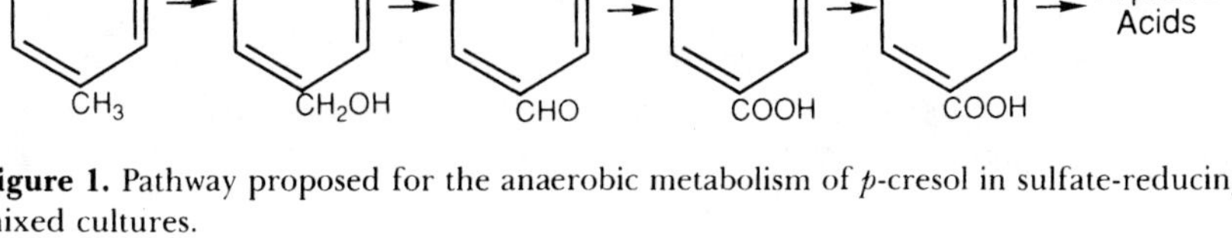

**Figure 1.** Pathway proposed for the anaerobic metabolism of $p$-cresol in sulfate-reducing mixed cultures.

The anaerobic metabolism of benzoate and phenol under sulfate-reducing conditions has been noted with a variety of inocula (Table II). These compounds are recognized as central intermediates common to the anaerobic biodegradation pathways of many aromatic compounds under a variety of conditions. The anaerobic degradation of benzoate and phenol have been extensively reviewed in recent years (Evans, 1977; Young, 1984; Sleat and Robinson, 1984; Berry et al., 1987; Evans and Fuchs, 1988; Haddock and Ferry, 1990; Schink et al., 1992) and will not be reiterated here. The generalizing features of the anaerobic decomposition of benzoate appear to involve the initial intracellular conversion of the molecule to benzoyl-CoA. The latter undergoes a six-electron-ring reduction step(s) to form cyclohexanecarboxyl-CoA. The cycloaliphatic compound is the precursor to ring cleavage. The types of intermediates produced as products of ring cleavage reactions are varied and depend on the organism or enrichment and the type of incubation conditions employed. In general, the ring cleavage products most often detected include a number of carboxylic acids like pimelate, adipate, heptanoate, and caproate. The subsequent anaerobic metabolism of these compounds appears to involve their conversion to central metabolic intermediates through $\beta$-oxidation reactions.

In comparison, the pathways for phenol decomposition by anaerobic microorganisms have received much less attention. At least two major pathways have been proposed (see reviews cited above). A ring reduction mechanism of phenol to cyclohexanone was suggested, and subsequent transformations of this intermediate resulted in the formation of caproate or adipate. However a convergence of the phenol pathway with the benzoate pathway via a 4-hydroxybenzoate intermediate has recently gained strong support. This reaction apparently proceeds via a carboxylation of the pheno ring with $CO_2$.

The formation of $p$-hydroxybenzoate as an intermediate in phenol biodegradation pathway is now well recognized. An examination of Table II will show that the transformation of this compound has frequently been observed under sulfate-reducing conditions. 4-Hydroxybenzoate can be metabolized via a direct ring reduction without the prior conversion to benzoic acid; it can be dehydroxylated and converted to benzoate, or it can sometimes be decarboxylated to phenol (see reviews above). The environmental significance of the latter bioconversion has yet to be fully elucidated.

Pure cultures of sulfate-reducers have been obtained using benzoate and phenol as electron donors (Table I). However the specific reaction mechanism that these cells employ for the metabolism of these compounds can now only be presumed similar to the pathways outlined above.

The susceptibility of nitrogen-substituted and -sulfonated benzenes to anaerobic decay under sulfate-reducing conditions has been evaluated by Kuhn and Suflita (1989a) (Table II). They tested a total of 24 model compounds including anilines, benzamides, benzene sulfonic acids, and benzenesulfonamides. The amino-substituted benzenes were relatively easily biodegraded when the aromatic nucleus was also substituted with a carboxyl group. This is in contrast to aniline and the toluidines, which proved much more recalcitrant. Only a weak transformation of the former compound was noted in these experiments and only the *m*-toluidine isomer was amenable to anaerobic decay under sulfate-reducing conditions. Benzamide as well as the aryl methyl and N-methyl derivatives were also metabolized under sulfate-reducing conditions. However, the addition of a second N-methyl group to the molecule completely precluded the anaerobic transformation of the resulting chemical. Thus, the number, position, and type of substituents on the molecule influenced its susceptibility to anaerobic biodegradation. Of the 12 model benzenesulfonic acids and sulfonamides, only *p*-benzenesulfonate metabolism was suggested under sulfate-reducing conditions. Notably, this chemical possesses an aryl carboxy group. As with other classes of substrates (anilines, phenols, etc.), the presence of an aryl carboxy group makes the resulting molecule much more susceptible to anaerobic microbial attack. Such results help illustrate the types of compounds that are likely to persist in anaerobic environments and the types of chemical substitution patterns that tend to favor anaerobic biotransformations.

An examination of Table 2 will also reveal that the biodegradation of a variety of nitrogen and oxygen heterocyclic chemicals has also been detected under sulfate-reducing conditions. Kuhn and Suflita (1989b) found that nicotinic acid was much more susceptible to anaerobic metabolism in sulfate-reducing aquifer slurries than the methylated (picoline isomers) or unsubstituted (pyridine) analogs. However, pyridine, 2-picoline, and nicotinic acid were all relatively easily degraded under sulfate-reducing conditions as verified by the loss of the parent substrate and the depletion of the near-expected amounts of the terminal electron acceptor. The 3- and 4-picoline isomers were more recalcitrant than the other test compounds in these experiments. Only the partial removal of 3-picoline was evident, while no substantial evidence for the removal of 4-picoline could be obtained. Similarly, 2-furoic acid was much more readily degradable than either furan or 2-methylfuran under sulfate-reducing conditions. Only partial degradation of the latter compound could be demonstrated while the former two substrates were completely removed (Kuhn and Suflita, 1989b).

Similar results were obtained by Kaiser and Bollag (1991) with sewage sludge as an inoculum. They found that pyridine and 3-hydroxypyride could be transformed when sulfate was provided as a terminal electron acceptor. However, no transformation of 2- or 4-hydroxypyridine was evident in these experiments.

In a study comparing the metabolism of pyridine derivatives in surface and subsurface soils, the presence of sulfate was evaluated as an experimental variable (Kaiser and Bollag, 1992). Under these conditions the picoline isomers were at least partially transformed in both surface and subsurface soils. Microorganisms in subsurface soil that was polluted by the pyridine derivatives were also adept at the transformation of hydroxypyridine isomers under sulfate-reducing conditions. However, anaerobic surface soil incubations only exhibited 3-hydroxypyridine transformation in the presence of sulfate. There is also little doubt that the several lutidine isomers were transformed, at least to a limited extent, under the imposed redox conditions. However, the picoline isomers were transformed more rapidly than di- or trimethylated pyridines exposed to contaminated subsurface soil under sulfate-reducing conditions (Kaiser and Bollag, 1992). Bak and Widdel (1986a) also demonstrated that sulfate-reducing bacterial enrichments could be obtained with several nitrogen heterocyclic compounds including quinoline, pyridine, indole, and skatole (3-methylindole).

Knezovich et al. (1990) studied the anaerobic microbial transformation of the polynuclear nitrogen heterocycle acridine under sulfate-reducing conditions. With acridine, a large number of oxidation products were detected, including 2-methylquinoline, phenyl-delta-oxopentanoic acid, N-methylanthranilic acid, 3-hydroxyanthranilic acid, phenylpropionic acid, benzyl alcohol, benzaldehyde, benzoic acid, 1,2-dimethoxyethylbenzene, phenylacetic acid, $p$-cresol, 4-methoxybenzoic acid, and phenyl-2-pentenoic acid. In addition, ethyl-, propyl- and hexylbenzene were also detected in experimental incubations, but not in controls.

Very little information is available on the anerobic biodegradation of sulfur heterocyclic compounds under sulfate-reducing conditions. The microbial metabolism of thiophene has been reported by Kurita et al. (1971). Hydrogen sulfide was detected as a product during this metabolism, but lower $C_1$–$C_4$ hydrocarbons could not be found. The bacterial cultures in this study were obtained from oily sludges and grown with polypeptone, lactic acid, glucose, meat extract, and yeast extract.

Kuhn and Suflita (1989b) found no evidence for the microbial removal of thiophene, thiophenecarboxylate, or 2-, or 3-methylthiophene following a three-month incubation period in sulfate-reducing aquifer-sediment incubations.

Other organosulfur compounds appear to be somewhat susceptible to anaerobic microbial attack. Kurita *et al.* (1971) found that benzothiophene could yield hydrogen sulfide as an anaerobic endproduct as well as several unknown carbon compounds. Köhler *et al.* (1984) reported the anaerobic destruction of several organosulfur compounds. In their study, the compounds were dissolved in paraffin oil and added to a sulfate-reducing mixed culture. Hydrogen gas was added to the headspace of some cultures. Under these conditions, a limited amount of desulfurization of dibenzylsulfide, dibenzothiophene, benzothiophene, dibenzyldisulfide, butylsulfide, and octylsulfide was evident after six days of incubation. The degree of desulfurization of these compounds ranged from only 6.5 to 16%. Similarly, Eckart *et al.* (1986) examined the anaerobic biodesulfurization of a Romashkino crude oil containing 1.8% sulfur. Mixed cultures grown with lactate and sulfate were employed in these experiments. *Desulfovibrio* spp. were the predominant members of the microbial community. When the pH was controlled to maintain circumneutral conditions, desulfurization of 26 to 40% was noted in a few days.

The biodegradation pathways for such materials are not known with certainty. A primary degradative event for many heterocyclic compounds likely involves the hydroxylation of the ring with $H_2O$ at a position adjacent to the heteroatom. Such an initial transformation route has been described for a number of heterocyclic chemicals amenable to anaerobic decay. Indeed, this transformation process has been demonstrated for the nitrogen heterocycles 4-picoline (Kaiser *et al.*, 1993) and indole (Shanker and Bollag, 1990) with SRB enrichments. Knezovich *et al.* (1990) proposed a pathway for acridine biodegradation under sulfate-reducing conditions that involved an initial hydroxylation of the parent molecule at the 3, 4, or 9 position. However, the initial oxidation products were not detected in their studies.

Often, the anaerobic microbial metabolism of nonhalogenated aromatic compounds proceeds via mechanisms that do not necessarily involve the destruction of the aromatic nucleus. The picture that emerges from such studies is that initial modification reactions function to generate intermediates that may then undergo a variety of substituent removal reactions. For instance, the former bioconversions include methyl group oxidations, enoate reductions, O-demethylations, ether cleavages, the alteration of alkanoic acid substituents, carboxylations, and several others. Often, initial modification reactions serve to convert the parent substrates to hydroxylated, carboxylated, or amino derivatives. Subsequent substituent removal reactions include decarboxylations, dehydroxylations, deaminations, and dehalogenations. The general presumption is that contaminant chemicals undergo the same type of bioconversion when incubated under sulfate-reducing conditions. Once these initial transformation have taken place, the intermediates may then enter the anaerobic biodegradation pathways associated with benzoate, phenol, or p-hydroxybenzoate as noted above, or other metabolic routes like the resorcinol or phloroglucinol pathway.

The hydroxybenzoate isomers are frequently found as intermediates in the anerobic biodegradation pathway of a variety of environmental contaminants. The biotransformation of these chemicals has been observed under sulfate-reducing conditions (Table II). Kuhn *et al.* (1989) examined the fate of hydroxybenzoate isomers in anoxic aquifer slurries. They found that benzoate was detected as an intermediate of all three hydroxybenzoate isomers suggesting that aryl dehydroxylation reactions are independent of the relative substitution pattern. However, while benzoate was the major degradation product detected with the *ortho* and *meta* isomers, it proved to be only a trace metabolite when aquifier slurries were amended with p-hydroxybenzoate. Rather, phenol proved to be the major metabolite of this substrate when sulfate served as the terminal electron acceptor. Several pure cultures of SRB are known to metabolize hydroxybenzoate isomers (Table I).

The biodegradation of more complicated phenolic structures like catechol, hydroquinone, resorcinol, phloroglucinol, and hydroxybiphenyl have also been detected in sulfate-reducing enrichments (Table II) and sometimes with pure cultures of SRB (Table I). The pathways employed for the destruction of these hydroxybenzene derivatives are far from clear. Reductive dehydroxylation of catechol and hydroquinone and convergence on the phenol pathway has been previously suggested for the anaerobic biodegradation of these substrates (Balba and Evans, 1980b; Szewzyk *et al.*, 1985; Young and Rivera, 1985). However, the carboxylation of hydroquinone and conversion of the parent substrate to gentisate by a fermenting bacterium has also been described (Szewzyk and Schink, 1989). Several resorcylic acids can be decarboxylated and converted to resorcinol. The latter can be directly reduced by fermenting bacteria to dihydroresorcinol and then hydrolyzed to 5-oxocaproic acid as summarized by Schink *et al.* (1992). Other pathways for the anaerobic microbial metabolism of resorcinol and related compounds are actively under investigation (Schink *et al.*, 1992). Another direct ring reduction followed by hydrolysis is used by some anaerobes to degrade phloroglucinol. Phloroglucinol is reduced to dihydrophloroglucinol, which is the immediate ring cleavage precursor. The latter is then converted to 3-hydroxy-5-oxocaproic acid (as reviewed by Schink *et al.*, 1992). Whether SRB employ these or still other as yet undiscovered pathways for the metabolism of these compounds remains to be determined. Similarly Liu *et al.* (1991) were able to demonstrate the anaerobic bioconversion of 2-hydroxybiphenyl under sulfate-reducing conditions. Sulfate consumption data suggested that the compound was completely mineralized.

The potential for the anaerobic biodegradation of saturated alkanes with sulfate as a terminal electron acceptor was evaluated by Aeckerberg *et al.* (1991) using marine sediment as an inoculum. In a dithionite-reduced seawater medium, more sulfide was formed in incubations containing *n*-hexadecane and *n*-dodecane than in alkane-free control incubations suggesting that the former two compounds were metabolized to some extent. However, no sulfide levels above controls could be measured when heptane, butane, ethane, or methane were the substrates of choice. In similar experiments using the particulate fraction taken from the bottom of an oil–water separator, the same authors could enrich cells capable of sulfide formation when hexadecane or oil was used as a source of carbon and energy (Aeckerberg *et al.*, 1991). However, the predominant organisms enriched in these experiments differed depending on the nature of the hydrocarbon source.

## 3. METABOLISM OF NONHALOGENATED COMPOUNDS BY PURE CULTURES

As noted above, benzoate is a compound common to the anaerobic biodegradation pathway of many aromatic compounds. Pure cultures of both spore forming and nonsporeforming SRB capable of degrading benzoate have been isolated (Table I). These include members of the genera *Desulfobacterium, Desulfotomaculum, Desulfococcus, Desulfosarcina, Desulfomonile, Desulfonema, Desulfoarculus,* and *Desulfobacula.* Benzoate-degrading SRB were enriched from soils using fatty acids as a sole carbon source and tested against a number of aromatic compounds and long-chain fatty acids (Cord-Ruwisch and Garcia, 1985). These cultures, as exemplified by *Desulfotomaculum sapomandens,* could mineralize phenylacetic and phenylpropionic acids and certain branched-chain fatty acids. Many benzoate degrading sulfate-reducers can also degrade phenylacetate, phenylpropionate as well as one of several of the hydroxybenzoate isomers (Table I; Widdel and Pfennig, 1984). Benzoate-utilizing

*Desulfobacterium* sp. are among some of the most nutritionally versatile SRB known. For instance, *D. anilini* isolated from marine sediment for its ability to anaerobically degrade aniline (Schnell *et al.*, 1989) can metabolize up to 17 different aromatic compounds including several homocyclic, heterocyclic, and nitrogen substituted aromatic chemicals.

SRB have also been isolated at the expense of the environmental contaminant phenol. Organisms with this ability include members of the genera *Desulfobacterium, Desulfotomaculum*, and *Desulfoarculus* (Table I). A strain of *Desulfobacterium phenolicum* has been isolated that will oxidize phenol, benzoate, hydroxybenzoates, *p*-cresol and indole (Bak and Widdel, 1986b) (Table I). This organism, enriched from marine sediment samples with phenol as the sole source of carbon and energy, exhibited a stoichiometry of phenol oxidation corresponding to slightly more than three moles of $H_2S$ formed per mole of phenol oxidized. This finding suggests that phenol was completely oxidized although no biochemical pathway for its metabolism was described. Putative intermediates in phenol degradation, such as cyclohexanol, cyclohexanone, or adipic acid would not support the growth of this organism and were not degraded. This may indicate either lack of a suitable uptake mechanism for these compounds or an alternate degradation pathway for phenol by this organism.

Recently, an SRB has been isolated and identified as a new species of *Desulfoarculus* that oxidizes benzoate, 4-hydroxybenzoate, and phenol. This marine isolate was enriched using anaerobic marine sediments with benzoate as a sole carbon source and sulfate as the electron acceptor. The stoichiometry of benzoate and 4-hydroxybenzoate oxidation with sulfate-reduction was measured and indicated that both compounds were completely oxidized to $CO_2$ (Drzyzga *et al.*, 1993).

Methylated phenols are also known environmental contaminants. Several pure cultures of sulfate-reducers are known to metabolize cresol isomers (Table I). For instance, Bak and Widdel (1986b) reported on pure cultures of SRB that catalyze *p*-cresol metabolism, while Schnell *et al.* (1989) isolated several strains of SRB that could use either *p*- and/or *m*-cresol. However, while there is no doubt that *o*-cresol can be degraded under sulfate-reducing conditions, pure cultures that grow at the expense of this compound are not known.

The cresol-degrading sulfate-reducing bacteria referred to in Table I could also degrade a variety of other aromatic compounds. Enrichments by Schnell *et al.* (1989) with the three dihydroxybenzene isomers as the substrates resulted in the isolation of three different strains of sulfate-reducers, each of which could grow only with the dihydroxybenzene used for enrichment. That is, cultures isolated on either catechol, resorcinol, or hydroquinone could not metabolize the other dihydroxybenzenes of interest. The catechol-degrading sulfate-red ucer (Strain Cat 2, Table I) obtained in this manner could grow with as many as 18 other aromatic compounds and many aliphatic acids and alcohols. The resorcinol- and hydroquinone-metabolizing isolates (Re10 and Hy5, respectively) (Table I) were not as versatile. The resorcinol-degrader could utilize 2,4-dihydroxybenzoate, formate, and 1,3 cyclohexanedione. The hydroquinone-degrader resembled *Desulfococcus multivorans* with the additional ability to grow with hydroquinone and 2,5-dihydroxybenzene (Schnell *et al.*, 1989).

Interesting results have been reported by Szewzyk and Pfennig (1987) in their studies of catechol degradation by an SRB that can use nitrate as well as a variety of oxidized sulfur species as electron acceptors. *Desulfobacterium catecholicum* can metabolize catechol, resorcinol, and hydroquinone but not phenol. Since this organism degraded all the dihydroxybenzenes, it is possible that it possesses different metabolic pathways for the various substrates or that the previously suggested pathways exist and the organism simply lacks a phenol uptake mechanism. This organism also utilized phloroglucinol, pyrogallol, benzoate, *p*-hydroxybenzoate, protocatechuate, and 2-aminobenzoate as electron donors to support dissimilatory sulfate-reduction (Table I).

Even more extensive characterization of an aromatic degrading sulfate-reducer has been published (Kuever *et al.*, 1993). This new species of the genus *Desulfotomaculum* strain Groll oxidized an extremely wide range of substituted aromatic compounds such as benzoate, catechol, phenol, vanillate, and protocatechuate among others (Table I). All of these compounds were oxidized with stoichiometries of $H_2S$ formed per mole of substrate oxidized that indicate complete oxidation of the parent molecules to $CO_2$. In addition to the aromatic compounds mentioned above, this metabolically diverse organism could utilize a wide range of fatty acids, 3- and 4-hydroxybenzoate, amino benzoate, cresols, and aromatic alcohols and aldehydes. This strain of *Desulfotomaculum* was also capable of *o*-demethylating several substituted aromatics, and the decarboxylation of protocatechuate to catechol. Numerous SRB have been described that can metabolize hydoxybenzoate isomers (Table I). Several are also able to metabolize benzoic acid but not known to biodegrade phenol. These include *Desulfobacterium catecholicum*, *Desulfococcus multivorans*, *Desulfosarcina variablis*, *Desulfonema magnum*, *Desulfotomaculum sapomandens*, and *Desulfobacula toluolica* (Table I). It may be that these organisms are able to reductively dehydroxylate the hydroxybenzoate isomer to form benzoate, which would then be anaerobically metabolized via ring reduction and cleavage reactions. Alternate explanations for such observations could involve the initial decarboxylation of the hydroxybenzoate isomer to phenol followed by the conversion of the latter compound to benzoate. The precise mechanism(s) of hydroxybenzoate decomposition by the SRB have yet to be established. Several other organisms like *Desulfobacterium phenolicum, anilini,* and strain Cat 2, as well as *Desultotomaculum* sp. strain Groll and *Desulfoarculus* sp. strain SAX are able to metabolize hydroxybenzoate, phenol and benzoate (Table I).

Recently a pure culture SRB has been isolated that will oxidize toluene with sulfate as the electron acceptor (Rabus *et al.,* 1993). This new culture with the proposed name *Desulfoarcula toluolica* (Table I) would oxidize toluene when supplied in an inert hydrophobic carrier phase. The toluene was completely oxidized to $CO_2$. Several putative intermediates of toluene degradation were tested in growth experiments. This organism will grow with *p*-cresol, benzaldehyde, phenylacetate, 4-hydroxybenzaldehyde, and benzoate as sole sources of carbon and energy. *O*- and *m*-cresol and *p*-hydroxybenzyl alcohol did not support growth. Attempts to elucidate the biochemical pathway for toluene degradation were made by measuring degradation rates of various compounds, such as benzoate, toluene, or *p*-cresol after growth in the presence of the test compound. Cells growing on benzoate displayed very low activities for toluene or *p*-cresol decomposition. Toluene grown cells oxidized benzoate immediately and *p*-cresol slowly. Cells grown on *p*-cresol were active against both benzoate and toluene. The authors suggest that since *p*-cresol is a good growth substrate but was utilized only slowly by toluene growing cells, toluene is not oxidized in this organism through ring hydroxylation mechanism. However, the addition of oxygen to a molecule such as toluene may markedly alter the biodegradation mechanism as well as the rate of uptake. A difference in toluene and cresol uptake rates could explain the low rate of *p*-cresol oxidation by toluene grown cells, while *p*-cresol may still be an intermediate in toluene oxidation by this organism.

As already noted, Schnell *et al.* (1989) described *D. anilini,* a metabolically versatile SRB enriched for its ability to grow with aniline as a substrate. Substrate induction experiments with this organism suggested a common pathway for the degradation of aniline and 4-aminobenzoate (Schnell and Schink, 1991). Like several other anaerobic bioconversions (phenol, *m*-cresol and hydroquinone; see above), the degradation of aniline by this organism was $CO_2$ dependent; but this dependency was not evident with 4-aminobenzoate. Extracts of aniline or 4-aminobenzoate grown cells exhibited 4-aminobenzoyl-CoA-synthetase activity and benzoate was detected as an intermediate during 4-aminobenzoate metabolism. The same authors (Schnell and Schink, 1991) also found that 4-aminobenzoyl-CoA was converted to benzoyl-CoA by cell extracts in the presence of low-potential electron donors. Thus, aniline was degraded by this organism via carboxylation to 4-aminobenzoate, which was subsequently activated to the CoA derivative and further metabolized by reductive dreamination to benzoyl-CoA. The latter compound then entered the benzoate degradation pathway.

Pure culture SRB are also known that metabolize a variety of N-, O-, or S-substituted heterocyclic compounds. Bak and Widdel (1986a) enriched SRB that metabolized skatole, indole, quinoline, and pyridine (Table I). They obtained an isolate, *Desulfobacterium indolicum,* that was able to use indole and quinoline as electron donors and carbon sources. Anthranilic acid and several other nonaromatic compounds were also utilized by this organism. It may be that such materials are intermediates of indole metabolism. *Desulfococcus niacini* is known for its ability to use nicotinic acid as a carbon and energy source and several pure cultures are known to metabolize oxygen heterocyclic compounds like furfural and furoic acid (Table I).

The situation with respect to pure cultures of SRB capable of transforming sulfur heterocyclic chemicals is worth noting. Kim *et al.* (1990a) studied the degradation of organosulfur compounds including several nonaromatic thiols, aromatic sulfides, and both benzothiophene and dibenzothiophene. The desulfurization reaction system consisted of harvested *Desulfovibrio desulfuricans* M6 cells suspended in a sulfate-free medium to which methyl viologen and a 2% organosulfur substrate in dimethylformamide was added. The greatest amount of removal was observed with benzothiphene (96%), although dibenzothiophene was also metabolized (42% removal). Ethanethiol, butaneithiol, phenylsulfide, benzylsulfide, and dibenzylsulfide were also transformed in these experiments, but the extent ranged from 7% removal to 73% removal depending on the particular compound. The reaction product associated with dibenzothiophene was identified as biphenyl. Thus, the organism apparently only cleaved the carbon sulfur bond and left intact the two aromatic nuclei. This has important implications, since the fuel value of the chemical would largely be unaffected by such a process. The same authors were also able to modify this system to demonstrate the desulfurization of a Kuwait oil (as evidenced by $H_2S$ evolution) using electrochemically supplied electrons (Kim *et al.,* 1990b).

Similarly, Köhler *et al.,* (1984) observed the anaerobic degradation of dibenzylsulfide by a *Desulfovibrio* sp. Toluene and benzylmercaptan were detected as intermediate metabolites and [35]S-sulfide was produced from [35]S-dibenzylsulfide. The former compound is sulfur-free and retains the bulk of the fuel value of the molecule. The same three compounds were also detected as products of the reductive desulfurization of dibenzyldisulfide degradation under methanogenic conditions (Miller, 1992).

Not all compounds that are transformed by SRB represent suitable electron donors. *Desulfovibrio desulfuricans* could be grown on a medium supplemented with either Kraft lignin or lignosulfonate (Ziomek and Williams, 1989). Interestingly, lignosulfonate contributed to the growth of the organism by replacing sulfate as an electron acceptor. Both lignin preparations could be transformed by this organism. This transformation included the partial depolymerization of the polyphenolic backbone as well as the lignin functional groups.

Boopathy and Kulpa (1993) describe a *Desulfovibrio* sp. (strain B; Table I) capable of using several nitroaromatic compounds as suitable nitrogen sources. Moreover, some of these compounds could also function as electron acceptors when the cell was cultivated in the absence of sulfate. Compounds with reduced aryl nitro groups (i.e., aniline) could be used as nitrogen sources but not as electron acceptors. None of the nitroaromatic compounds served as the sole sources of carbon and energy for the organism. The aryl nitro groups were reductively deaminated to ammonia, which was used as the nitrogen source for the cell, while the aromatic ring was left intact. Thus, toluene, phenol and benzene were the organic endproducts of 2,4,6-trinitrotoluene, 2,4-dinitrophenol and aniline decomposition, respectively (Boopathy and Kulpa, 1993; Boopathy *et al.*, 1993). Similar nitroaromatic reduction reactions of 2,4,6-trinitrotoluene were catalyzed by a pure culture sulfidogen as reported by Preuss *et al.* (1993). The organism was able to reduce the parent compound in a series of discrete steps to triaminotoluene and to transform the latter compound to unknown products. Pyruvate, hydrogen, and carbon monoxide served as the electron donors for the nitro-group reduction reactions. Inhibitor studies and the decrease in sulfide formation from sulfite in the presence of partially reduced intermediates were taken as evidence by the authors for the involvement of a sulfite reductase in the overall bioconversion process.

Insight into the possible subsequent transformations of triaminotoluene can be gleaned from the work of Funk *et al.* (1993). They found 2,4,6-trihydroxytoluene and *p*-cresol as intermediates of 2,4,6-trinitrotoluene decomposition in anaerobic soils. The authors note that these observations can only be understood when the chemistry of the triaminotoluene intermediate is taken into account. In aqueous solutions, the latter compound is unstable and will be in equilibrium with the trihydroxy derivative and the corresponding triketomethylcyclohexane.

The trihydroxytoluene may possibly get rearranged to methylpyrogallol or dehydroxylated to *p*-cresol. An noted above, these compounds are known or likely to be amenable to subsequent anaerobic microbial decay. Whether the same or similar bioconversions are operative in the SRB remains to be determined.

## 4. METABOLISM OF HALOGENATED COMPOUNDS BY ISOLATES AND MIXED CULTURES

In addition to the metabolism of nonhalogenated organic pollutants by SRB, other mechanism(s) exist for the biotransformation of halo-organic compounds. One such mechanism is reductive dehalogenation. Reductive dehalogenation involves the transfer of electrons, usually from an endogenous electron donor, to a haloorganic molecule. The reduction of the latter results in the removal of a halogen from an organic moiety and its replacement by a proton. The lesser halogenated product generally has reduced toxicity associated with it and a greater susceptibility to other biodegradative processes.

Many studies document reductive dehalogenation of halogenated aromatic and nonaromatic compounds and they have been reviewed previously (Kuhn and Suflita, 1989c; Mohn and Tiedje, 1992; Vogel *et al.*, 1987; Colberg, 1990; Suflita and Townsend, 1995). The picture that emerges from such studies is that the tendency for a haloorganic compound to be reduced is dependent on the redox chemistry of the compound itself. Thus, many halogenated pollutants are oxidized relative to their nonhalogenated counterparts and tend to accept electrons and be reductively dehalogenated. This process occurs more readily in reducing environments than when oxidizing conditions prevail. However, there is a considerable range in the tendency of these compounds to be oxidized or reduced. Some haloorganic pollutants are more electronegative than oxygen and can be reductively dehalogenated even under aerobic conditions. Thus, highly reducing conditions are not necessary for the reductive dehalogenation of nonaromatic compounds like tetrachloroethylene (PCE), trichloroethylene (TCE), hexachloroethane (HCE), and other haloaliphatic chemicals. However, comparable transformations of haloaromatic compounds often require highly reducing conditions and occur most readily when methanogenic conditions prevail. The differences between the anaerobic biotransformation of halogenated aliphatic and aromatic chemicals has been previously summarized (Kuhn and Suflita, 1989c).

Reductive dehalogenation reactions offer an alternate route for the treatment of halogenated contaminants. Such bioconversion reactions ultimately reduce the environmental lifetime of some of the most persistent chemicals. While this type of bioconversion does not provide a source of carbon to the requisite microorganisms, it may provide energy to the cells. Dehalogenating microorganisms transfer electrons, perhaps by a series of intermediate carriers, to the halogenated compound. This process may be coupled with energy conservation with the haloorganic compound serving as the terminal electron acceptor.

Thermodynamic calculations suggest that the standard free energy change for the oxidation of hydrogen using a haloaromatic compound as an electron acceptor can yield more energy than some of the more common electron acceptors available in anaerobic environments (Dolfing and Harrison, 1992). The opportunity to use an alternate electron acceptor, particularly one for which there may be only a limited degree of competition in energy-limited environments, may provide dehalogenating cells with some selective advantage.

Evidence supporting the hypothesis that reductive dehalogenation may result in some energetic benefit for a dehalogenating bacterium has been suggested in studies of resting cell suspensions of the SRB *D. tiedjei* (Mohn and Tiedje, 1991). Upon the addition of 3-chlorobenzoate to cell suspensions, an increase in ATP pool size concomitant with dehalogenation of the substrate was observed. Additionally, when 3-chlorobenzoate was coupled with a suitable electron donor, proton translocation was observed as evidenced by a decrease in pH of the medium, while in separate experiments, an artificially imposed pH gradient led to increased ATP pools. This evidence is consistent with the chemiosmotic coupling of reductive dehalogenation of 3-chlorobenzoate with ATP synthesis by *D. tiedjei*. Thus, the reductive dehalogenation of chlorinated aromatic compounds is not a gratuitous reaction in *D. tiedjei;* the cells can apparently derive some benefit from transferring electrons to halogenated aromatic compounds. However, a clear demonstration of an energetic benefit to *D. tiedjei* as evidenced by an increase in growth yield when cells were grown with pyruvate alone and compared with cells grown with pyruvate and 3-chlorobenzoate could not be obtained (personal observation). That is, the growth rate and yield were identical under these two cultivation conditions even though dehalogenation was easily evident when 3-chlorobenzoate was present.

This finding contrasts somewhat with those of Dolfing (1990). He suggests that the strain *D. tiedjei* can couple ATP synthesis to the transfer of electrons from formate and acetate to 3-chlorobenzoate. Growth yield and ATP concentrations in starved cells of *D. tiedjei* were dependent on the presence of 3-chlorobenzoate under defined conditions (Dolfing,

1990). This study, as well as Mohn and Tiedje (1991), suggests that anaerobic microorganisms cannot only transfer electrons to chlorinated compounds that may be pollutants but that they can use these electron transfer reactions to generate energy for growth. The reasons why a definitive growth yield increase linked to aryl dehalogenation reactions has not yet been conclusively demonstrated remain enigmatic.

Not surprisingly then, halogenated compounds must compete with other available electron acceptors in order to be incorporated into microbial redox reactions. Thus, the effectiveness of reductive dehalogenation reactions is often influenced by the presence of other electron acceptors. This is particularly important in the context of this review, since sulfate as well as other sulfur oxyanions, are known to have a dramatic influence on reductive dehalogenation reactions. However, for the reasons noted above, the influence of these electron acceptors is more pronounced when haloaromatic compounds are being metabolized, as opposed to haloaliphatic contaminants.

The anaerobic oxidation of natural halophenols such as 2,4-dibromophenol by SRB enrichments has been observed (King, 1988). In this study, the halogenated aromatic compound was not introduced into the environment by man but rather by a hemichordate marine worm. Sulfate-reducing cultures enriched from the burrows of this worm would oxidize and dehalogenate dibromophenol. It was postulated that dehalogenation occurred before the resulting phenol was oxidized by SRB. Dibromophenol inhibited the production of $H_2S$ in the presence of an added electron donor initially, but $H_2S$ production was eventually markedly enhanced. It was also observed that the addition of molybdate (an inhibitor of sulfate-reduction) to sediments abolished phenol decomposition, but not the reduction of dibromophenol. In this case, it appeared that the reductive dehalogenation process occurred in the presence of the alternate electron acceptor sulfate, but was not necessarily inhibited by it.

This is in contrast to many observations on the influence of sulfate on reductive dehalogenation reactions. Several of these observations emanate from experiments designed to examine the fate of haloaromatic compounds in samples from an anoxic aquifer polluted by municipal landfill leachate. To date, over 30 haloaromatic compounds have been examined with such samples, including a variety of halogenated benzoates, phenols, phenoxyacetates, anilines, and several heterocyclic molecules (Gibson and Suflita, 1986, 1990; Kuhn et al., 1990; Suflita et al., 1988; Ramanand et al., 1993). A rather consistent pattern emerged when the biodegradation of these material was compared with such samples. In all cases, the reductive dehalogenation of the substrates was detected as the primary degradative event under methanogenic conditions. However, the same compounds proved extremely recalcitrant when assayed in sulfate-reducing aquifer slurries or when sulfate was added to the previously methanogenic incubations. Such findings do not specifically implicate the involvement of any particular microbial group (e.g., methanogens, acetogens, sulfate-reducing bacteria etc.) in the dehalogenation of these substrates. The inhibition or delay of aryl reductive dehalogenation by the presence of alternate electron acceptors has also been confirmed by many other investigators (Madsen and Aamand, 1991; Genthner *et al.*, 1989; Allard *et al.*, 1992; Hale *et al.*, 1991; Kohring *et al.*, 1989; May *et al.*, 1992a,b; Morris *et al.*, 1992.

This point notwithstanding, several reports indicate that once acclimation periods are overcome, the amendments of environmental samples with alternate electron acceptors may or may not influence dehalogenation processes (Kohring *et al.*, 1989; May *et al.*, 1992a; Allard *et al.*, 1992). Reductive dehalogenation of 2,4 dichlorophenol in anaerobic sediments occurred in the presence of added sulfate, although this amendment significantly reduced the rate of dehalogenation (Kohring *et al.*, 1989). Sulfate-reducing conditions were maintained during the entire experiment, as evidenced by a lack of methane formation. During these experiments the concentration of sulfate was reduced from an initial 25 mM to between 6–8 mM. This experiment demonstrates that anaerobic dehalogenation can occur simultaneously with sulfate-reduction. In these same experiments, the addition of nitrate as an electron acceptor also inhibited dehalogenation of 2,4 dichlorophenol. In other cases, the microbial metabolism of a haloaromatic compound in unacclimated environmental samples may be either unaffected or even dependent on another electron acceptors (Haggblom and Young, 1990; Haggblom *et al.*, 1992; King, 1988). For instance, chlorinated phenols and benzoic acids are degraded in estuarine and fresh water sediments under sulfate-reducing conditions. Oxidation of 3- and 4-chlorobenzoate and all three monochlorophenol isomers (Haggblom *et al.*, 1993) was accompanied by the reduction of sulfate in excess of background levels. Although reductive dechlorination of these compounds was presumed to occur, it was not specifically demonstrated. Therefore, the relationship of several of these studies to aryl dehalogenation *per se* remains to be clarified.

At least partial insight into the effect of sulfur oxyanions on reductive dehalogenation reactions can be obtained by examining the metabolism of the electron donor hydrogen by *D. tiedjei* in the presence of the potential electron acceptors (DeWeerd *et al.*, 1990). The rate of hydrogen consumption was about six times faster when sulfite was provided as an electron acceptor, compared to the rate observed with 3-chlorobenzoate as the electron acceptor. Therefore, a competition for reducing equivalents when both electron acceptors were present is possible. In consistent fashion, when both sulfite and 3-chlorobenzoate were available as potential electron acceptors, an intermediate rate of hydrogen consumption was observed. The same effect on hydrogen consumption was also observed with the other sulfate and thiosulfate. That is, that rate of hydrogen consumption in the presence of both electron acceptors was less than that observed with the sulfur oxyanions as the only acceptor. It is therefore tempting to speculate that a mutual electron carrier for dehalogenation and sulfur oxyanion reduction could possibly account for the intermediate hydrogen consumption rates and that sulfur oxyanions should inhibit aryl dehalogenation reactions. However, when 3-chlorobenzoate metabolism was examined in these experiments, it was found that dehalogenation was inhibited only in the presence of sulfite and thiosulfate, but not by sulfate. Comparable results were obtained with aryl dehalogenation in cell-free extracts as well (DeWeerd and Suflita, 1990).

The environmental implications of such findings are worth noting. The kinetics of hydrogen metabolism in the presence of sulfur oxyanions and halobenzoates suggest that in environments predominated by sulfate-reduction, the levels of hydrogen may be too low to support significant amounts of aryl dehalogenation. This suggestion is consistent with the bulk of the literature. The ability of dehalogenating bacteria to scavenge reducing equivalent in anaerobic environments may be dependent on other microbial electron accepting processes present in the particular habitats. Thus, the regulation of dehalogenation in anaerobic habitats may ultimately depend upon the availability of alternate electron acceptors and the ability of the dehalogenating cells inhabiting such environments to compete for electron donors with other bacteria.

Much more is known about the anaerobic biotransformation of haloaliphatic contaminants. Among the most persistent and important environmental pollutants are the halogenated alkanes and alkanes, such as carbon tetrachloride, PCE, TCE, and trichloroethane. Many of these contaminants are readily amenable to anaerobic decay. For example, the complete transformation of TCE and partial degradation of other chlorinated organic chemicals by a proteolytic *Clostridium* has been reported (Galli and McCarty, 1989) and dehalogenation reactions by other organisms has been frequently observed. The production of ethane, ethylene, and acetylene from brominated or chlorinated alkanes by methanogens has been well documented (Belay and Daniels, 1987). Since many environments are anaerobic or can be made so, reductive dehalogenation of persistent contaminants is an important remedial option.

Although the anaerobic transformation of halogenated alkanes and alkanes has primarily been studied under methanogenic conditions (Bouwer and McCarty, 1983), some microcosms enriched with PCE have shown little involvement of methanogenic bacteria in dehalogenation. In one study, the population of SRB in dehalogenating sediments was measured at 1% of the total population, while the methanogen population was negligible (Major *et al.*, 1991). Indeed, it would appear that many different types of anaerobes are involved in the dehalogenation of aliphatic molecules. Different electron donors may enrich different types of microbial populations in subsurface sediments and several types of organisms may be capable of dehalogenation. This study also demonstrated the complete dehalogenation of PCE to ethene and ethane under anaerobic conditions, an important observation because many previous studies have measured the accumulation of vinyl chloride, a potent carcinogen formed as an intermediate during the dehalogenation of PCE and TCE.

There is significant evidence accumulating in the literature that TCE dehalogenation is associated with sulfate-reduction. The simultaneous reductive dehalogenation of TCE and sulfate-reduction in anaerobic Vero beach sediments has been reported (Barrio-Lage *et al.*, 1987). Although methane was produced in these sediments after 18 months of incubation, reductive dehalogenation of TCE was essentially complete by that time. These microcosms produce cis-1,2 dichloroethylene (DCE) as a dehalogenation product. In some samples DCE was also observed to decrease over time, possibly due to further dehalogenation. As with previous studies, the link between sulfate-reduction and reductive dehalogenation was circumstantial; other anaerobes besides sulfate-reducers may have been using the halogenated organic compounds as electron acceptors. However, the data presented showed that methanogens were not abundant or active during the dehalogenation process.

The reductive dehalogenation of PCE to TCE and cis-1,2-DCE by sulfate enriched cultures has also been observed (Bagley and Gossett, 1990). Cultures from sewage sludge rapidly dehalogenated PCE to TCE and cis-1,2-DCE. The authors speculated that sulfate-reducing organisms were responsible for dechlorination of PCE in these experiments. The percentage of PCE removed was also highest in those cultures generating the most hydrogen sulfide; further indirect evidence of a relationship between sulfate-reduction and dehalogenation of chloroalkenes.

Degradation of halogenated organic compounds has also been observed in biofilm reactors. A column reactor containing a biofilm with sequential zones of aerobic respiration dentrification and sulfate-reduction was exposed to concentrations of halogenated methane, ethane, and other halogenated organic compounds. Transformation of the halogenated aliphatic compounds occurred with the onset of sulfate-reduction, and depletion of sulfate in the column feed decreased degradation of the halogenated compounds. These results suggested that, under the operating conditions of this reactor, SRB were at least indirectly responsible for the removal of the chlorinated aliphatics from the contaminated stream. This experiment also demonstrated that it may be possible to operate a commercial anaerobic biofilm reactor to catalyze the degradation of persistent environment contaminants such as halogenated aliphatics (Cobb and Bouer, 1991).

Strong evidence for the role of SRB in reductive dehalogenation has been provided by studies involving pure cultures. The organism *Desulfobacterium autotrophicum* has been shown to dehalogenate a number of chlorohydrocarbons, including tetrachloromethane conversion to trichloromethane and dichloromethane, and trichloroethane reduction to dichloroethane (Egli *et al.*, 1987). Trichloromethane was also dehalogenated by this pure culture. This work confirms earlier speculation that sulfate-reducers can dehalogenate persistent environmental contaminants. These single-step dehalogenation reactions did not completely remove the contaminants from the environment. However, under long incubation times or different conditions, subsequent dehalogenation may occur, since tetrachloromethane, trichloromethane and dichloromethane were all substrates for dehalogenation. It is possible that dehalogenation (at lower rates) could result in the complete reduction of molecules such as PCE to ethane or ethane by SRB. A later study using the same culture demonstrated the transformation of tetrachloromethane to trichloromethane and dichloromethane over short periods of time (less than 20 days) under heterotrophic growth conditions (Egli *et al.*, 1988). Other anaerobes such as acetogens also dehalogenated these compounds even more rapidly. This report also provides some indication of a biochemical pathway for reductive dehalogenation. The acetogens and the sulfate-reducing microorganisms contained cobamides, whereas another pure culture of a sulfur-reducer that does not contain cobamides lacked dehalogenation activity. This information suggests that cobamides may be involved in the reductive dehalogenation of tetrachloromethane and other chloroorganic compounds by SRB.

There is some evidence for a route of electron transfer from anaerobic bacteria to halogenated hydrocarbons through cobalamin and co-

balamin derivatives. Pure aquocobalamin catalyzes the reductive dehalogenation of carbon tetrachloride, chloroform, methylene chloride, and
chloromethane with titanium citrate as the electron donor (Krone et al,
1989a). Other cobalamin derivatives such as methylcobalamin and co-
bimamide and bacterial transition metal coenzymes such as vitamin
B-12, coenzyme F430, and hematin can also catalyze reductive dehalogenation of chlorinated hydrocarbons (Gantzer and Wackett, 1991).
With titanium citrate as an electron donor, vitamin B-12 can reductively
dehalogenate PCE. It has been shown that vitamin B-12 and coenzyme
F430 dehalogenate all of the intermediates between PCE and ethylene
(TCE, DCE, and vinyl chloride). In addition, coenzyme F430 catalyzes
the reductive dehalogenation of carbon tetrachloride with titanium citrate as the electron donor and reduced intermediates, including chloroform, methylene chloride, chloromethane, and methane were detected
(Krone et al., 1989b).

The nonspecific transfer of electrons to halogenated hydrocarbons
by transition metal coenzymes may not benefit the cell since it can bypass
ATP generating steps involved in electron transfer to natural terminal
electron acceptors. This may also not be the only route to reductive
dehalogenation of chlorocarbons by anaerobes. A recent paper describ-
ing the dehalogenation of chlorophenols by an anaerobic sulfate-
reducing microorganism (Mohn and Kennedy, 1992), showed that live
intact cells were necessary for reductive dehalogenation of this chloroaromatic chemical.

An extremely promising report has recently appeared that illustrates the use of enrichment of sulfate-reducers in the field for dehalogenation of compounds such as PCE and TCE. (Beeman et al., 1993).
An aquifer contaminated with PCE, TCE, and DCE was treated in situ by
pumping water to the surface, amending the site water by the addition
of either sodium benzoate or magnesium sulfate to obtain sulfate-
reducing conditions, and recharging the water into the aquifer. Groundwater flow carried the amended water towards recovery wells. This sulfate enrichment process resulted in the dehalogenation of PCE and TCE
to the extent that the concentration of PCE decreased in the recovery
wells by at least 98% and TCE concentrations decreased between 85 and
89% over a four-month period. The concentrations of DCE and vinyl
chloride increased significantly during the first year of treatment. By the
end of the second year under sulfate-reducing conditions, the concentrations of PCE, TCE, DCE, and vinyl chloride had decreased in recovery wells to below the limits of detection. Ethylene appeared in samples
from all of the treated wells during the second year, indicating the
complete dehalogenation of PCE and TCE through daughter products

to ethylene. The concentration of sulfate in the feed water were maintained from between 0.4 mM to 1 mM. A set of control wells run in parallel without sulfate-reducing conditions displayed no decrease in the concentrations of PCE or TCE, indicating that sulfate-reducing conditions were necessary for reductive dehalogenation.

One of the best-characterized pure cultures of a sulfate-reducing dehalogenating microorganism is *D. tiedjei*. The organism was originally isolated from a methanogenic consortium that was able to use 3-chlorobenzoic acid as a sole carbon and energy source (Shelton and Tiedje, 1984). This consortium contained seven species of bacteria, including a dechlorinating stain that was designated DCB-1. As isolated this pure culture would dechlorinate 3-chlorobenzoate in the presence of pyruvate under anaerobic conditions. It was later discovered that this organism would dechlorinate tetrachloroethylene (PCE). *D. tiedjei* dehalogenated PCE to TCE at a rate significantly higher than any of the other anaerobes tested including two methanogens (Fathepure *et al.*, 1987). Under the conditions of this experiment PCE was converted stoichiometrically to TCE; further dechlorination was not observed. However, a mixed consortium containing *D. tiedjei*, a benzoate degrader, and *Methanospirillium* sp. would apparently dehalogenate both PCE and TCE. As of this writing, it is unknown if the biochemistry for aryl dehalogenation is the same that is involved in PCE dehalogenation. Given the differences in the dehalogenation processes mentioned above, it is not inconceivable that individual cells could possess multiple dehalogenation mechanisms. The complete dehalogenation of PCE results in the formation of either ethene or ethane and removal of this persistent and toxic contaminant from anaerobic environments. Complete metabolism is important since it avoids the transitory accumulation of vinyl chloride. The reductive dehalogenation of PCE to ethane has been observed (De Bruin *et al.*, 1992), but until recently not under sulfate-reducing conditions. Further analysis of the dehalogenation products from PCE by *D. tiedjei* may reveal that this organism is also capable of complete dehalogenation.

The introduction of *D. teidjei* into an anaerobic granular sludge reactor caused this reactor to display new catabolic potential. Anaerobic granular sludge blanket reactors are efficient and rapid methods for treating contaminated waste water, but recalcitrant pollutants such as haloorganic compounds are not efficiently removed in these reactors. When a granular sludge reactor was inoculated with a pure culture of *D. tiedjei*, the microorganism established itself as a stable member of the reactor microflora, became incorporated into reactor granules, and efficiently catalyzed the reductive dechlorination of 3-chlorobenzoate

(Ahring *et al.*, 1992). This is the first report of a pure or defined microbial mixture added to a relatively poorly characterized granulated sludge system to improve performance. In this case, the biochemical activities of *D. tiedjei* were incorporated into the reactor and permitted the system to treat waste containing recalcitrant haloorganic compound.

## 5. PERSPECTIVES

There have been many recent discoveries about the role of SRB in the degradation of environmental pollutants. New biochemical pathways, new organisms, and new substrate chemicals are appearing in the scientific literature with greater frequency. It would not be surprising if eventually it was discovered that the sulfate-reducers display a similar metabolic versatility to that of the denitrifying bacteria or perhaps the iron-reducing bacteria; there is already a long list of compounds regarded as environmental pollutants that can be metabolized by anaerobes with sulfate or another sulfur oxyanion as an electron acceptor. This information will be used to develop approaches that stimulate the environmental degradation of pollutants by sulfate-reducers. Both the rate and extent of hydrocarbon oxidation by SRB can be enhanced, at least theoretically, if the appropriate environmental conditions are created or the proper microbial culture is enriched. It may also be possible to introduce a specific organism into an anaerobic environment to stimulate degradation of a pollutant.

The literature also shows that certain SRB can dehalogenate chloroorganics. Some extremely important environmental pollutants such as pentachlorophenol, perchloroethylene, and trichloroethylene can be at least partially metabolized by sulfate-reducing bacteria. The physiology of these dehalogenation reactions is still poorly understood, but identifying the right environmental conditions for complete dehalogenation will be important for field applications. The conversion of haloorganic compounds must be done with some sensitivity toward avoiding the accumulation of undesirable intermediates. Of course, the risks associated with the transient accumulation of such intermediates must be weighed against the risks and expense associated with other treatment options, or the 'do nothing' scenario. The sulfate-reducers are known to possess metabolic activity against some of the most persistent and troublesome environmental contaminants known. These organisms certainly play a role in a variety of transformations and in many anaerobic environments the activities of SRB may be the major route of detoxification for certain contaminants.

# REFERENCES

Aeckerberg, F., Bak, F., and Widdel, F., 1991, Anaerobic oxidation of saturated hydrocarbons to $CO_2$ by a new type of sulfate-reducing bacterium, *Arch. Microbiol.* **156**:5–14.

Ahring, D., Christiansen, N., Mathrani, I., Hendrickson, H., Macairo, A., and McCario, E., 1992, Introduction of a *de novo* bioremediation ability, aryl reductive dechlorination, into anaerobic granular sludge by inoculation of sludge with *Desulfomonile tiedjei, Appl. Environ. Microbiol.* **58**:3677–3682.

Allard, A.-S., Hynning, P.-A., Remberger, M., and Neilson, A. H., 1992, Role of sulfate concentration in dechlorination of 3,4,5-trichlorocatechol by stable enrichment cultures grown with coumarin and flavanone glycones and aglycones, *Appl. Environ. Microbiol.* **58**:961–968.

Bagley, D. M., and Gossett, J. M., 1990, Tetrachloroethylene transformation to trichloroethylene and cis-1,2-dichloroethylene by sulfate reducing enrichment cultures, *Appl. Environ. Microbiol.* **56**:2511–2516.

Bak, F., and Widdel, F., 1986a, Anaerobic degradation of indolic compounds by sulfate-reducing enrichment cultures, and description of *Desulfobacterium indolicum* gen. nov. sp. nov., *Arch. Microbiol.* **146**:170–176.

Bak, F., and Widdel, F., 1986b, Anaerobic degradation of phenol and phenol derivatives by *Desulfobacterium phenolicum* sp. nov., *Arch. Microbiol.* **146**:177–180.

Balba, M. T., and Evans, W. C., 1980a, The anaerobic dissimilation of benzoate by *Pseudomonas aeruginosa* coupled with *Desulfovibrio vulgaris*, with sulphate as terminal electron acceptor, *Biochem. Soc. Trans.* **8**:624–627.

Balba, M. T., and Evans, W. C., 1980b, The methanogenic biodegradation of catechol by a microbial consortium: evidence for the production of phenol through cis-benzenediol, *Biochem. Soc. Trans.* **8**:452–454.

Barrio-Lage, G. A., Parsons, F. Z., Nassar, R. J., and Lorenzo, P. A., 1987, Biotransformation of trichloroethylene in a variety of subsurface materials, *Environ. Tox. Chem.* **6**:571–578.

Beeman, R. E., Howell, J. E., Schumaker, S. H., Salazar, E. A., and Buttram, J. R., 1983, A field evaluation of *in-situ* microbial reductive dehalogenation by the biotransformation of chlorinated ethynes. in: *In-Situ and On-Site Bioremediation.* (R. E. Hinchee ed.). Lewis Publishers, Boca Raton, Florida, pp. 14–27.

Belay, N., and Daniels, L., 1987, Production of ethane, ethylene, and acetylene from halogenated hydrocarbons by methanogenic bacteria, *Appl. Environ. Microbiol.* **53**:1604–1610.

Beller, H. R., Grbić-Galić, D., and Reinhard, M., 1992, Microbial degradation of toluene under sulfate reducing conditions and the influence of iron on the process, *Appl. Environ. Microbiol.* **58**:786–793.

Berry, D. F., Francis, A. J., and Bollag, J.-M., 1987, Microbial metabolism of homocyclic and heterocyclic aromatic compounds under anaerobic conditions, *Microbiol. Rev.* **51**:43–59.

Boopathy, R., and Daniels, L., 1991, Isolation and characterization of a furfural degrading sulfate-reducing bacterium from an anaerobic digester, *Curr. Microbiol.* **23**:327–332.

Boopathy, R., and Kulpa, C. F., 1993, Nitroaromatic compounds serve as nitrogen source for *Desulfovibrio* sp. (B. strain), *Can. J. Microbiol.* **39**:430–433.

Boopathy, R., Kulpa, C. F., and Wilson, M., 1993, Metabolism of 2,4,6-trinitrotoluene (TNT) by *Desulfovibrio* sp. (B strain), *Appl. Microbiol. Biotechnol.* **39**:270–275.

Bouwer, E. J., and McCarty, P. L., 1983, Transformations of 1- and 2-carbon halogenated aliphatic organic compounds under methanogenic conditions, *Appl. Environ. Microbiol.* **45**:1286–1294.

Brune, G., Schoberth, S. M., and Sahm, H., 1983, Growth of a strictly anaerobic bacterium on furfural (2-Furaldehyde), *Appl. Environ. Microbiol.* **46**:1187–1192.

Cobb, G., and Bouer, E., 1991, Effects of electron acceptors on halogenated organic compound biotransformations in a biofilm column, *Environ. Sci. Technol.* **25**:1068–1074.

Colberg, P.J.S., 1990, Role of sulfate in microbial transformations of environmental contaminants: chlorinated aromatic compounds, *Geomicrobiol. J.* **8**:147–165.

Cord-Ruwisch, R., and Garcia, J. L., 1985, Isolation and characterization of an anaerobic benzoate-degrading spore-forming sulfate-reducing bacterium, *Desulfotomaculum sapomandens* sp. nov., *FEMS Microbiol. Lett.* **29**:325–330.

De Bruin, W. P., Kotterman, M. J., Posthumus, M. A., Schraa, G., and Zehnder, A.J.B., 1992, Complete biological reductive transformation of tetrachloroethylene to ethane, *Appl. Environ. Microbiol.* **58**: 1996–2000.

DeWeerd, K. A., Mandelco, L., Tanner, R. S., Woese, C. R., and Suflita, J. M., 1990, *Desulfomonile teidjei* gen. nov. and sp. nov., a novel anaerobic, dehalogenating, sulfate-reducing bacterium, *Arch. Microbiol.* **154**:23–30.

DeWeerd, K. A., and Suflita, J. M., 1990, Anaerobic aryl reductive dehalogenation of halobenzoates by cell extracts of "*Desulfomonile tiedjei*", *Appl. Environ. Microbiol.* **56**:2999–3005.

Dolfing, J., 1990, Reductive dechlorination of 3-chlorobenzoate is coupled to ATP production and growth in an anaerobic bacterium strain DCB-1, *Arch. Microbiol.* **153**:249–266.

Dolfing, J., and Harrison, B. K., 1992, The Gibbs free energy of formation of halogenated aromatic compounds and their potential role as electron acceptors in anaerobic environments, *Environ. Sci. Technol.* **26**:2213–2218.

Drzyzga, O., Kuver, J., and Blotevogel, K.-H., 1993, Complete oxidation of benzoate and 4-hydroxybenzoate by a new sulfate-reducing bacterium resembling *Desulfoarculus*, *Arch. Microbiol.* **159**:109–113.

Eckart, V., Köhler, M., and Hieke, W., 1986, Microbial desulfurization of petroleum and heavy petroleum fractions. 5. Anaerobic desulfurization of Romashkino petroleum, *Zbl. Mikrobiol.* **141**:291–300.

Edwards, E. A., and Grbić-Galić, D., 1992, Complete mineralization of benzene by aquifer microorganisms under strictly anaerobic conditions, *Appl. Environ. Microbiol.* **58**:2663–2666.

Edwards, E. A., Wills, L. E., Reinhard, M., and Grbić-Galić, D., 1992, Anaerobic degradation of tolulene and xylene by aquifer microorganisms under sulfate reducing conditions, *Appl. Environ. Microbiol.* **58**: 794–800.

Egli, C., Scholtz, R., Cook, A. M., and Leisinger, T., 1987, Anaerobic dechlorination of tetrachloromethane and 1,2-dichloroethane *FEMS Microbiol. Lett.* **43**:257–261.

Egli, C., Tschen, T., Scholtz, R., Cook, A. M., and Leisinger, T., 1988, Transformation of tetrachloromethane to dichloromethane and carbon dioxide by *Acetobacterium woodii*, *Appl. Environ. Microbiol.* **54**:2819–2824.

Evans, W. C., 1977, Biochemistry of the bacterial catabolism of aromatic compounds in anaerobic environments, *Nature* **270**:17–22.

Evans, W. C., and Fuchs, G., 1988, Anaerobic degradation of aromatic compounds, *Ann. Rev. Microbiol.* **42**:289–317.

Fathepure, B. Z., Nengu, J. P., and Boyd, S. A., 1987, Anaerobic bacteria that dechlorinate perchloroethene, *Appl. Environ. Microbiol.* **53**:2671–2674.

Folkerts, M., Ney, U., Kneifel, H., Stackebrandt, E., Witte, E. G., Forstel, H., Schobert, S. M., and Sahm, H., 1989, *Desulfovibrio furfuralis* sp. nov., a furfural degrading strictly anaerobic bacterium, *Syst. Appl. Microbiol.* **11**:161–169.

Funk, S. B., Roberts, D. J., Crawford, D. L., and Crawford, R. L., 1993, Initial-phase optimization of munition compound-contaminated soils, *Appl. Environ. Microbiol.* **59**:2171–2177.

Galli, R., and McCarty, P. L., 1989, Biotransformation of 1,1,1-trichloroethane, trichloromethane, and tetrachloromethane by a Clostridium sp., *Appl. Environ. Microbiol.* **55**:837–844.

Galusko, A. S., and Rosanova, E. P., 1991, *Desulfobacterium cetonicum*, new species, a sulfate-reducing bacterium oxidizing fatty acids and ketones, *Mikrobiologiya* **60**:102–107.

Gantzer, C. J., and Wackett, L. P., 1991, Reductive dechlorination catalyzed by bacterial transition-methyl coenzymes, *Environ. Sci. Technol.* **25**:715–722.

Genthner, B.R.S., Price, W. A., II, and Pritchard, P. H., 1989, Anaerobic degradation of chloroaromatic compounds under a variety of enrichment conditions, *Appl. Environ. Microbiol.* **55**:1466–1471.

Gibson, S. A., and Suflita, J. M., 1986, Extrapolation of biodegradation results to groundwater aquifers: reductive dehalogenation of aromatic compounds, *Appl. Environ. Microbiol.* **52**:681–688.

Gibson, S. A., and Suflita, J. M., 1990, Anaerobic biodegradation of 2,4,5-trichlorophenoxyacetic acid in samples from a methanogenic aquifer: stimulation by short-chain organic acids and alcohols, *Appl. Environ. Microbiol.* **56**:1825–1832.

Grbić-Galić, D., and Vogel, T. M., 1987, Transformation of toluene and benzene by mixed methanogenic cultures, *Appl. Environ. Microbiol.* **53**:254–260.

Haag, F., Reinhardt, M., and McCarthy, P. L., 1991, Degradation of toluene and paraxylene in anaerobic microcosms: evidence for sulfate as a terminal elecron acceptor, *Environ. Tox. Chem.* **10**:1379–1389.

Haddock, J. D., and Ferry, J. G., 1990, Anaerobic metabolism of aromatic chemicals, in: *Bioprocessing and Biotreatment of Coal.* (D. Wise, eds.) Marcel Dekker, Inc. New York, pp. 393–416.

Haggblom, M. M., Rivera, M. D., Oliver, D., and Young, L. Y., 1992, Anaerobic degradation of halogenated phenols by a sulfate reducing consortium., abstr. Q-198, p. 368. Abstr. 92nd Gen. Meet. Am. Soc. Microbiol. 1992. American Society for Microbiology, Washington, D.C.

Haggblom, M., Rivera, M., Young, L. Y. 1993, Influence of alternative electron acceptors on the anaerobic biodegradability of chlorinated phenols and benzoic acids, *Appl. Environ. Microbiol.* **59**:1162–1167.

Hale, D. D., Rogers, J. E., and Wiegel, J., 1991, Environmental factors correlated to dichlorophenol dechlorination in anoxic freshwater sediments, *Environ. Toxicol. Chem.* **10**:1255–1265.

Imhoff-Stuckle, D., and Pfennig, N., 1983, Isolation and characterization of a nicotinic acid-degrading sulfate-reducing bacterium, *Desulfococcus niacini* sp. nov., *Arch. Microbiol.*, **136**:194–198.

Kaiser, J.-P., and Bollag, J.-M., 1991, Metabolism of pyridine and 3-hydroxypyridine under aerobic, denitrifying and sulfate-reducing conditions, *Experientia* **47**:292–296.

Kaiser, J.-P., and Bollag, J.-M., 1992, Influence of oil inoculum and redox potential on the degradation of several pyridine derivatives, *Soil Biol. Biochem* **24**:351–357.

Kaiser, J.-P., Minard, R. D., and Bollag, J.-M., 1993, Transformation of 3- and 4-picoline under sulfate-reducing conditions, *Appl. Environ. Microbiol.* **59**:701–705.

Kim, H. Y., Kim, T. S., and Kim, B. H., 1990a, Degradation of organic sulfur compounds and the reduction of dibenzothiophene to biphenyl and hydrogen sulfide by *Desulfovibrio desulfuricans* M6, *Biotechnol. Lett.* **10:**761–764.

Kim, T. S., Kim, H. Y., and Kim, B. H., 1990b, Petroleum desulfurization by *Desulfovibrio desulfuricans* M6 using electrochemically supplied reducing equivalent, *Biotechnol. Lett.* **12:**757–760.

King, G. M., 1988, Dehalogenation in marine sediments containing natural sources of halophenols, *Appl. Environ. Microbiol.* **54:**3079–3085.

Knezovich, J. P., Bishop, D. J., Kulp, T. J., and Grbić-Galić, D., 1990, Anaerobic microbial degradation of acridine and the application of remote fiber spectroscopy to monitor the transformation process. *Environ. Toxicol. Chem.* **9:**1234–1244.

Köhler, M., Genz, I.-L., Schicht, B., and Eckart, V., 1984, Microbial desulfurization of petroleum and heavy petroleum fractions. 4. Anaerobic degradation of organic sulfur compounds of petroleum, *Zbl. Mikrobiol.* **139:**239–247.

Kohring, G.-W., Zang, X., and Wiegel, J., 1989, Anaerobic dechlorination of 2,4-dichlorophenol in fresh water sediments in the presence of sulfate, *Appl. Environ. Microbiol.* **55:**2735–2737.

Krone, U. E., Thauer, R. K., and Hogencamp, H.P.C., 1989a, Reductive dehalogenation of chlorinated C1-hydrocarbons mediated by corrinoids, *Biochem.* **28:**4908–4914.

Kuever, J., Kulmer, J., Jannsen, S., Fischer, U., and Blotevogel, K-H., 1993, Isolation and characterization of a new spore-forming sulfate-reducing bacterium growing by complete oxidation of catechol, *Arch. Microbiol.* **159:**282–288.

Kuhn, E. P., and Suflita, J. M., 1989a, Anaerobic biodegradation of nitrogen-substituted and sulfonated benzene aquifer contaminants, *Haz. Waste Haz. Mater.* **6:**121–133.

Kuhn, E. P., and Suflita, J. M., 1989b, Microbial degradation of nitrogen, oxygen, and sulfur heterocyclic compounds under anaerobic conditions: studies with aquifer samples, *Environ. Toxicol. Chem.* **8:**1149–1158.

Kuhn, E. P., and Suflita, J. M., 1989c, Dehalogenation of pesticides by anaerobic microorganisms in soils and groundwater—a review, in: *Reactions and Movements of Organic Chemicals in Soils* (B. L. Sawhney, and K. Brown, eds.), SSSA Special Publication No 22., Soil Science Society of America and American Society of Agronomy, Madison, Wisconsin, pp. 111–180.

Kuhn, E. P., Suflita, J. M., Rivera, M. D., and Young, L. Y., 1989, Influence of alternate electron acceptors on the metabolic fate of hydroxybenzoate isomers in anoxic aquifer slurries, *Appl. Environ. Microbiol.* **55:**590–598.

Kuhn, E. P., Townsend, G. T., and Suflita, J. M., 1990, Effect of sulfate and organic carbon supplements on reductive dehalogenation of chloroanilines in anaerobic slurries, *Appl. Environ. Microbiol.* **56:**2630–2637.

Kurita, S., Endo, T., Nakamura, H., Yagi, T., and Tamiya, N., 1971, Decomposition of some organic sulfur compounds in petroleum by anaerobic bacteria, *J. Gen. Appl. Microbiol.* **17:**185–198.

Liu, Shi, Liang, L.-N., and Suflita, J. M., 1991, The metabolism of the ground water contaminant 2-hydroxybiphenyl under sulfate-reducing conditions, *Cur. Microbiol.* **22:**69–72.

Love, C. A., Patel, B.K.C., Nichols, P.D., and Stackerbrandt, E., 1993, *Desulfotomaculum australicum,* sp. nov., a Thermophilic sulfate-reducing bacterium isolated from the Great Artesian Basin of Australia, *System. Appl. Microbiol.* **16:**244-251.

Madsen, T., and Aamand, J., 1991, Effects of sulfuroxy anions on degradation of pentachlorophenol by a methanogenic enrichment culture, *Appl. Environ. Microbiol.* **57:**2453–2458.

Major, D. W., Hodgins, E. W., and Butler, B. J., 1991, Field and laboratory evidence of *in situ* biotransformation of tetrachloroethylene to ethene and ethane at a chemical transfer facility in North Toronto, in: *On Site Bioreclamation*, (R. E. Hinchee and R. F. Olfenbuttel, eds.), Butterworth-Heinemann, Stonehand, Massachusetts, pp. 539.

May, H. D., Boyle, A. W., Price II, W. A., and Blake, C. K., 1992a, Declorination of PCBs with Hudson River bacterial cultures under anaerobic conditions: effect of sulfate, sulfite, and nitrate., abstr. Q-28, p. 340. Abstr. 92nd Gen. Meet. Am. Soc. Microbiol. 1992. American Society for Microbiology.

May, H. D., Boyle, A. W., Price, II, W. A., and Blake, C. K., 1992b, Subculturing of a polychlorinated biphenyl-dechlorinating anaerobic enrichment on solid media, *Appl. Environ. Microbiol.* **58**:4051–4054.

Miller, K. W., 1992, Reductive desulfurization of dibenzyldisulfide, *Appl. Environ. Microbiol.* **58**:2176–2179.

Mohn, W. W., and Kennedy, K. J., 1992, Reductive dehalogenation of chlorophenols by *Desulfomonile tiedjei* DCB-1, *Appl. Environ. Microbiol.* **58**:1367–1370.

Mohn, W. W., and Tiedje, J. M., 1992, Microbial reductive dehalogenation, *Microbiol. Rev.* **56**:482–507.

Morris, P. J., Mohn, W. W., Quensen III, J. F., Tiedje, J. M., and Boyd, S. A., 1992, Establishment of a polychlorinated biphenyl-degrading enrichment culture with predominantly meta dechlorination, *Appl. Environ. Microbiol.* **58**:3088–3094.

Preuss, A., Fimpel, J., and Diekert, G., 1993, Anaerobic transformation of 2,4,6-trinitrotoluene (TNT), *Arch. Microbiol.* **9**:345–353.

Rabus, R., Nordhaus, R., Ludwig, W., Widdel, F., 1993, Complete oxidation of toluene under strictly anoxic conditions by a new sulfate-reducing bacterium, *Appl. Environ. Microbiol.* **59**:1444–1451.

Ramanand, K., Nagarajan, A., and Suflita, J. M., 1993, Reductive dechlorination of the nitrogen heterocyclic herbicide picloram, *Appl. Environ. Microbiol.* **59**:2251–2256.

Ramanand, K., and Suflita, J. M., 1991, Anaerobic degradation of *m*-cresol in anoxic aquifer slurries: carboxylation reactions in a sulfate-reducing bacterial enrichment, *Appl. Environ. Microbiol.* **57**:1689–1695.

Ramanand, K., and Suflita, J. M., 1993, Carboxylation and mineralization of *m*-cresol by a sulfate-reducing bacterial enrichment, *Curr. Microbiol.* **26**:327–332.

Roberts, D. J., Fedorak, P. M., and Hrudey, S. E., 1990, $CO_2$ incorporation and 4-hydroxy-2-methylbenzoic acid formation during anaerobic metabolism of *m*-cresol by a methanogenic consortium, *Appl. Environ. Microbiol.* **56**:472–478.

Shanker, R., and Bollag, J.-M., 1990, Transformation of indole by methanogenic and sulfate-reducing microorganisms isolated from digested sludge, *Microb. Ecol.* **20**:171–183.

Shanker, R., Kaiser, J.-P., and Bollag, J.-M., 1991, Microbial transformation of heterocyclic molecules in deep subsurface sediments, *Microb. Ecol.* **22**:305–316.

Schink, B., Brune, A., and Schnell, S., 1992, Anaerobic degradation of aromatic compounds, in: *Microbial Degradation of Natural Products* (G. Winkelmann, ed.), VCH Publishers, Weinheim, Germany pp. 220–242.

Schnell, S., Bak, F., and Pfennig, N., 1989, Anaerobic degradation of aniline and dihydroxybenzenes by newly isolated sulfate-reducing bacteria and description of *Desulfobacterium anilini*, *Arch. Microbiol.* **152**:556–563.

Schnell, S., and Schink, B., 1991, Anaerobic aniline degradation via reductive deamination of 4-aminobenzoyl-CoA in *Desulfobacterium anilini*, *Arch. Microbiol.* **155**:183–190.

Shelton, D. R., and Tiedje, J. M., 1984, Isolation and partial characterization of bacteria in an anaerobic consortium that mineralizes 3-chlorobenzoic acid, *Appl. Environ. Microbiol.* **48**:840–848.

Sleat, R., and Robinson, J. P., 1984, The bacteriology of anaerobic degradation of aromatic compounds, *J. Appl. Bacteriol.* **57**:381–394.

Smolenski, W. J., and Suflita, J. M., 1987, Biodegradation of cresol isomers in anoxic aquifers, *Appl. Environ. Microbiol.* **53**:710–716.

Suflita, J. M., Liang, L., and Saxena, A., 1989, The anaerobic biodegradation of *o*-, *m*-, and *p*-cresol by sulfate-reducing enrichment cultures obtained from a shallow anoxic aquifer, *J. Ind. Microbiol.* **4**:255–266.

Suflita, J. M., Gibson, S. A., and Beeman, R. E., 1988, Anaerobic biotransformations of pollutant chemicals in aquifers, *J. Indust. Microbiol.* **3**:179–194.

Suflita, J. M., and Townsend, G. T., 1995, The microbial ecology and physiology of aryl dehalogenation reactions and implications for bioremediation, in: *Microbial Transformation and Degradation of Toxic Organic Chemicals*, L. Y. Young and C. Cerniglia (Eds.), John Wiley and Sons, Inc., N.Y., pp. 237–262.

Szewzyk, R., and Pfennig, N., 1987, Complete oxidation of catechol by the strictly anaerobic sulfate-reducing *Desulfobacterium catecholicum* sp. nov., *Arch. Microbiol.* **147**:163–168.

Szewzyk, U., and Schink, B., 1989, Degradation of hydroquinone, gentisate, and benzoate by a fermenting bacterium in pure or defined mixed culture, *Arch. Microbiol.* **151**:541–545.

Szewzyk, U., Szewzyk, R., and Schink, B., 1985, Methanogenic degradation of hydroquinone and catechol via reductive dehydroxylation to phenol, *FEMS Microbiol. Ecol.* **31**:79–87.

Tanaka, K., 1992, Anaerobic oxidation of isobutyl alcohol, 1-pentanol, and 2-methoxyethanol by *Desulfovibrio vulgaris* strain marburg, *J. Ferment. Bioeng.* **73**:503–504.

Tasaki, M., Kamagata, Y., Nakamura, K., and Mikami, E., 1991, Isolation and characterization of a thermophilic benzoate-degrading, sulfate-reducing bacterium, *Desulfotomaculum thermobenzoicum* sp. nov., *Arch. Microbiol.* **15**:348–352.

Vogel, T. M., and Grbić-Galić, D., 1986, Incorporation of oxygen from water into toluene and benzene during anaerobic fermentatative transformation, *Appl. Environ. Microbiol.* **52**:200–202.

Vogel, T. M., Criddle, C. S., and McCarthy, P. L., 1987, Transformations of halogenated aliphatic compounds, *Environ. Sci. Technol.* **21**:722–736.

Widdel, F., 1988, Microbiology and ecology of sulfate- and sulfur-reducing bacteria, in: *Biology of Anaerobic Microorganism* (A.J.B. Zehnder ed.), John Wiley and Sons, New York, pp. 469–585.

Widdel, F., and Bak, F., 1992. Gram-negative mesophilic sulfate-reducing bacteria, in: *The Prokaryotes*, Second Edition. (A. Balows, H. G. Truper, M. Dworkin, W. Harder, and K-H. Schleifer eds.), Springer-Verlag, New York, pp. 3352–3378.

Widdel, F., Kohring, G.-W., and Mayer, F., 1983, Studies on dissimilatory sulfate reducing bacteria that decompose fatty acids, *Arch. Microbiol.* **134**:286–294.

Widdel, F., and Pfennig, N., 1984, Dissimilatory sulfate- or sulfur-reducing bacteria, in: *Bergey's Manual of Systematic Bacteriology*, Vol. 1. (N. R. Krieg ed.), Williams and Wilkins, Baltimore, pp. 663–679.

Young, L. Y., and Rivera, M. D., 1985, Methanogenic degradation of four phenolic compounds, *Water Res.* **19**:1325–1332.

Ziomek, E., and Williams, R. E., 1989, Modification of lignins by growing cells of the sulfate-reducing anaerobe *Desulfovibrio desulfuricans*, *Appl. Environ. Microbiol.* **55**:2262–2266.

# Index

Adenine ribonucleotide deaminase, 13
Alanine dehydrogenae, 11
Aldehyde oxidoreductase, 10
Antibodies
  to elongation factors, 35
  to RNA polymerase, 37
  to SRB, 8
APS reductase, 11, 42, 74, 90, 114, 165,
  171, 175
Apurinic endodeoxyribonuclease, 13
*Archaeoglobus*
  *fulgidus*, 33, 34, 38–40, 43, 44, 60, 69,
    70, 228
  *profundus*, 33, 34, 38–40, 60, 228
Aspartate decarboxylase
ATPase, 13
ATP sulfurulase, 11, 42, 90, 175
Autotrophic growth, 38, 123

Biocides
  effectiveness, 275
  performance, 281
Biofilms, 248, 286, 287
Biophotolysis, 18
Bioremediation
  acid mine drainage, 15
  gas desulfurization, 15
  organic compounds, 50, 295
  paper industry, 20
  precipitation of heavy metals, 16
  smelter wastes, 16
  tannery effluent, 16
Bisulfite reductase, 11, 74, 92, 93, 96,
  100, 101, 104 105, 107 130

Catalase, 10
Chemoreceptor A, 70
*Chloropseudomonas ethylica*, 232
Coenzyme F420, 36, 41

Commercial problems
  biocorrosion of concrete, 20
  biocorosion of iron, 20, 247
  cutting emulsions, 20
  food processing, 20
  olive brines, 20
Cell structures, 4, 37, 232
CO dehydrogenase, 10, 41, 44, 70, 119
Corrosion
  abiotic, 244
  analysis, 245, 246
  control of, 266
  iron concentration, 253
  microbial influenced, 247, 285
  mechanisms, 249–257
  role of oxygen, 266
Cysteine synthase, 11
Cytochromes, 9, 14, 17, 19, 58, 72, 74, 99,
  105, 114, 116–118, 120, 123, 126,
  127, 129–135, 185, 200, 204–207

*Desulfoarculus baarsii*, 54, 57, 63, 65, 75,
  76, 226
*Desulfobacter*
  *curvatus*, 54, 62, 226
  *hydrogenophilus*, 54, 62, 226
  *latus*, 54, 226
  *postgatei*, 54, 62, 226
*Desulfobacterium*
  *anilini*, 14, 54, 226, 296, 312, 314
  *autotrophicum*, 43, 54, 62, 65, 226, 323
  *catecholicum*, 54, 226, 296, 313
  *cetonicum*, 297
  *indolicum*, 54, 226, 297, 315
  *macestii*, 54, 226
  *niacini*, 54, 57, 62, 65, 226, 299
  *oleovorans*, 297
  *phenolicum*, 54, 226, 295, 312, 314
  *vacuolatum*, 54, 57, 62, 65, 225, 226, 297

**333**

*Desulfobacula toluolica*, 300, 313

*Desulfobulbus*
  *elongatus*, 70, 126, 225
  *marinus*, 57, 62, 225
  *propionicus*, 11, 14, 21, 54, 62, 157, 175, 225

*Desulfobotulus sapovorans*, 54, 57, 63, 225

*Desulfococcus*
  *biacutus*, 54, 226
  *multivorans*, 54, 62, 65, 157, 160, 226, 299, 313
  *nianini*, 315

Desulfoferrodoxin, 120, 129

*Desulfohalobium retbaense*, 54, 225

*Desulfomicrobium*
  *aspheronum*, 54, 57, 225
  *baculatum*, 57, 121–123, 125, 128, 134, 232
  *baculatus*, 54, 56, 72, 195, 196, 199, 201, 205

*Desulfomonile tiedjei*, 54, 62, 65, 226, 299, 318, 320, 325

*Desulfonema*
  *limicola*, 54, 65, 226
  *magnum*, 54, 226, 299, 313

Desulforedoxin, 70, 117, 120, 121, 129

Desulforubidin: *see* Bisulfite reductase

*Desulfosarcina variabilis*, 54, 62, 65, 226, 299, 313

*Desulfotomaculum*
  *acetoxidans*, 21, 43, 58, 59, 66–68, 227
  *antarticum*, 58, 59, 227
  *australicum*, 58, 60, 67–69, 229
  *geothermicum*, 58, 67, 68, 227
  *guttoideum*, 58, 59, 227
  *kuznetsovii*, 58, 227
  *nigrificans*, 11, 13, 53, 58, 60, 66–68, 90, 94, 98–100, 105, 227
  *orientis*, 15, 58, 67, 68, 163, 166, 174, 227, 298
  *ruminis*, 21, 58, 67, 68, 227
  *sapomandens*, 58, 67, 68, 227, 298, 311, 313
  *thermoacetoxidans*, 58, 59, 67, 227
  *thermobenzoicum*, 58, 67, 68, 227, 299

*Desulfovibrio*
  *africanus*, 9, 12, 54, 56, 62–64, 93, 121, 124, 224
  *alcholovorans*, 54, 57, 74, 224

*Desulfovibrio (cont.)*
  *baarsii*, 54, 57, 62, 63, 75
  *baculaatus*, 12, 53, 56, 57, 62, 64, 73
  *carbionolicus*, 54, 57, 74, 224
  *desulfuricans*, 3, 5, 6, 8–17, 19, 54, 56, 62–64, 72, 94, 100–103, 107, 116, 119–122, 127, 129, 134, 135, 155, 159, 161, 167, 169–171, 188–190, 192, 206, 244, 255, 258, 295, 315, 316
  *fructosovorans*, 54, 72, 74, 189–191, 202, 224
  *furfuralis*, 54, 224, 295
  *giganteus*, 21, 54, 74, 224
  *gigas*, 5, 9–14, 54, 62, 64, 70, 72, 73, 93, 98–100, 105, 116, 119–125, 128–135, 185, 187, 196, 199, 201, 205, 206, 295
  *halophilus*, 54, 64, 224
  *longus*, 54, 62, 224
  *multispirans*, 10, 12, 132
  *piger*, 57, 224
  *sapovorans*, 14, 56, 57, 62, 63, 169
  *salexigens*, 3, 12, 54, 56, 62–64, 70, 124, 125, 135, 160, 188, 224
  *simplex*, 54, 64, 224
  *sulfodismutans*, 54, 74, 224
  *termitidis*, 21, 54, 64, 224
  *thermophilus*, 56, 58
  *vulgaris*, 3, 5, 7, 9–14, 53, 54, 62, 63, 70, 72, 73, 76, 90–94, 98, 99, 106, 116, 117, 119, 120, 123, 125, 127–129, 134–137, 166, 169–171, 178, 185, 187, 189–195, 197, 199, 201, 203, 205, 224, 295

Desulfoviridin: *see* Bisulfite reductase

*Desulfuromonas*
  *acetoxidans*, 18, 129, 232
  *pigra*, 21, 54, 57, 62, 63

Detection, 8, 39, 74, 189, 273, 287

Dissimilatory sulfate reduction, 222

Dissimilaatory sulfur reduction, 222

Disproportionation of sulfur compounds, 174

Dithionate as a intermediate, 104, 105

Ecology, 8, 38, 218, 230, 232

Energetics of sulfate activation, 164

Energetics of sulfite reduction, 166, 167

Enzymes: *see* Specific enzyme
  from *A. fugidus*, 43
  immobilized, 9, 14
  isolated form Eubacteria, 10–13
Evolution, 77

Fatty acids, 7, 73
Fermentation, 174
Fe-S clusters, 72, 115, 118, 120, 121, 122,
  124, 125, 133, 191, 193, 203
Ferredoxin, 42, 71, 117, 118, 120–122,
  125, 128, 131, 133, 134
Flavodoxin, 70, 117, 118, 125, 131, 133–
  135
Flavoredoxin, 117, 123, 131, 133
FMN, 114, 117, 129, 195
Formate dehydrogenase, 10
N-Formylmethanofuran dehydrogenase,
  41, 44
N-Formylmethanofuran: H₄MPT formyl-
  transferase, 41, 44
Fumarate dehydrogenase, 10

*Geobacter metallireducens*, 62, 75
Genes
  for cytochrome c, 126, 128, 136, 137,
    185, 204
  for flavodoxin, 125, 136
  for hydrogenase, 136, 185, 200
  for prismane proteins, 204
  for rubredoxin, 136
  for rubrerythrin, 136
  *rbo*, 120, 121
Genome size, 70, 187
Growth
  characteristics, 34, 38, 189, 228
  media, 3

Habitats, 224–227
Halogen compounds, 317–326
Hydrogenase, 9, 12, 14, 72, 76, 105, 123,
  125, 128–133, 168, 171, 185, 192–202
Hydrogen
  cycling hypothesis, 129, 130, 132, 133,
    168
  interspecies transfer, 233
  production, 9, 18, 131, 169

Identification, 7, 22, 33
Inhibition of growth, 7, 33

Interactions
  algae, 20, 232
  cattle, 20
  clams, 21
  cyanobacteria, 20, 232
  humans, 21, 224
  methanogens, 233
  sheep, 21
  swine, 50
  termites, 21, 224
Iron metabolism, 17, 77, 125, 304
Isolation, 7, 189, 224–227

Lactate dehydrogenase, 10, 41, 44
Lipids, 38, 74, 286

Magnetotactic bacteria, 78
Malate:NADP oxidoreductase, 10
Membranes, 101, 103, 130, 131, 132, 158,
  173
Mercury methylation, 15, 17
Metabolism
  endproducts, 9, 34, 40, 59, 102, 103, 167,
    227
  of lignin, 15
Methane production, 61, 218, 228
*Methanobacterium thermoautrophicum*, 39, 43
*Methanosarcina barkeri*, 39, 43
Methanofuran b, 36, 41, 61
N⁵N¹⁰-Methylene₄MPT dehydrogenase,
  41, 44
Microbial mats, 76, 232
Molecular biology, 188

NAD:rubredoxin oxidoreductase, 10, 114,
  119
NAD(P)H₂:menadione oxidoreductase, 10
Nigerythrin, 120
Nitrate reduction, 175
Nitrite reductase, 12
Nitrogenase, 12, 70, 224
Nutrition, 6, 38, 224–228

Oil reservoir souring, 283
Oil well treatment, 19
Oxygen metabolism, 6, 127, 129, 157, 169,
  177, 178, 230, 231, 232, 257–260

P582, 95, 96, 98, 100, 104
Phylogenic characteristics, 34, 35, 43, 50,
  61, 222

Phototrophic associations, 232
Plasmids, 187, 188
Polyglucose, 119, 132
Prismane protein, 202–204
delta-1-proline-5-carboxylate reductase, 11
Proton potential, 159, 171, 173
Protoporphyrinogen oxidase, 10, 70
Pyrophosphatase, 13, 90, 165
Pyruvate:ferredoxin oxidoreductase, 10, 41, 44

Restriction endonuclease, 13
Rhodanese, 11
RNA
  23S, 34
  16S, 21, 33, 34, 56, 61, 62, 66, 76
  7S, 35, 37
RNA polymerase, 37
Rubredoxin, 70, 116, 118–120, 135
Rubredoxin:oxygen oxidoreductase, 10, 120
Rubrerythrin, 120

Salinity range, 230
Serine transacetylase, 11
Succinic dehydrogenase, 70, 126
Sulfate transport, 156, 173
Sulfite reductase, 11, 42, 45, 70, 104, 115, 123, 125, 202, 203

Sulfur cycle, 218
Sulfuretum, 220
Superoxide dismutase, 10
Selenium, 5, 17

Tar sands, 19
*Thermodesulfobacterium*
  *commune*, 57, 58, 65, 66, 94, 228
  *mobile*, 8, 56–58, 73, 228
*Thermodesulfovibrio yellowstonii*, 66
Tetrahydromethanopterin, 36, 41, 61
Tetrathionate pathway, 92
Thiosulfate reductase, 11, 99
Thiosulfate transport, 161
Transformations
  chemical, 3, 295
  distilleries, 19
  fish canning industry, 19
  furfural wastewater, 19
  methanol wastes, 19
  molasses wastes, 20
  starch producing factory, 19
Trithionate reactions, 94, 98, 103, 107, 167
Trithionate reductase, 11, 98

Uranium, 17

Virus, 5, 188